Friction Stir Welding and Processing

Friction Stir Welding and Processing

Fundamentals to Advancements

Edited By

Dr. Sandeep Rathee
Department of Mechanical Engineering,
National Institute of Technology Srinagar
Jammu and Kashmir, India-190006

Dr. Manu Srivastava
Department of Mechanical Engineering,
PDPM Indian Institute of Information Technology,
Design and Manufacturing Jabalpur,
Madhya Pradesh, India-482005

Dr. J. Paulo Davim
Department of Mechanical Engineering,
University of Aveiro,
Campus Santiago,
3810-193 Aveiro, Portugal

Library of Congress Cataloging-in-Publication Data:

Names: Rathee, Sandeep, editor. | Srivastava, Manu, editor. | Davim, J. Paulo, editor. | John Wiley & Sons, publisher.
Title: Friction stir welding and processing: fundamentals to advancements / edited by Dr. Sandeep Rathee, Dr. Manu Srivastava, Dr. J. Paulo Davim.
Description: Hoboken, New Jersey : Wiley, [2024] | Includes index.
Identifiers: LCCN 2023049870 (print) | LCCN 2023049871 (ebook) | ISBN 9781394169436 (cloth) | ISBN 9781394169450 (adobe pdf) | ISBN 9781394169443 (epub)
Subjects: LCSH: Friction stir welding.
Classification: LCC TS228.9 .F758 2024 (print) | LCC TS228.9 (ebook) | DDC 671.5/2—dc23/eng/20231205
LC record available at https://lccn.loc.gov/2023049870
LC ebook record available at https://lccn.loc.gov/2023049871

Cover Design: Wiley
Cover Image: © NASA/MSFC/D. Stoffer (public domain)

Set in 9.5/12.5pt STIXTwoText by Straive, Chennai, India

Contents

About the Editors

Dr. Sandeep Rathee is currently serving the Department of Mechanical Engineering, National Institute of Technology Srinagar, Jammu & Kashmir, India, as an Assistant Professor. His previous assignment was as a Post-Doctoral Fellow at Indian Institute of Technology Delhi (IIT Delhi). He is the recipient of the prestigious National Post-Doctoral Fellowship by SERB (Govt. of India). He was awarded PhD degree from Faculty of Technology, University of Delhi. His field of research mainly includes friction stir welding/processing, advanced materials, composites, additive manufacturing, advanced manufacturing processes, and characterization. He has authored over 60 publications in various international journals of repute and refereed international conferences. He has been awarded 10 industrial design patents and 1 copyright. He has authored/edited seven books in the field of advanced manufacturing. He is working as an editor-in-chief of a book series titled "Advanced Manufacturing of Materials" with CRC Press, Taylor & Francis group. He has also worked as managing guest editor for a special issue of a Scopus Indexed Elsevier journal. He is the editor of International Journal of Experimental Design & Process Optimization, Inderscience Publishers. He is associated with several reputed journals in the capacity of editorial member. Additionally, he is serving as a reviewer in more than 30 Journals. He has completed/is handling externally funded projects of more than 07 Million INR in the field of additive manufacturing and friction stir welding. He has supervised/is supervising 4 PhD scholars, 2 MTech, and 20 BTech students. His works have been cited more than 1600 times (as per google scholar).

He has a total teaching and research experience of more than ten years. He has delivered invited lectures, chaired scientific sessions in several national and international conferences, STTPs, and QIP programs. He is a life member of the Additive Manufacturing Society of India (AMSI), and Vignana Bharti (VIBHA).

Orcid id: https://orcid.org/0000-0003-4633-7242

Dr. Manu Srivastava is presently serving PDPM Indian Institute of Information Technology, Design and Manufacturing Jabalpur, India, in the Department of Mechanical Engineering. Her previous assignment was as a Prof. & Head, Department of Mechanical Engineering and Director Research, Faculty of Engineering and Technology, MRIIRS, Faridabad, India. She has completed her PhD in the field of additive manufacturing from Faculty of Technology, University of Delhi. Her field of research is additive manufacturing, friction-based AM, friction stir processing, advanced materials, manufacturing practices, and optimization techniques. She has authored/ coauthored around 100 publications in various technical platforms of repute. She has been awarded 18 industrial design patents and 1 copyright.

She has authored/edited seven books in the field of advanced manufacturing. Out of these, three have already been published with CRC press, Taylor & Francis group, and others are either in press or in an accepted stage. Two of them are with Wiley, one with Elsevier, and one with Springer publishing house.

She is working as an editor-in-chief of a book series titled "Advanced Manufacturing of Materials" with CRC Press, Taylor & Francis Group. She has also worked as guest editor for a special issue of a Scopus Indexed Elsevier journal. She is on the editorial board of several reputed journals in the capacity of editorial member. She is also serving as the regional editor of International Journal of Experimental Design & Process Optimization, Inderscience Publishers. Additionally, she is serving as a reviewer in more than 30 Journals. She is working on various projects of around 5 million INR and has completed several consultancy works funded by Govt. of India in the field of Hybrid Additive Manufacturing and rehabilitation robotics. She has a total teaching and research experience of around 15 years. She has won several proficiency awards during the course of her career including merit awards, best teacher awards, etc. One special award that needs mention is the young leader award. She has delivered invited lectures, chaired scientific sessions in several national and international conferences, STTPs, and QIP programs. She is a life member of Additive Manufacturing Society of India (AMSI), Vignana Bharti (VIBHA), The Institution of Engineers (IEI India), Indian Society for Technical Education (ISTE), Indian society of Theoretical and Applied Mechanics (ISTAM), and Indian Institute of Forging (IIF).

Orcid id: https://orcid.org/0000-0001-6513-7882

Dr. J. Paulo Davim is a Full Professor at the University of Aveiro, Portugal. He is also distinguished as honorary professor in several universities/colleges/institutes in China, India, and Spain. He received his PhD degree in Mechanical Engineering in 1997, MSc degree in Mechanical Engineering (materials and manufacturing processes) in 1991, Mechanical Engineering degree (5 years) in 1986, from the University of Porto (FEUP), the Aggregate title (Full Habilitation) from the University of Coimbra in 2005, and the DSc (Higher Doctorate) from London Metropolitan University in 2013. He is Senior Chartered Engineer by the Portuguese Institution of Engineers with an MBA and Specialist titles in Engineering and Industrial Management as well as in Metrology. He is also Eur Ing by Engineers Europe FEANI-Brussels and Fellow (FIET) of IET-London. He has more than 35 years of teaching and research experience in Manufacturing, Materials, Mechanical, and Industrial Engineering, with special emphasis in Machining & Tribology. He has also interest in Management, Engineering Education, and Higher Education for Sustainability. He has guided large numbers of postdoc, PhD, and master's students as well as has coordinated and participated in several financed research projects. He has received several scientific awards and honors. He has worked as evaluator of projects for ERC-European Research Council and other international research agencies as well as examiner of PhD thesis for many universities in different countries. He is the Editor in Chief of several international journals, Guest Editor of journals, books Editor, book Series Editor, and Scientific Advisory for many international journals and conferences.

   Orcid: https://orcid.org/0000-0002-5659-3111

List of Contributors

Mahmoud Abbasi
Department of Materials and Metallurgical
Engineering
Amirkabir University of Technology
Tehran
Iran

Bommana B. Abhignya
Department of Mechanical Engineering
Hybrid Additive Manufacturing Laboratory
PDPM Indian Institute of Information
Technology
Design and Manufacturing
Jabalpur
India

Shanmugam Arun Kumar
Department of Mechatronics Engineering
Kongu Engineering College
Erode
Tamil Nadu
India

Mohankumar Ashok Kumar
Department of Mechanical Engineering
Government College of Engineering
Anna University
Coimbatore
Tamil Nadu
India

Behrouz Bagheri
Department of Materials and Metallurgical
Engineering
Amirkabir University of Technology
Tehran
Iran

Umashankar Bharti
Department of Mechanical Engineering
Hybrid Additive Manufacturing Laboratory
PDPM Indian Institute of Information
Technology
Design and Manufacturing
Jabalpur
India

Narendra B. Dahotre
Center for Agile and Adaptive Additive
Manufacturing
University of North Texas
Denton
TX
USA

Department of Materials Science and
Engineering
University of North Texas
Denton
TX
USA

Kolar Deepak
Department of Mechanical Engineering
Vardhaman College of Engineering
Jawaharlal Nehru Technological University
Hyderabad
Telangana
India

Melaku Desta
Department of Mechanical Engineering
CEME
Addis Ababa Science and Technology
University
Addis Ababa
Ethiopia

Devasri Fuloria
School of Mechanical Engineering
Vellore Institute of Technology
Vellore
India

Vijay Shivaji Gadakh
Department of Mechanical Engineering
Dr. Vithalrao Vikhe Patil College of
Engineering
Savitribai Phule Pune University
Ahmednagar
Maharashtra
India

Department of Automation and Robotics
Engineering
Amrutvahini College of Engineering
Savitribai Phule Pune University
Ahmednagar
Maharashtra
India

Raja Gunasekaran
Department of Mechanical Engineering
Velalar College of Engineering and Technology
Erode
Tamil Nadu
India

Yogesh Ramrao Gunjal
Department of Mechanical Engineering
Dr. Vithalrao Vikhe Patil College of
Engineering
Savitribai Phule Pune University
Ahmednagar
Maharashtra
India

Department of Mechanical Engineering
Amrutvahini College of Engineering
Savitribai Phule Pune University
Ahmednagar
Maharashtra
India

Velu Kaliyannan Gobinath
Department of Mechatronics Engineering
Kongu Engineering College
Erode
Tamil Nadu
India

Nikhil Jaiswal
Department of Mechanical Engineering
Hybrid Additive Manufacturing Laboratory
PDPM Indian Institute of Information
Technology
Design and Manufacturing
Jabalpur
India

Yuqi Jin
Center for Agile and Adaptive Additive
Manufacturing
University of North Texas
Denton
TX
USA

Department of Physics
University of North Texas
Denton
TX
USA

Durairaj Raja Joseph
Department of Aerospace Engineering
School of Aeronautical Sciences
HITS
Hindustan Institute of Technology & Science
Chennai
India

Satish V. Kailas
Department of Mechanical Engineering
IISc
Bangalore
Karnataka
India

Narayan Sahadu Khemnar
Department of Mechanical Engineering
Dr. Vithalrao Vikhe Patil College of
Engineering
Savitribai Phule Pune University
Ahmednagar
Maharashtra
India

Department of Automation and Robotics
Engineering
Amrutvahini College of Engineering
Savitribai Phule Pune University
Ahmednagar
Maharashtra
India

Atul Kumar
School of Mechanical Engineering
Vellore Institute of Technology
Vellore
India

Gaurav Kumar
Department of Mechanical Engineering
VCE
Dr. A.P.J. Abdul Kalam Technical University
Lucknow
Uttar Pradesh
India

Mukesh Kumar
Department of Mechanical Engineering
VCE
Dr. A.P.J. Abdul Kalam Technical University
Lucknow
Uttar Pradesh
India

Vinayak Malik
Department of Mechanical Engineering
IISc
Bangalore
Karnataka
India

Department of Mechanical Engineering
KLS Gogte Institute of Technology
Visvesvaraya Technological University
Belagavi
Karnataka
India

Neelam Meena
Department of Metallurgical Engineering and
Material Science
Indian Institute of Technology Bombay
India

Husain Mehdi
Department of Mechanical Engineering
Meerut Institute of Engineering and
Technology
Dr. A.P.J. Abdul Kalam Technical University
Lucknow
Uttar Pradesh
India

Hitesh Mhatre
Department of Mechanical Engineering
Sardar Vallabhbhai National Institute of
Technology
Surat
Gujarat
India

Amrut Shrikant Mulay
Department of Mechanical Engineering
Sardar Vallabhbhai National Institute of
Technology
Surat
Gujarat
India

Palaniappan Muthukumar
Department of Mechanical Engineering
Kongunadu College of Engineering and
Technology
Thottiyam
Tamil Nadu
India

Shazman Nabi
Department of Mechanical Engineering
National Institute of Technology Srinagar
Jammu & Kashmir
India

Farooz A. Najar
Department of Mechanical Engineering
National Institute of Technology Srinagar
Jammu & Kashmir
India

R. Ganesh Narayanan
Department of Mechanical Engineering
Indian Institute of Technology Guwahati
Guwahati
India

Nagarajan Nithyavathy
Department of Mechatronics Engineering
Kongu Engineering College
Erode
Tamil Nadu
Indi

Jinu Paul
Department of Mechanical Engineering
National Institute of Technology Calicut
Calicut
India

Ardula G. Rao
Naval Materials Research Laboratory
Defence Research and Development
Organization
Thane
India

Sandeep Rathee
Department of Mechanical Engineering
National Institute of Technology,
Srinagar
Jammu & Kashmir
India

Kazi Sabiruddin
Department of Mechanical Engineering
Indian Institute of Technology Indore
Indore
India

Divya Sachan
Department of Mechanical Engineering
Indian Institute of Technology Guwahati
Guwahati
India

Prem Sagar
Department of Mechanical Engineering
The Technological Institute of Textile Sciences
Maharshi Dayanand University
Rohtak
Haryana
India

Baidehish Sahoo
School of Mechanical Engineering
MIT World Peace University
Pune
India

Mohd Sajid
Department of Mechanical Engineering
VCE
Dr. A.P.J. Abdul Kalam Technical University
Lucknow
Uttar Pradesh
India

Sushma Sangwan
Department of Mechanical Engineering
The Technological Institute of Textile Sciences
Maharshi Dayanand University
Rohtak
Haryana
India

Abhishek Sharma
Research Division of Materials Joining
Mechanism
Joining & Welding Research Institute
Osaka University
Osaka
Japan

Arshad N. Siddiquee
Department of Mechanical Engineering
Jamia Millia Islamia
New Delhi
India

Tanvir Singh
Department of Mechanical Engineering
St. Soldier Institute of Engineering &
Technology
Punjab Technical University
Jalandhar
Punjab
India

Palani Sivaprakasam
Department of Mechanical Engineering
CEME
Addis Ababa Science and Technology
University
Addis Ababa
Ethiopia

Murugan Srinivasan
Department of Mechanical Engineering
Mahendra Engineering College
Anna University
Namakkal
India

Manu Srivastava
Department of Mechanical Engineering
Hybrid additive manufacturing Laboratory
PDPM Indian Institute of Information
Technology
Design and Manufacturing
Jabalpur
India

Kandasamy Suganeswaran
Department of Mechatronics Engineering
Kongu Engineering College
Erode
Tamil Nadu
India

Setu Suman
Department of Mechanical Engineering
Indian Institute of Technology Indore
Indore
India

Putti Venkata Siva Teja
Department of Mechanical Engineering
Dhanekula Institute of Engineering &
Technology
Jawaharlal Nehru Technological University
Kakinada
Ganguru
Andhra Pradesh
India

Tianhao Wang
Energy and Environment Directorate
Pacific Northwest National Laboratory
Richland
WA
USA

Ashish Yadav
Department of Mechanical Engineering
Hybrid additive manufacturing Laboratory
PDPM Indian Institute of Information
Technology
Design and Manufacturing
Jabalpur
India

Teng Yang
Center for Agile and Adaptive Additive
Manufacturing
University of North Texas
Denton
TX
USA

Department of Materials Science and
Engineering
University of North Texas
Denton
TX
USA

Preface

"This book is dedicated to the relationships of trust and respect that stay strong like an anchor to ward off all obstacles and are a divine intervention to create an oasis of hope in the desert of aimless souls"

Manu

Morally, every researcher and academician is bound to effectively share, exchange, and communicate ideas, knowledge, and experience with the global technical society. The unpredictable nature of life encourages us as an investigator of scientific truths to make every effort to not only share our learnings but also bring together a group of eminent researchers of this technical society to come up with this edited book. This book has contributions from researchers in the field of friction stir welding, processing (FSW/P), and their variants from across the world. The intent is also to facilitate our future generation of researchers with the knowledge gained so far to provide them a consolidation of the accomplished research. This will most definitely empower them with a vision to make a technologically and socially strong community based on deep-rooted foundations. With this thought process in place, the editors have come together to disseminate information gathered from years of experience in the field of FSW/P and their variants with the technical society. Today, a wide variety of high-quality literature in the form of a few monographs and a multitude of journal articles are available in the field of FSW/P, but most of these are confined only to some focused areas. A resource that presents the overall picture in FSW/P and their variants is very much required. This book is a novel venture toward the said direction. It is ensured to present details in simple yet precise language with clarity to cater to a wide variety of readers globally.

FSW is the art and science of joining materials in solid state using a nonconsumable tool with the application of frictional heat. It is frequently utilized for obtaining high-strength welds and join a wide range of materials including but not limited to aluminum and its alloys, copper and its alloys, titanium and its alloys, stainless steel variants, magnesium and its alloys, and so on. When FSW is applied for processing applications or for fabricating composites, the FSW technology is called FSP. With slight modifications, today this technology has more than 25 different variants, each dedicated to some specific applications. At present, FSW/P technology is been increasingly utilized in shipbuilding, aircraft and space applications, welding and processing a variety of exotic and specially engineered structures like shape memory alloys, honey comb, metal matrix, polymer matrix composites, and so on.

This book has 20 chapters each of which is dedicated to a different aspect of FSW/P. The chapters included in this book have been briefly introduced here to make the reader well versed with the overall content.

Chapter 1 presents an overview of friction stir welding technique as a sustainable alternative to conventional metal joining and welding techniques. This chapter also explains the principle and working of FSW, its different variants, and the kind of tool variations incorporated in these variants. Some common defects encountered during FSW are also mentioned. Some of FSW's advantages and limitations are also added in the end.

Chapter 2 presents an introduction to friction stir welding and the single-point incremental forming procedure on friction stir welded blanks, recent developments in tool design, tool materials, parameter optimization, mechanical properties, etc. The chapter offers quick idea about selection of FSW and SPIF processes to prepare customized components in real-time applications. Two similar or dissimilar materials can be joined and formed with this hybrid manufacturing process. The combined process has versatile applications in automotive and defense sector. SPIF has special advantages over conventional forming processes which includes heterogeneity, quicker lead time, versatility, etc. The chapter concludes with several recommendations for future research.

Chapter 3 presents an overview of friction stir brazing and its variants. Two case studies are presented viz. joining of low carbon steels by application of friction stir brazing and Sn–Pb alloy as filler and intermetallic compound formation and mechanical characteristics of brazed samples made by friction stir vibration brazing with Sn–Pb filler material and SiC reinforcing particles. Applications of friction stir brazing in different sectors are discussed followed by summary and future directions of the chapter.

Chapter 4 introduces and discusses details of friction stir processing (FSP) as a comprehensive microstructure tailoring tool. This chapter starts with the history behind the evolution of the FSP from the conventional friction stir welding processes. The second section of the chapter gives an overview of the working principle of FSP followed by its comparison with other SPD techniques in third section. The fourth section of the chapter elaborates on the factors affecting the process such as tool rotational speed, traverse speed, and in-process cooling. It is followed by discussion related to mechanisms of microstructural evolution. The last section of the chapter describes the challenges and opportunities associated with FSP to resonate with recent trends and become industry ready.

Chapter 5 gives a fundamental overview of friction stir processing along with the process parameters that affect surface integrity. Discussion on the basic mechanism of thermal spray technique along with their classification and applications is provided. Further, discussed the role of surface engineering as a modification technique and how it affects surface morphologies. In the end, the inappropriate parameters that affect the surface modification technique have been assessed.

Chapter 6 deals with surface composites manufacturing by solid-state friction stir processing method. A detailed discussion on the factors affecting the microstructure and mechanical properties of FSPed processed surface composites is provided. This chapter also explains the factors promoting the dominance of various strengthening mechanisms in the surface composites processed through FSP.

Chapter 7 presents the issues in friction stir welding (FSW) of dissimilar material joining such as dissimilar aluminum (Al) alloys, aluminum to copper alloys (Al–Cu), aluminum to titanium (Al–Ti), and aluminum to steel (Al–Fe), recent developments in tool design, tool materials, parameter optimization, microstructure, mechanical properties, common defects occurred, etc. The chapter concludes with several recommendations for future research.

Chapter 8 covers the friction stir welding process, particularly the joining of aluminum and its alloys, which includes the introduction, advancements, applications, and conclusion. Emphasis is given to critical factors that will affect the FSW method in the joining of aluminum alloys.

The current state of the art for FSW of aluminum and its alloy, as well as the basics of the process and its influences, are examined. Additionally, additive mixed FSW of Al alloy, testing and characterization like tensile, hardness, and microstructure analysis, as well as industrial applications, are covered in the FSW mechanical characteristics section.

Chapter 9 presents an experimental study on the mechanical characterization of FSWed joints of dissimilar aluminum alloys of AA7050 and AA6082. This work analyzed the effect of processing parameters on the mechanical characterization of the friction stir welded joint (FSWed) of AA7050 and AA6082. Well-material mixing of FSWed joints on high rotational tool speed (RTS) was observed. The fracture of the welded joint at HAZ reveals the excellent weld quality and bonding between the dissimilar metals.

Chapter 10, In this chapter, standard microstructural characterization techniques are presented. Various aspects such as the microstructure sample preparation procedure with brief introduction of equipment used in microstructural characterization are covered. A few illustrative examples are added toward the end of each section of this chapter with an aim to facilitate understanding of concepts.

Chapter 11 focuses on the microstructural characterization and mechanical testing of samples subjected to friction stir welding (FSW) and friction stir processing (FSP) techniques. The aim is to investigate the resulting microstructure and evaluate the mechanical properties of the welded/processed samples. The chapter begins by describing the methodology used for microstructural analysis, which includes techniques such as optical microscopy, scanning electron microscopy (SEM), and X-ray diffraction (XRD). The microstructural features, such as grain structure, grain boundaries, and phase composition, are examined and documented. The findings provide valuable insights into the relationship between process parameters, microstructure, and mechanical behavior, aiding in the optimization of FSW and FSP techniques for desired material properties. In conclusion, this chapter provides a comprehensive analysis of the microstructural characterization and mechanical testing of FSWed/FSPed samples, shedding light on the effects of the welding/processing techniques on the resulting material properties.

Chapter 12 provides state-of-the-art information on the joining of metal matrix-reinforced (MMR) welds by employing the FSW technique. The results are critically evaluated with more emphasis on the reinforcement particles and plasticized material flow behavior that affects the metallurgical properties of MMR welds. In addition, the mechanical performance of reinforced FSW welds is evaluated directly related to welding specifications. Fractography and wear behavior characteristics of reinforced FSW welds are also evaluated based on the materials' combination of reinforcement particles and base material.

Chapter 13 attempts to summarize the applications, and manufacturing methods of sandwich sheet structures, specifically the solid-state joining methods such as friction stir welding, friction stir spot welding, accumulative roll bonding, and adhesive bonding. Applications in trains, marine sector, turbines, aerospace, and ship construction are presented. From the scarce literature, the problems associated with the manufacturing and joining methods is also highlighted at last.

Chapter 14 provides a bird's eye view of different types of defects noticed in friction stir welding (FSW). It elaborates on major defects and discusses the possible causes of the emergence of certain types of flaws and the ways to mitigate them. It further exposes the readers to discrete friction stir variants and typical defects observed in these processes. The chapter concludes with a few solutions to avoid defects in friction stir processes.

Chapter 15 explains ultrasound wave propagation behaviors associated with the macrostructure, microstructure, and residual stresses. Then, the explanations of the methodologies are discussed that involved such principles into ultrasonic inspections, evaluations, and monitoring.

Furthermore, the cases studies regarding the recent ultrasonic NDT/Es in friction stir-based manufacturing processes are introduced and discussed.

Chapter 16 covers the significance of friction stir welding comparing with other solid-state metal joining processes. The application of FSW on different materials such as aluminum alloys, magnesium alloys, copper alloys, titanium alloys, steels, composite materials, polymers, and plastics is discussed. The recent developments and applications of FSW metal joining process in various industries were also discussed. The industrial applications such as aerospace, automobile, ship building, railways, and other industries were reviewed.

Chapter 17 discusses on the development of an indigenous method of friction stir process (FSP) fabrication technique. A discussion is made on the impetus development of FSP setup to fabricate surface composites with fortified properties for specific applications. Various shoulder and pin profiles have been designed and their effects on the microstructural evolution and mechanical properties were also discussed. The fundamental understanding and critical thinking of the tool modifications and the manufacturing process paves way for the development of new surface composites.

Chapter 18 presents a case study that provides a comprehensive analysis of the effects of various pin profiles on the performance of the FSW tool.

Chapter 19 presents a case study that provides static analysis of honey comb structure (HCS) fabricated by FSP. Several researchers have carried out fabrication of HCS by using FSP method with different materials, but there is meager reported research on its analysis.

Chapter 20 provides a comprehensive review of the friction stir additive manufacturing (FSAM) technique, highlighting its advantages over other gas-/liquid-based technologies. FSAM offers clean and green technology, defect-free parts, cost-effectiveness, and the ability to process intricate designs with excellent mechanical properties. The chapter explores the application of FSAM in the field of additive manufacturing (AM), particularly in complex structural design. The chapter concludes by discussing the microstructural development, recent advancements, and future prospects of FSAM.

The quantum of information related to different friction stir welding/processing techniques, fast degree of obsolescence, and extremely high levels of ongoing technical as well as technological advances puts a restraint on presenting details of every aspect of each related topic. Editor group has, however, put in their best efforts in making this book informative and interesting. This book is a result of dedicated research in the field of FSW/P as well as collaboration with different peer groups and an in-depth literature review. Editors most sincerely hope that the book is a valued knowledge source for upcoming research groups, academia, and industry. It is advised to apply the information in this book for promoting research and development in the field of FSW/P and related processes.

All queries, advice, and observations regarding the book are most welcome.

Dr. Sandeep Rathee
Dr. Manu Srivastava
Dr. J. Paulo Davim

January 2024

Acknowledgments

At the start, our editors Becky Cowan and Lauren Poplawski for being such a huge pillar of strength need a wholehearted acknowledgment. The editors profusely thank the entire team of Wiley Press for their support toward the initiative of bringing forth this book.

The editors thank all the contributing authors of different chapters for sharing their expertise and helping us in bringing the book to its present form. The editors wholeheartedly thank their respective institutions, National Institute of Technology Srinagar, Jammu & Kashmir and PDPM Indian Institute of Information Technology, Design and Manufacturing Jabalpur, Madhya Pradesh. We express deep gratitude to them for their valuable support toward the endeavor to come up with the present edited book.

Dr. Sandeep Rathee wishes to thank his mentors. He also acknowledges the support of his students and family, especially his parents Shri Raj Singh Rathee and Smt. Krishna Rathee, for their constant support.

Dr. Manu Srivastava wishes to thank her mentors. She acknowledges the support of her students for their inspiration and support. Projects like this need a lot of dedicated effort which often comes at the cost of time kept apart for the family. No words of gratitude would suffice to acknowledge the support of her family.

Authors devote and dedicate this work to the Divine Creator based upon their belief that the strength to bring any thoughtful and noble endeavor into being emerges from the Almighty with a wish that this work makes a valuable addition for its readers.

Dr. Sandeep Rathee
Dr. Manu Srivastava
Dr. Paulo Davim

List of Figures

List of Tables

1

Friction Stir Welding: An Overview

Farooz A. Najar[1], Shazman Nabi[1], Sandeep Rathee[1], and Manu Srivastava[2]

[1] Department of Mechanical Engineering, National Institute of Technology Srinagar, Jammu & Kashmir, India
[2] Department of Mechanical Engineering, PDPM Indian Institute of Information Technology, Design and Manufacturing, Jabalpur, India

1.1 Introduction

Friction stir welding (FSW) is a solid-state welding method patented in 1991 by The Welding Institute (TWI). It is a solid-state joining process that, at its inception, was used to weld relatively soft materials such as aluminum, magnesium, and their alloys. Today's global competition has necessitated producing objects with versatile functions, compact size, lightweight, aesthetics, durability, and cost-effectiveness, making manufacturing more complex. Most products are usually complex and consist of an assembly of smaller components that are simple in shape, small in size, and often made of different materials. Until recently, most of the industry's joining and assembly requirements were achieved by fusion welding processes, mostly arc welding techniques. Fusion welding has its own problems, and to name some: requirement of skill, weld defects, microstructural and mechanical property changes, inability to join different materials with distant properties, and solidification-related defects.

Unlike typical welding processes, which employ a heat source to melt and fuse two materials together, FSW generates heat and plasticizes the material using a nonconsumable tool. One of the primary advantages of FSW is that it yields a high-quality, defect-free joint with outstanding mechanical characteristics such as high strength and fatigue resistance. FSW is also a comparatively low-heat input solid-state procedure that produces lesser fumes and spatter, making it a safer and more eco-friendly alternative to fusion-based welding processes. FSW is now utilized in several sectors, including aerospace, automotive, robotics, shipbuilding, offshore energy production, and rail transportation. FSW is a cost-friendly welding technology as it could be done using the readily available vertical milling machine in any workshop with minimal modifications.

This chapter covers FSW in detail starting with FSW machine setup and its working principle. Then, the various weld zones developed during FSW will be discussed. Various FSW variants will be then discussed succeeded by some of the defects encountered during the FSW processes. In the end, the advantages and limitations of FSW are mentioned as well.

1.2 FSW Working Principle

The FSW tool consists of tool shoulder of suitable shoulder diameter and smaller diameter pin that is also known as probe. The axis of the rotating tool is tilted at a small angle of the order of $0°-3°$, and during FSW, the tool shoulder penetrates slightly below the base metal (BM) surface by a distance known as plunge depth. As the rotating tool is inserted in the workpiece, heat generation starts owing to the frictional resistance between the tool (shoulder and pin) and workpiece, and plastic deformation of BM occurs around the pin. The generated heat is sufficient to soften BM under the shoulder. The extent of heat generation increases as plunge depth further increases. Tool rotation plastically deforms material adjacent to the pin by an action of stirring. The stirring causes severe plastic deformation (SPD) which also contributes to heat input. Under the combined effect of friction and plastic deformation, material being stirred softens, and its temperature may rise above its recrystallization temperature. With the directional traverse of the rotating tool softened material next to pin, it experiences complex material flow and undergoes extrusion ahead of the tool and forging action by shoulder behind the tool. The space vacated by the traversing tool is replenished by the material transported from the front to the back as a result of extrusion and forging action. During extrusion, material being processed rotates in the direction of tool and simultaneously rises up; the rising material is pressed by the shoulder to forge it back behind the tool, replenishing the space vacated by the tool which has already moved ahead. This complex transport mixes the material and consolidates it behind the tool in the form of processed region.

FSW setup generally includes FSW machine, FSW tool, base materials, clamping system, and so on. Figure 1.1 illustrates some of these elements. Subsequent paragraph briefly discusses these elements.

FSW tool: FSW tool is a specially designed tool with a shoulder and pin. The diameter of the pin is generally kept as one-third to one-fourth of the shoulder diameter. However, different researchers have utilized different shoulder-to-pin diameter ratios.

Workpiece: The materials to be welded together are referred to as the workpiece. Metals such as aluminum, copper, and titanium alloys, as well as polymers are frequently joined via FSW.

Clamping system: During the FSW process, the clamping system keeps the workpiece in place. It must sustain enough force to prevent the workpiece from moving or distorting during welding.

Figure 1.1 Friction stir welding setup.

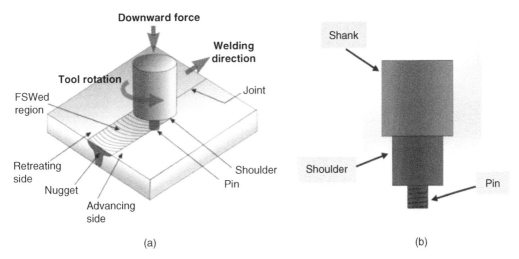

Figure 1.2 (a) Schematic of FSW process. Source: Elatharasan and Kumar [2]/with permission of Elsevier; AS: Advancing side, RS: Retreating side. (b) FSW tool.

Welding Machine: The welding machine supplies the necessary power and control to operate the FSW tool. It usually consists of a motor that rotates the tool, a mechanical, hydraulic, or pneumatic system that provides force to the tool, and a controller that adjusts welding settings.

Cooling System: Because FSW creates a substantial quantity of heat, a cooling system is essential to regulate the workpiece's temperature and avoid deformation or damage.

FSW utilizes frictional heat and plastic deformation to join two metal surfaces without the need of additional consumables like filler material or shielding gases [1]. The FSW process involves a nonconsumable rotating tool (Figure 1.2b), that is plunged into the workpiece and moved along the joint line (Figure 1.2a), generating frictional heat that softens the material and makes it easier to deform. As the tool rotates and moves, it creates a plasticized zone that mixes the material from both sides of the joint, which is then allowed to cool and solidify to form a seamless weld [1].

FSW poses numerous advantages compared to fusion welding methods, namely elimination of filler materials during welding process, no requirement of shielding gases are required, and less heat generation during the process. Thus, the formation of oxide layer, high coefficient of thermal expansion, and loss of mechanical properties can be overcome by FSW [3]. The heat developed during the process is localized to the welding area, which helps to minimize distortion and reduce the risk of thermal damage to the workpieces. The heat dissipation can be studied in terms of the weld zones created by the friction between tool and workpiece(s).

1.3 Weld Zones

During the FSW process, following different zones appear:

- Stir zone (SZ)/Nugget zone (NZ)
- Thermo-mechanical affected zone (TMAZ)
- Heat affected zone (HAZ)
- Base material (BM)

Figure 1.3 illustrates different zones formed during FSW.

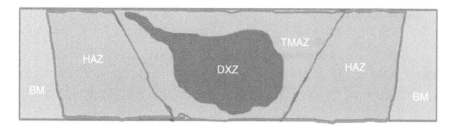

Figure 1.3 Various welding zones associated with FSW. Source: El-Sayed et al. [1]/with permission of Elsevier.

The portion of workpiece where welding is to be done (tool pin stirs the material) is termed as stir zone, and it experiences heavy plastic deformation due to rotating and advancing movement of tool, which leads to dynamic recrystallization (DRX) of grains. The grains thus obtained are severely elongated and reoriented and are generally smaller than BM [4], the strength of joint enhances as the grain size decreases. Due to the smaller size of the grains, they are packed very closely within the structure. TMAZ is closer to the NZ, where both heat and plastic deformation occur to microstructure but without the grain refinement due to insufficient deformation rate; however, some dissolution of precipitates is observed [3].The microstructure of the material is also altered in this zone owing to the development of thermal and mechanical stresses. HAZ is the region after the TMAZ and is far from the NZ, where the material has experienced comparatively lesser increase in temperature. The microstructure of the material in this zone may have been altered due to the heating, but not to the extent of the TMAZ. The BM zone in FSW refers to the region of the material that is not affected by the frictional heat and the stirring action of the working tool. This zone is situated on either sides of the joint and remains relatively undeformed by the welding process [5]. There has been a huge improvement in terms of various techniques used for joining materials using FSW. These different techniques or variants are described in the following section in detail.

1.4 Variants of FSW

Invention of FSW has revolutionized the metal joining by virtue of it being a solid-state process, its automation and independence from welder's skill, environment friendliness, and high productivity. Ever since its invention, there has been a spurt of research output on FSW and it did not remain merely a welding process, but it gave birth to a magnificent fabrication technology. There are various variants of FSW, some of them are illustrated (refer to Figure 1.4) and discussed in subsequent subsections.

1.4.1 Friction Stir Spot Welding (FSSW)

Friction stir spot welding (FSSW) is a type of FSW process used to create strong and high-quality joints between metal workpieces. The process involves using a specialized tool that is rotated at high speed and forced into the workpieces to be joined. This generates frictional heat that softens the workpieces and allows the tool to move along the joint line to create a bond. In FSSW, no traverse/linear motion of tool takes place as in the case of FSW that leads to localized spot weld. Unlike conventional spot-welding techniques, FSSW is suitable for welding dissimilar metals, thin materials, or materials with high melting points. The resulting welds are stronger and more fatigue-resistant

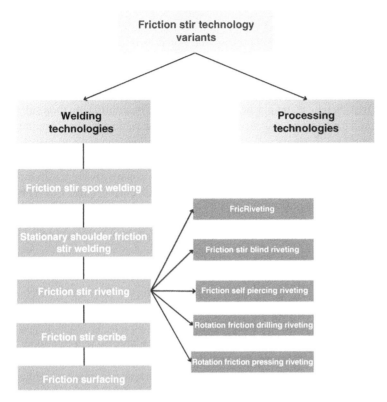

Figure 1.4 Different variants of FSW.

than conventional spot welds, making FSSW a promising alternative method. FSSW is generally completed in three different stages:

Stage 1: Plunging, in which the rotating tool is plunged into the workpiece in a controlled manner. At the joint, heat gets generated owing to friction which softens the work material while the applied axial downward force and stirring action of tool plasticizes the material.

Stage 2: When the stirred tool reaches up to a certain required depth, known as plunge depth, the tool is kept there for some time known as the dwell time.

Stage 3: After the completion of dwell period, tool is retracted which leaves a key hole or void replica of tool on the specimen and this hole can be compensated by additional filler material [3]. The schematic illustration of these stages can be seen in Figure 1.5.

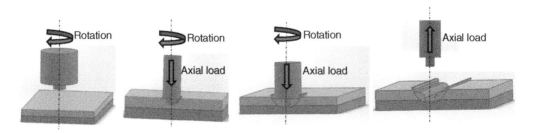

Figure 1.5 Schematics diagram of FSSW: (a) Tool rotation, (b) Plunge, (c) Dwell (stirring action), (d) Retraction.

1.4.2 Stationary Shoulder FSW

Stationary shoulder FSW (SS-FAW) is a type of FSW process that generally does not require shoulder rotation. Initially, works were reported by Russell and Blignault [6]. In SS-FSW, the tool shoulder remains stationary while tool probe rotates. The probe of tool and shoulder are connected by some mechanisms such as bearing assembly, while the workpiece is moved under the tool. As the tool probe rotates, it generates frictional heat and plasticizes the material. The rotating pin then stirs the material, causing it to mix and form a solid-state bond [7].Owing to the stationary shoulder, the range of material to be stirred and heat input are reduced. That further reduces the HAZ and levels of distortion. In SS-FSW, the stationary shoulder provides additional support to the workpiece and helps to reduce deformation during the welding process. This technique is particularly useful for welding materials with high melting points or for welding dissimilar materials (Figure 1.6). The weld produced by SS-FSW is strong, defect-free, and has good fatigue properties [7].

1.4.3 Friction Stir Riveting

Friction stir riveting (FSR) is a type of FSW process that creates a bond between two or more metal sheets without melting the material. Specially designed tool with a rotating shoulder and a nonrotating probe is used to apply pressure and stir the material, resulting in a strong and durable joint that offers benefits over conventional riveting methods. FSR produces high-quality joints having good mechanical strength and eliminates the need for consumables. It is suitable for joining similar or dissimilar materials. The different steps of this process include (i) Rivet advance or feed, wherein the samples are secured on the die and the rivet advances toward the workpiece for contact while rotating at a predetermined speed; (ii) Hot riveting stage, where the revolving rivet penetrates the softened top specimen; (iii) Friction stage, where the downward motion is ceased but the rotation of rivet continues for generating a high-temperature environment for solid-state joining of specimens; and (iv) Off stage, where the rotation of rivet is abruptly stopped and the static contact is created with the surrounding specimens that results in the formation of a solid-state joint.

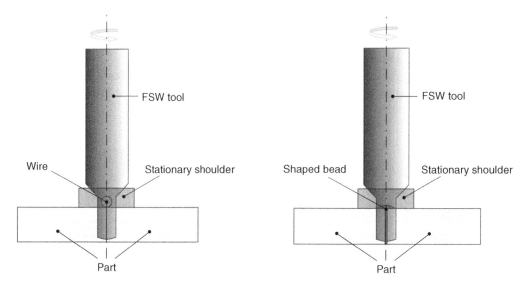

Figure 1.6 Stationary shoulder FSW. Source: Martin [8]/with permission of Elsevier.

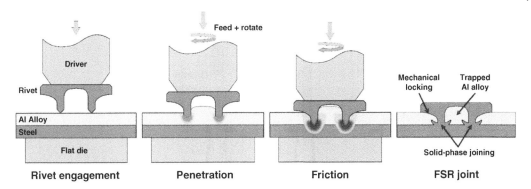

Figure 1.7 Illustration of FSR for two dissimilar metals: (a) holding of thin sheets, (b) plunging of rivet, (c) generation of heat due to friction, (d) retraction of tool, eventually the above surface rivet is trimmed off from the material surface. Source: Shan et al. [9]/with permission of Elsevier.

Enhanced fatigue and static strength are provided by the FSR in comparison to the traditional spot welding [9]. Figure 1.7 illustrates FSR of two dissimilar metals.

Depending upon the tool type used in FSR, it can be classified into various techniques viz. FricRiveting, friction stir blind riveting, friction self-piercing riveting, rotation friction drilling riveting, and rotation friction pressing riveting.

1.4.3.1 FricRiveting

Friction riveting (FricRiveting) is a solid-state joining process that creates a bond between two or more metal components without melting the material usually joining the polymer–metal hybrid structures [10]. The schematic view of FricRiveting is shown in Figure 1.8. The mechanical interference and adhesion between a metallic rivet and polymeric connecting partners helps the structures to be joined. Friction welding and mechanical anchoring are the basic principles of the procedure. The rotating metallic rivet provides the joining energy in the form of frictional heat [10]. The process involves a rotating tool that applies axial pressure to a cylindrical rivet, resulting in plastic deformation and flowability of the material around the rivet shank to develop a strong and stable bond.

1.4.3.2 Friction Stir Blind Riveting

Friction stir blind riveting (FSBR) is a solid-state joining process that creates a bond between two or more metal sheets without melting the material. It is specifically designed for joining materials that cannot be accessed from both sides, such as blind joints. The process uses a tool with a nonrotating probe and a rotating shoulder that penetrates and stirs the material while applying pressure, creating a durable bond around the rivet shank (Figure 1.9). FSBR offers several benefits over conventional blind riveting methods, such as improved joint strength and suitability for dissimilar

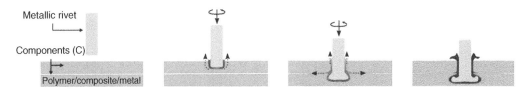

Figure 1.8 Schematic view of FricRiveting process. Source: Borba et al. [11]/with permission of Elsevier.

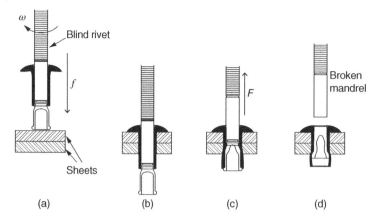

(a)   (b)   (c)   (d)

Figure 1.9 Friction stir blind riveting. Source: Min et al. [12]/with permission of Elsevier.

materials. However, FSBR requires specialized equipment and is still a new process. Despite this, it has promising applications in various industries where blind joints are common.

Some other variations of FSR which result by modifying the tool geometry or tool type are friction self-piercing riveting, rotation friction drilling riveting, and rotation friction pressing riveting.

1.4.4 Friction Stir Scribe

Friction stir scribe process is an advanced and unique style of joining dissimilar materials. It develops exceptionally strong bond and is faster and less costly than other processes. In this technique, two dissimilar materials with varying melting points held one over the other are welded. The scribe is attached to the tool, which is about the size of spark plug, and is rotated and plunged onto the stacking materials (Figure 1.10). The scribe cuts the bottom sheet while the pin deforms the top sheet all along the joint line. The cutting scribe forms the trough in the bottom material and leaves two hooks which interlocks with the top material plate on either side.

1.4.5 Friction Surfacing

Friction surfacing is a solid-state joining process used to create metallurgical bonds between two or more metallic components. It is a form of friction welding where a consumable rod or wire is

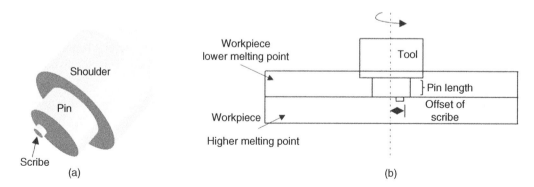

(a)   (b)

Figure 1.10 Friction stir scribe: (a) Tool, (b) process.

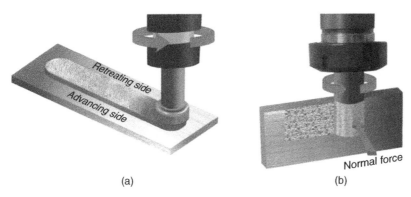

(a)　　　　　　　　　　　　　　　　　　(b)

Figure 1.11 Friction surfacing: (a) the end of the consumable tool, (b) the radial surface of the consumable tool. Source: Seidi et al. [13]/with permission from Elsevier.

rotated against the surface of a BM, generating frictional heat. This heat softens the BM and the consumable, forming a plasticized region. As the consumable material is continuously fed into the contact zone, it undergoes SPD and is gradually bonded to the BM. The frictional heat and the pressure applied during the process cause intermixing of the materials, promoting metallurgical bonding and diffusion at the interface.

The friction surfacing technique is commonly used for repair, reclamation, and surface modification applications. It can be employed to enhance the wear resistance, corrosion resistance, or other properties of the BM by depositing a suitable consumable material. The consumable can be selected to have specific properties, such as higher hardness or improved chemical resistance, depending on the desired outcome. Figure 1.11 illustrates the working of friction surfacing.

1.4.6 Friction Stir Processing

Friction stir processing is regarded as a FSW derivative. It can be coined as a generic surface modification process. In 1999, the word FSP was used as keyword for the first time in the work of Mishra et al. [14]. FSP works on a similar principle as of FSW. In its simplest form, a specially designed rotating tool is inserted into a substrate BM plate secured in a specially designed fixture and then traversed in the desired direction, as shown in Figure 1.12.

The specifically designed tool, often composed of a hard material, such as tungsten carbide and tool steel, is fixed in the spindle of the machine utilized for FSW which rotates around its own axis while being traversed along the region to be processed. Although the working principle for FSW and FSP is the same, yet both are used for different purposes in industrial applications. Primary aim of FSP is to modify properties at the surface of workpiece while FSW aims at joining of two plates/jobs together. Need behind the invention of FSP can be justified as, in many industrial applications, the service life of components is dictated by surface properties, such as hardness, corrosion resistance, and tribological properties. FSP is proving to be very useful in improving the structure of cast surfaces, impregnating the BM surface with exotic reinforcement materials such as piezoelectric materials, carbon nanotubes (CNTs), and the like. These reinforcement materials induce extraordinary characteristics on the friction stir-processed surfaces. Of late, FSP is also being used to fabricate the functionally structured surfaces by embedding such extraordinary materials as reinforcement. In the times to come, we may expect tremendous growth in applications of such materials which find FSP more efficient to fabricate, resulting in a parallel growth in the FSP process as well.

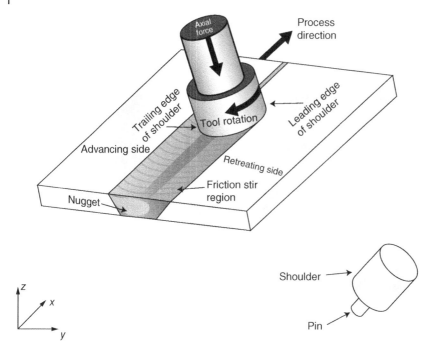

Figure 1.12 Schematic arrangement of FSP. Source: Rathee et al. [15]/Taylor & Francis.

Also, derivatives of the FSW/P process such as friction stir channeling, friction stir alloying, friction stir brazing, friction stir additive manufacturing, and additive friction stir processes are gaining popularity in developing functional materials from a wide variety of raw materials.

1.5 Defects

FSW is a solid-state welding process that produces high-quality, defect-free welds in many cases. However, like any welding process, certain defects occur during FSW. Some of the common defects associated with FSW include:

Incomplete or lack of penetration: This defect occurs when the tool fails to adequately penetrate the entire thickness of the workpiece. It can result in weak or incomplete welds, compromising the joint strength.

Wormhole defects: These defects are characterized by a void or cavity within the weld zone. They are typically caused by the entrapment of gas or foreign material during the welding process.

Flash or excess material: Flash refers to the excess material that gets extruded out from the weld joint during the process. It can occur when the welding parameters are not properly controlled especially tool plunge depth, leading to excessive heat or pressure. Flash can create an uneven surface and may require post-weld machining or trimming.

Defects at the weld start or end: FSW can sometimes exhibit defects at the start or end of the weld seam. These defects can be caused by issues with tool alignment, insufficient material flow, or improper welding technique. They can result in weak or incomplete welds at the joint edges.

Lack of fusion or incomplete bonding: In certain cases, FSW may result in incomplete bonding between the workpiece and the weld material. This defect can occur if the tool does not adequately stir and mix the material or if the welding parameters are not optimized. Lack of fusion can lead to weak joints and reduced mechanical properties.

Residual stress and distortion: FSW can induce residual stresses and distortion in the welded components. These effects are primarily influenced by the welding parameters, material properties, and design considerations. Residual stresses and distortion can affect the dimensional accuracy, stability, and long-term performance of the welded structure.

1.6 Advantages and Limitations of FSW

FSW offers several advantages over traditional fusion welding techniques, some of them are discussed below:

1.6.1 Advantages

Strong and high-quality welds: FSW produces strong, defect-free welds with excellent mechanical properties. The process works on solid-state principles, resulting in reduced susceptibility to defects such as porosity, solidification cracking, or lack of fusion.

Joining of dissimilar materials: FSW enables the joining of dissimilar materials that are difficult to weld using conventional fusion welding methods. It can join materials with significantly different melting points, such as aluminum and steel, while maintaining the properties of both materials.

No fumes or arc radiation: FSW is a solid-state process that does not involve the use of filler metals, shielding gases, or an arc. As a result, it does not produce fumes, spatter, or arc radiation, making it a safer and environmentally friendly welding technique.

In addition to the above-discussed advantages, a list of established FSW strengths is provided as below:

- Energy efficient;
- Ease of operation on retrofitted milling machines amounting to lower costs;
- Can be completely automated and operated in both horizontal and vertical positions;
- Reduced expensive post-process machining owing to improved workpiece aesthetics after welding;
- And so on.

1.6.2 Limitations

Despite several advantages of FSW, it suffers from limitations also, some of them are discussed below:

Limited joint access: The FSW process requires access to only one side of the workpieces, making it challenging to weld components with restricted access or complex geometries. Joint designs and fixturing may need to be modified to accommodate FSW.

Limited joint configurations: FSW is best suited for butt and lap joint configurations. It may not be as effective for other joint types, such as T-joints or corner joints. Specialized techniques or modifications may be required for welding complex joint geometries.

Limited to solid materials: FSW is specifically designed for joining solid-state materials, and it may not be applicable for welding materials such as polymers or nonmetallic composites.

Despite these limitations, FSW has proven to be a valuable joining technique for a wide range of applications in various industries, including aerospace, automotive, and shipbuilding, due to its many advantages and its ability to produce high-quality, reliable welds.

1.7 Conclusion and Future Prospectus

This chapter has provided a comprehensive overview of FSW and its various aspects. We discussed the principle of FSW, the process parameters, tool design and working in different variants of FSW, some common defects that are encountered during FSW, and the advantages and limitations of this welding technique. Through the examination of research studies and practical applications, it is evident that FSW offers numerous benefits, such as improved joint strength, reduced defects, and enhanced material properties.

In the area of FSW, there are several promising avenues for research and development. FSW can be applied to new and challenging materials, such as high-temperature alloys, composites, and dissimilar material combinations, to extend its industrial applications. Enhancing the weld quality, productivity, and cost-effectiveness of FSW processes involves optimizing rotational speed, traverse speed, and tool design. Process control, defect detection, and quality assurance can be improved through the development of real-time monitoring techniques, such as temperature, force, and acoustic emission sensors. The increasing concern regarding sustainability makes it necessary for future research to investigate how FSW affects the environment, including its energy consumption, its material use, and its recyclability. Research directions that are important to the future of FSW include exploring and optimizing alternative heat sources, developing eco-friendly variants of FSW, and designing processes for reduced environmental impact.

Acknowledgments

The authors, Dr Manu Srivastava and Dr Sandeep Rathee thank the Science & Engineering Research Board (SERB) for its financial assistance under the project (vide sanction order no. SPG/2021/003383) to perform this work.

References

1 El-Sayed, M.M., Shash, A.Y., Abd-Rabou, M., and ElSherbiny, M.G. (2021). *Welding and processing of metallic materials by using friction stir technique: a review. Journal of Advanced Joining Processes* **3**: 100059.

2 Elatharasan, G. and Kumar, V.S.S. (2013). *An experimental analysis and optimization of process parameter on friction stir welding of AA 6061-T6 aluminum alloy using RSM. Procedia Engineering* **64**: 1227–1234.

3 Prabhakar, D.A.P., Shettigar, A.K., Herbert, M.A. et al. (2022). *A comprehensive review of friction stir techniques in structural materials and alloys: challenges and trends. Journal of Materials Research and Technology* **20**: 3025–3060.

4 Mishra, D., Roy, R B., Dutta, S., Pal, S. K., Chakravarty, D. (2018). *A review on sensor based monitoring and control of friction stir welding process and a roadmap to Industry 4.0. Journal of Manufacturing Processes* **36**: 373–397.

5 Singh, V.P., Patel, S.K., and Kuriachen, B. (2021). *Mechanical and microstructural properties evolutions of various alloys welded through cooling assisted friction-stir welding: a review. Intermetallics* **133**: 107122.

6 Russell, M.J. (2006). Recent developments in friction stir welding of ti alloys. *Proc. 6th Int. Symp. on FSW*, St. Sauveur, Canada, 2006.

7 Li, D., Yang, X., Cui, L., He, F., Zhang, X. (2015). *Investigation of stationary shoulder friction stir welding of aluminum alloy 7075-T651. Journal of Materials Processing Technology* **222**.

8 Martin, J.P. (2013). *Stationary shoulder friction stir welding.* In: *Proceedings of the 1st International Joint Symposium on Joining and Welding* (ed. H. Fujii), 477–482. Woodhead Publishing.

9 Shan, H., Yunwu, M., Sizhe, N., Bingxin, Y., Ming, L., Yongbing, L., Zhongqin, L. (2021). *Friction stir riveting (FSR) of AA6061-T6 aluminum alloy and DP600 steel. Journal of Materials Processing Technology* **295**: 117156.

10 Blaga, L., dos Santos, J.F., Bancila, R., and Amancio-Filho, S.T. (2015). *Friction riveting (FricRiveting) as a new joining technique in GFRP lightweight bridge construction. Construction and Building Materials* **80**: 167–179.

11 Borba, N.Z., Blaga, L., dos Santos, J.F., and Amancio-Filho, S.T. (2018). *Direct-friction riveting of polymer composite laminates for aircraft applications. Materials Letters* **215**: 31–34.

12 Junying, M., Jingjing, L., Yongqiang, L., Carlson, B.E., Jianping, L., Wang, W.M. (2015). *Friction stir blind riveting for aluminum alloy sheets. Journal of Materials Processing Technology* **215**: 20–29.

13 Seidi, E. and Miller, S.F. (2021). *Lateral friction surfacing: experimental and metallurgical analysis of different aluminum alloy depositions. Journal of Materials Research and Technology* **15**: 5948–5967.

14 Mishra, R.S., Mahoney, M.W., McFadden, S.X., Mara, N.A., Mukherjee, A.K. (1999). *High strain rate superplasticity in a friction stir processed 7075 Al alloy. Scripta Materialia* **42** (2): 163–168.

15 Rathee, S., Maheshwari, S., Siddiquee, A.N., Srivastava, M. (2018). *Issues and strategies in composite fabrication via friction stir processing: a review. Materials and Manufacturing Processes* **33** (3): 239–261.

2

Friction Stir Welding and Single-Point Incremental Forming: *State-of-the-Art*

Hitesh Mhatre[1], Amrut Mulay[1], and Vijay Gadakh[2]

[1] *Department of Mechanical Engineering, Sardar Vallabhbhai National Institute of Technology, Surat, Gujarat, India*
[2] *Department of Automation and Robotics Engineering, Amrutvahini College of Engineering, Ahmednagar, India*

2.1 Introduction

Solid-state joining and processing techniques are increasingly being employed by industries to link softer metals that are hard to join using regular fusion welding operations. Friction stir welding (FSW) is beneficial since it does not require filler, resulting in a large weight decrease. FSW has successfully connected high-strength aluminum alloys, as well as other metallic alloys, utilized in the automotive and aerospace sectors [1]. FSW's remarkable achievements have led to the refinement, enhancement, and modernization of the friction stir concept, resulting in several innovative techniques for joining and processing materials. These advances are gradually allowing the transfer of technological competence to higher strength and stiffness materials and cutting-edge applications [2].

This chapter's goal is to provide an introduction to FSW and the single-point incremental shaping procedure on friction stir welded blanks. The two fundamental processes in friction stir are joining and processing. While processing is intended to enhance material properties, welding is often used to combine different materials [3]. FSW involves the rotation of the tool in contact with the workpiece and uses a nonconsumable tool. This circular motion generates the required heat for the process through the frictional interaction between the tool and the workpiece.

A "third body region" is one that is physically separate from the workpiece and the tool. This region is formed either on the tool (in cases where the tool is expendable) or on the workpiece (in cases where the tool is nonexpendable). As shown in Figure 2.1, the third region of the material, despite its solidity, exhibits a fluidity that extends into three dimensions, enabling it to combine and merge with other materials at the interface. This zone, which is located between the material's melting point and crystallization temperature, develops as a result of frictional heat produced at the operating surfaces. It has a low flow stress and a high viscosity, and it resembles the deformed or plasticized material used in friction stir operations [4, 5].

Friction stir techniques often involve the appearance of a third body region that is not observed in fusion technologies, as frictional heat is not generated near the material's melting point. The third body region facilitates material intermixing and interatomic diffusion at elevated temperatures, creating a strong bond between similar or dissimilar materials. Yet, at greater pressures or lower temperatures, interatomic bonding may become the primary mechanism for material joining or processing.

Friction Stir Welding and Processing: Fundamentals to Advancements, First Edition.
Edited by Sandeep Rathee, Manu Srivastava, and J. Paulo Davim.
© 2024 John Wiley & Sons, Inc. Published 2024 by John Wiley & Sons, Inc.

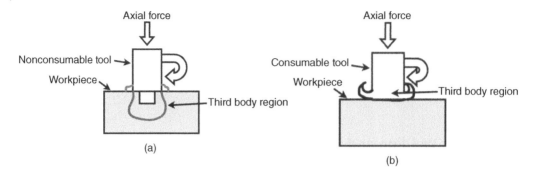

Figure 2.1 Third body area in friction stir welding using (a) a nonconsumable tool and (b) a consumable tool.

The fundamental goal of friction stir methods is to generate enough heat through friction to form a third body area in the material, allowing it to be joined or processed. The tool is important to the process, creating the required heat for the creation of the third body area and enclosing it when it is nonconsumable. As a result, the shape and substance of the tool are critical to the performance of friction stir procedures. A procedure must be efficient, successful, and economically viable in order to be considered sustainable.

2.2 Friction Stir Welding (FSW)

Despite the fact that the concept of friction stir has been employed in the past, FSW has significantly improved the success of combining low melting point materials. The FSW approach employs both pressure- and displacement-controlled devices. The nonconsumable tool used in FSW is made up of a shoulder and a pin, with the particular design and material of the tool dictated by variables including the size and joint configuration of the materials being welded, as well as other important criteria. Researchers have extensively recorded the many FSW tool types, including their designs, material selections, strengths, failure mechanisms, and failure preventive techniques [6].

2.2.1 Friction Stir Welding Process Features

The FSW method comprises of four main stages: plunging, dwelling, welding, and retracting/cooling, which are illustrated in Figure 2.2 along with various steps. During the plunging phase, the tool is rotated at a constant rotational rate and pushed downward with a vertical axial force to initiate contact with the joint line. The deformation process begins as soon as the tool's shoulder makes contact with the workpiece's face. The spinning tool is subjected to an axial force for 5–10 seconds when in contact with the workpiece surface during the dwell time. The material and thickness of the workpiece affect how long this stage takes. The material becomes deformable because to the fixed position's considerable friction heat generation at the tool–workpiece contact. This plastic deformation produces more heat, which contributes to the total heating process. During the welding step, the plastic substance that surrounds the tool promotes smooth movement.

During the welding process, a rotating tool applies an axial load and surrounds the plasticized volume with its shoulder. The tool travels with a constant traversing force and speed along the specified joint line. Friction and deformation generated by the tool's constant rotation generate

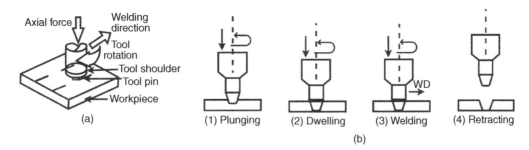

Figure 2.2 The FSW process (a) as a schematic and (b) at various phases.

heat, and the rotation also induces material stirring and flow. The plastic covering the tool pin flows from the side that the tool is moving towards the pin's rear as it does so. The material behind the pin is subsequently filled by the shoulder's forging action, which closes the hole left by the pin's forward motion. Depending on the temperature and pressure settings, the process can lead to either atomic diffusion or bonding, causing material mixing and forming a junction behind the tool. Once the welding is done, the tool is withdrawn and cooled, exit hole is created. Unwanted exit holes in the weld can be prevented by either refilling the hole or using a longer weld length than was originally anticipated. The tool has three main uses: to confine plastic material under the shoulder, to control material movement, and to generate heat through friction or deformation. When welding thin sheets, the shoulder generates the most of the heat; however, when welding larger materials, the pin generates the majority of the heat. The shoulder and pin surfaces might be given specialized characteristics to improve heat production and material flow.

2.2.2 FSW Process Parameters

FSW is defined as the combination of extrusion, forging, and stirring of the material, resulting in high strain rates and temperatures. The procedure involves complicated material movement and significant plastic deformation. Factors, such as tool geometry, weld variables, and joint design, have a significant impact on the temperature distribution and material flow pattern, which directly affect the microstructural changes. To pinpoint the crucial procedural elements affecting the standard of friction stir welds (FSWs), a cause–effect diagram, shown in Figure 2.3, was created. The following categories apply to these factors:

(1) The factors that go into tool design considerations include things like shoulder geometry, shape, and pin length in addition to tilt angle, pin geometry, plunge depth, and shoulder-to-pin diameter ratio (D/d).
(2) Factors including clamp force, plate thickness fluctuation, clamp form, and thickness mismatch are taken into account while designing a clamp.
(3) Machine-based variables include, among others, weld speed, weld gap (the separation between the upper and lower portions of the shoulder), tool rotating speed, and others.
(4) Workpiece parameters include things like material, composition, mechanical and physical characteristics, size, and form. Understanding and managing these variables can improve the quality of FSWs, assuring intended results and better process performance.

Weld speed, tool rotation speed, vertical force on the tool, tool tilt angle, and tool design are the main independent process factors used to regulate the FSW process. These factors directly affect the heat generation rate, temperature field, cooling rate, x-direction force, torque, and power of the

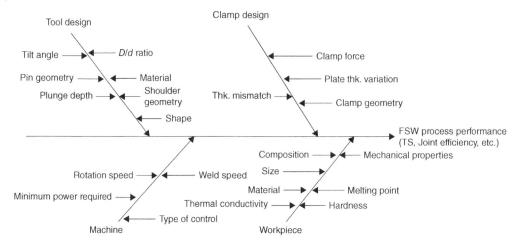

Figure 2.3 Cause–effect diagram.

FSW process. In general, increasing axial pressure and tool rotating speed cause peak temperatures to rise, although weld speed somewhat lowers peak temperatures. In the FSW process, the tool shape has a significant impact on the material flow, temperature history, grain size, and mechanical characteristics.

Reducing the time, the tool and workpiece are in contact at a specific point is a result of a higher weld speed, which ultimately leads to a decrease in heat input. An increase in rotating speed results in longer contact times. An optimal heat input amount is required to create plastic deformation, stirring, and material flow, which are crucial factors in influencing the formation and robustness of the welded connections. Therefore, maintaining control over these factors is essential for an effective FSW strategy.

2.2.3 Feasibility and Application of FSW

FSW is a room-temperature working technique that does away with fusion welding's drawbacks, such as the development of weld defects during the solidification of liquid metal. Al-alloy welded joints for commercial usage have been successfully produced on an industrial scale using FSW. As long as the tool material, tool design, and process variables match the essential parameters, FSW is a viable method for commercial joining of magnesium alloys as well as other high strength and high melting point alloys, such as Fe, Ti, Ni, and Cu-based alloys. Current developments are hastening the adoption of FSW for these alloys. Additionally, FSW is being researched as a technique for successfully mixing various materials, such as wood–plastic composites, metal matrix composites, ceramics, and polymers. [7]

2.3 Single-Point Incremental Forming (SPIF)

Jeswiet et al. defined incremental sheet forming (ISF) as a manufacturing method with various distinguishing characteristics. In contrast to previous procedures, ISF uses a small forming tool rather than big, specialized dies. The forming tool maintains continual contact with the sheet metal and may be moved in three dimensions. This method allows for the fabrication of both symmetric and asymmetric structures in sheet metal [8].

In ISF, deformation occurs gradually by applying pressure to the sheet's surface. To prevent displacement and material flow into the forming area, a blank holder is typically used to restrain the metal sheet. As a result, the entire forming process relies on thinning of the blank to achieve the desired shape.

The following process variables are associated with incremental sheet formation.

A component can be made using the single-point incremental forming (SPIF) process, which uses a CNC forming tool with just one point of contact with the sheet metal. A constant-height blank holder firmly holds the blank in place (as depicted in Figure 2.4). In some cases, this process can be entirely dieless, without the need for any additional supporting tools. To produce a distinct transition between the flange and the component, a specialized rig or backplate is usually used. Figure 4 depicts both variants of the SPIF procedure. It is worth noting that when a backplate is employed, the single-point contact characteristic is absent. Nevertheless, the acronym SPIF is widely recognized and commonly used to refer to the process variations depicted in Figure 2.4.

In order to expand the range of possible geometries and improve process accuracy, two-point incremental forming (TPIF), a forming process improvement, may require the use of a full or partial positive die (as illustrated in Figure 2.5). A complete support system becomes particularly crucial for components with distinct changes in curvature. It is critical to ensure that the parts of the component that have already developed stay undisturbed during the future stages. This may be accomplished by using a support tool. TPIF refers to process changes including the use of a positive die.

The sheet is formed in TPIF by a CNC forming tool while being held in a blank holder. In contrast to SPIF, which includes the sheet gradually approaching the final shape from the inside, TPIF involves the sheet gradually reaching the final form from the outside, assisted by the lowering action of the blank holder over the die. Unlike dieless SPIF, TPIF relies on the presence of tooling, although the tooling materials can be less expensive compared to stamping, providing a cost advantage. In some circumstances, a partial die (as seen in Figure 2.5) can be used to manufacture variations of a component design while maintaining the same setup by just changing the tool path.

The innovative method used by kinematic incremental sheet forming (KISF) involves simultaneously activating two forming tools that are placed on the opposite sides of the blank (as

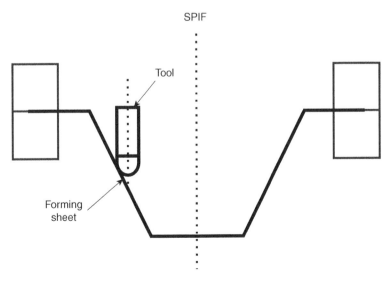

Figure 2.4 Single-point incremental sheet forming (SPIF).

TPIF

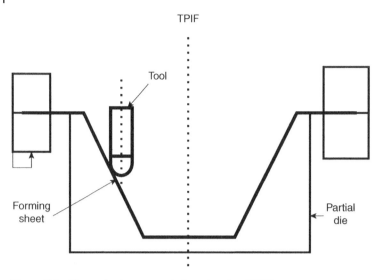

Figure 2.5 Two-point incremental sheet forming (TPIF).

KISF

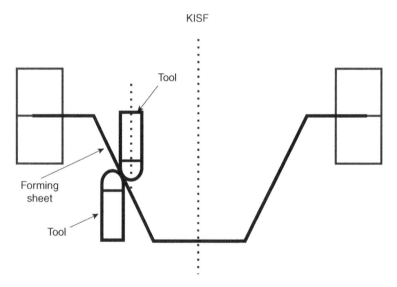

Figure 2.6 Kinematic incremental sheet forming using two forming tools.

illustrated in Figure 2.6). The nature of tool movement restricts the thinning of the sheet and fracture. The springback and geometrical accuracy are found to be better. This process type uses dieless sheet metal forming, which is more adaptable than SPIF. It is also known as roboforming or entirely kinematic ISF. In order to enable the slave tool to serve as a support tool during the forming process, the tool paths for the master and slave tools must be precisely defined and synchronized. The second major challenge is maintaining control over the dimensional accuracy of the parts, particularly when using two synchronized robots [9].

2.4   FSW and SPIF

Material variations necessitate the use of technologies, such as formability reduction and weld line movement, when producing friction stir welded blanks. As these blanks are welded as per the requirement like dissimilar material welded blanks, different thickness blanks is termed as tailor welded blanks (TWBs). Unequal stresses and strains during traditional forming processes might induce weld line displacement toward the stronger material. This is due to the fact that TWBs are made up of incompatible materials and welds, resulting in a heterogeneous blank in which the thinner or weaker material deforms preferentially and tears early. Aluminum TWBs have worse formability than its participating materials, according to studies. Also, increasing the thickness ratio of TWB materials might reduce formability. Unbalanced forces produced during traditional forming techniques, such as deep drawing, might lead to formability loss and weld line displacement. Despite the challenges, progress has been made in metal forming technologies. One such advancement is SPIF, a technique that involves localized deformation of a clamped blank. A diagram of the SPIF process is shown in Figure 2.7.

During the SPIF process, a tool is used to create deformation, whereas forming is performed by stretching and bending together [6]. Additionally, as compared to other traditional forming processes [10], the SPIF process provides greater formability [11] for homogenous blanks. The SPIF approach was utilized to analyze the formation behavior of TWBs due to this advantage. Although the combination of SPIF and TWB technologies has promise, further study is still needed.

To meet increasingly stringent fuel efficiency rules, the car industry has achieved important technical advances, one of which is the development of TWBs. These blanks are created by welding numerous sheets of metal with varied grades, gauges, or surface qualities (such as coated or uncoated) together, generally using laser or FSW processes. TWBs have the advantage of putting the right materials in the right places depending on functional or design requirements, resulting in weight savings, optimum material use, cost effectiveness, and other benefits. On the other hand, TWB formation may be difficult owing to the presence of several materials with different thicknesses or properties on each side of the weld.

Many studies on the evolution of formability in TWBs under various deformation modes have been done. Panda et al. [12] found that nonuniform deformation during the stretch-forming procedure resulted in failure on the weaker HSLA side, as well as loss in formability of DP980-HSLA

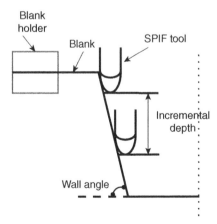

Figure 2.7   Schematic depiction of the SPIF process.

TWBs. As per the findings of Bandyopadhyay et al. [13], the formability of DP980-IFHS TWBs (Dual-Phase 980 – Intercritical Ferrite and Martensite Hot-Stamped Tailored Welded Blanks) gradually diminished during the deep drawing process. The failure was observed to transpire at the cup bottom in the weaker material and aligned parallel to the weld. Along with the reduction in formability, a common issue associated with the forming of TWBs is weld line displacement due to uneven deformation. In the context of deep drawing procedures, the displacement of the weld line was examined in DP980-IFHS (Dual-Phase 980 – Intercritical Ferrite and Martensite Hot-Stamped) and DP980-DP600 laser welded blanks. It was observed that, during the deep drawing process, the weld line exhibited movement toward the stronger material at the cup bottom. Conversely, in regions experiencing higher compressive stress, such as the flange or cup wall, the weld line migrated toward the weaker material. Several techniques have been proposed to enhance formability and regulate the movement of the weld line, including the split binder method [14], clamping pin technique [15], and optimal initial weld line placement [16].

The bulk of research into the formability of TWBs has focused on traditional sheet metal forming, with only a limited amount of literature available on TWB SPIF. In a preliminary investigation of SPIF of TWBs, Silva et al. [11] reported encouraging results. TWBs were produced by joining sheets of AA1050-H111 with varying thicknesses, which were subsequently subjected to the SPIF process through experimental benchmark testing. In a similar way, Tayebi et al. [17] conducted SPIF on TWBs joined through the FSW process, using AA6061 and AA5083. The study concluded that selecting appropriate welding and SPIF parameters can prevent the formability of TWBs from diminishing during the SPIF process.

2.5 Summary and Outlook

While the friction stir concept has been around for a while, it only gained widespread popularity with the introduction of FSW. This technique has proven highly effective in welding Al-alloys and is increasingly becoming the preferred method for joining other types of alloys as well. By modifying various aspects of the process, such as the machine, tool, workpiece, and tool and workpiece positions, several unique ways based on the friction stir concept have been developed. This shows a strong knowledge of the underlying physics and related mechanics. Friction stir joining and processing typically resulted in a refined and recrystallized region with superior mechanical and microstructural properties. Numerous ferrous and nonferrous materials have been subjected to the idea, and with just minor adjustments, a variety of materials may be generated for various uses. While several of these processes are still in the early phases of development, research indicates that, in terms of practicality and product quality, especially for lightweight metals, they are at least as good as, if not better than, their fusion-based equivalents.

Despite the advancements in processes utilizing friction stir equipment, certain challenges still require attention. These problems include the flexibility and affordability of such processes, as well as the requirement for modeling advances to enhance process monitoring and control and create higher quality products at a cheaper cost. Additionally, the immobility of the equipment and resulting tool wear due to the significant loads required for tool stability are issues that must be addressed. Consequently, efforts should be made to minimize the equipment's size for easier transportation and on-site use, ultimately realizing the advantages of lightweight design.

The combination of FSW with SPIF TWBs has demonstrated potential as a viable solution for forming intricate shapes while improving mechanical properties. The technology's future applications include enhanced formability for sheet metal parts, particularly for materials that are typically difficult to shape through traditional means. Additionally, FSW with SPIF TWBs can result in superior dimensional accuracy, surface finish, and reduced springback, all of which can improve the overall quality of parts and minimize scrap rates. The use of FSW with SPIF TWBs can minimize the amount of material used in the process, and the cheap tooling costs of SPIF can result in considerable cost savings for manufacturers. Additionally, the use of FSW with SPIF TWBs may lower waste production, energy use, and carbon emissions, making it a green and sustainable option for the industrial sector. Because FSW with SPIF TWBs enable the construction of complex geometries, it is perfect for the production of distinctive components in industries like aerospace, automotive, and medical.

References

1 Threadgill, P.L., Leonard, A.J., Shercliff, H.R., and Withers, P.J., (2009). International. *Metallurgical Research* 54: 49. https://www.scopus.com/inward/record.uri?eid=2-s2.0-85061234320&partnerID=40&md5=6f04c2bc58f6661b9f1ed4a08bd4acd4.

2 Watanabe, T., Takayama, H., and Yanagisawa, A. (2006). Joining of aluminum alloy to steel by friction stir welding. *Journal of Materials Processing Technology* 178 (1): 342–349. https://doi.org/10.1016/j.jmatprotec.2006.04.117.

3 Arbegast, W.J. (2007). Application of friction stir welding and related technologies. *Friction Stir Welding and Processing* 273–308. https://www.scopus.com/inward/record.uri?eid=2-s2.0-60849123025&partnerID=40&md5=c0d6aa1e4e4f27cd5bddd4d7db66177d.

4 Vilaça, P., Gandra, J., and Vidal, C. (2012). Linear friction based processing technologies for aluminum alloys: surfacing, stir welding and stir channeling. *Aluminium Alloys – New Trends in Fabrication and Applications* 159–197. https://www.scopus.com/inward/record.uri?eid=2-s2.0-84883858195&partnerID=40&md5=16f69f9f1b853996cab1d02837f44594.

5 Thomas, W.M., Staines, D.G., Norris, I.M., and de Frias, R. (2003). Friction stir welding tools and developments. *Welding in the World* 47 (11–12): 10–17. https://doi.org/10.1007/BF03266403.

6 Martins, P.A.F., Bay, N., Skjødt, M. et al. (2008). Theory of single point incremental forming. *CIRP Annals* 57 (1): 247–252. https://doi.org/https://doi.org/10.1016/j.cirp.2008.03.047.

7 Rahbarpour, R., Azdast, T., Rahbarpour, H., and Shishavan, S.M. (2014). Feasibility study of friction stir welding of wood–plastic composites. *Science and Technology of Welding and Joining* 19 (8): 673–681. https://doi.org/10.1179/1362171814Y.0000000233.

8 Jeswiet, J., Micari, F., Hirt, G. et al. (2005). Asymmetric single point incremental forming of sheet metal. *CIRP Annals* 54: https://doi.org/10.1016/s0007-8506(07)60021-3.

9 Hirt, G. and Bambach, M. (2020). Incremental sheet forming. In: *Sheet Metal Forming*, 273–287. ASM International https://doi.org/10.31399/asm.tb.smfpa.t53500273.

10 Mugendiran, V. and Gnanavelbabu, A. (2014). Comparison of FLD and thickness distribution on AA5052 aluminium alloy formed parts by incremental forming process. *Procedia Engineering* 97: 1983–1990. https://doi.org/https://doi.org/10.1016/j.proeng.2014.12.353.

11 Silva, M.B., Skjødt, M., Vilaça, P. et al. (2009). Single point incremental forming of tailored blanks produced by friction stir welding. *Journal of Materials Processing Technology* 209 (2): 811–820. https://doi.org/10.1016/j.jmatprotec.2008.02.057.

12 Panda, S.K., Hernandez, V.H.B., Kuntz, M.L. et al. (2009). Formability analysis of diode-laser-welded tailored blanks of advanced high-strength steel sheets. *Metallurgical and Materials Transactions A* 40 (8): 1955–1967. https://doi.org/10.1007/s11661-009-9875-4.

13 Bandyopadhyay, K., Panda, S.K., Saha, P. et al. (2015). Limiting drawing ratio and deep drawing behavior of dual phase steel tailor welded blanks: FE simulation and experimental validation. *Journal of Materials Processing Technology* 217: 48–64. https://doi.org/10.1016/j.jmatprotec .2014.10.022.

14 Seyam, M.S., Shazly, M., El-Mokadem, A., and Wifi, A.S. (2018). A binding scheme to minimize thinning of formed tailor welded blanks. *The International Journal of Advanced Manufacturing Technology* 96 (9): 3933–3950. https://doi.org/10.1007/s00170-018-1686-6.

15 Kinsey, B., Liu, Z., and Cao, J. (2000). A novel forming technology for tailor-welded blanks. *Journal of Materials Processing Technology* 99 (1–3, 153): 145. https://doi.org/10.1016/ S0924-0136(99)00412-4.

16 Mennecart, T., Güner, A., Khalifa, N.B. et al. (2014). Effects of weld line in deep drawing of tailor welded blanks of high strength steels. *Key Engineering Materials* 611–612: 955–962. https://doi.org/ 10.4028/www.scientific.net/KEM.611-612.955.

17 Tayebi, P., Fazli, A., Asadi, P. et al. (2019). Formability analysis of dissimilar friction stir welded AA 6061 and AA 5083 blanks by SPIF process. *CIRP Journal of Manufacturing Science and Technology* 25: 50–68. https://doi.org/10.1016/j.cirpj.2019.02.002.

3

Friction Stir Brazing and Friction Stir Vibration Brazing

Behrouz Bagheri and Mahmoud Abbasi

Department of Materials and Metallurgical Engineering, Amirkabir University of Technology, Tehran, Iran

3.1 Introduction to FSB

Brazing is known as a joining process used to join various kinds of materials and metals by molten brazing materials. Brazing includes various steps, namely inserting of brazing metal between joining specimens, heating the specimens, melting of brazing metal, capillary action of molten metal between the joining specimens, and molten solidification [1, 2]. After solidification, the joint is formed, and the joining specimens are attached. The joint is achieved without the melting of base samples. Zhang et al. [3] introduced friction stir brazing (FSB) initially for the joining of dissimilar metals, such as aluminum and steel. Joining specimens with filler metal between, are fixed on milling machine. Rotating tool with a cylindrical shape contacts the surfaces of joining specimens. Heat is produced due to friction between the tool and the joining specimens, and the filler metal is melted. As heat produced during FSB is not too much, it can be deduced that filler metals with high melting point temperature cannot be used in FSB.

Mechanical bonding as well as metallurgical bonding are two mechanisms for joining in FSB. Although the role of the preceding mechanism is lower than the former one because temperature is not too much during FSB and diffusion is insignificant. FSB originates from friction stir welding (FSW) which is a solid-state joining process. Rotating tool in contact with joining specimens makes heat due to friction. Temperature during FSW is higher than that during FSB and it is about 0.7–0.8 T_m. In this regard, the joining materials around the joint place are soft and they can be mixed when tool consisting of shoulder and pin rotates and movers [4]. In FSB, due to application of a pin-less tool, occurrence of hook, keyhole, and severe wear of the pin tip by strong parent materials is not observed. At the end of FSB process, a bond with a width up to shoulder diameter develops. Application of a preplaced solder/braze is a method to decrease oxide film formation when molten filler solidifies through a eutectic reaction because oxide film fragments are repelled by eutectic liquid phase [5, 6]. During FSB, the temperature does not increase largely. So, development of residual stresses in the joining specimens is not considerable.

Recently, improving the microstructure of the joint and enhancing the strength of it have been the point of focus of many researchers [7]. The first patent for ultrasonic brazing originated in Germany and was invented in 1939. Ultrasonic brazing, by the 1970s, was applied to join different materials, such as metals, composites, and ceramics [8]. The ultrasonic-assisted brazing process to

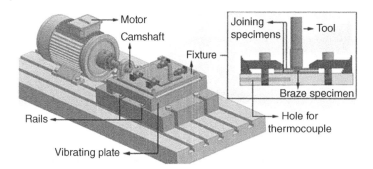

Figure 3.1 Detailed view of the machine used for friction stir vibration brazing.

study the different characteristics of the joint generated between aluminum matrix composites contained Si3Ni4 particles as reinforcement by a Zn–Al material as filler metal was investigated by Xu et al. [9]. It was found that the surface oxide layer between the base material and filler metal broke and a bond consisted of uniform distribution of the Si_3Ni_4 particles as reinforcements was produced. Zhang et al. [10] analyzed the effect of the ultrasonic-assisted brazing process and Al-5 wt%Si brazing filler for the joining of Ti-6Al-4V alloy and ZrO_2 ceramic. It was concluded that ultrasonic shows a significant role in the uniformity of microstructure and lack of formation of Si segregation areas both at the interface of Ti-6Al-4V and center of the joint. In addition, the application of ultrasonic enhanced the shear strength of the joint during the brazing process. Abbasi and Bagheri introduced "friction stir vibration brazing" (FSVB) as a modified version of FSB [11]. During FSVB, the joining materials fixed in a fixture and the fixture is vibrated in a direction normal to joining direction with low frequency and high vibration amplitude. The machine shown in Figure 3.1 was designed and fabricated to apply vibration during FSB. The joining specimens are fixed on the fixture and the fixture is installed on milling machine. A camshaft mechanism was applied to vibrate the fixture. In fact, the motor shaft rotation is transformed to a linear movement of the fixture which is reciprocating. The amplitude of vibration is 0.5 mm and vibration frequency is controlled by a driver connected to motor. A motor of 0.5 kW power was used for experiments.

During FSVB, as the tool touches the top surface of joining specimen, the motor is turned on, and the fixture vibrates, while during FSB, the fixture is immobile. During FSB, the rotating movement of the tool on the joining specimens leads to heat production and filler melting, while in FSVB, the rotating movement of the tool is accompanied by the reciprocating movement of the fixture. This results in higher heat production due to enhanced friction. Higher heat as well as fixture movement enhance the flow of molten filler between the joining specimens and correspondingly, a joint with high strength is developed. More details regarding the application of mechanical vibration during the brazing process will be presented in the following sections.

3.2 Variants of FSB

Despite various benefits of FSB, there are different parameters to obtain a sound joint, including rotational speed, plunge depth, pass numbering, work media, vibration frequency, the type of filler material, shoulder diameter, and so on [12–14]. Due to the formation of intermetallic compound (IMC) and its role in the strength of joints during the brazing of workpieces with filler material, the selection of adequate filler materials is important. Chang et al. [15] applied the Ni metal as an interlayer material during the joining of Al–Mg alloys. It was reported that the microstructure evolution and mechanical characteristics of the joint improved by introducing Ni as an interlayer. Abdollahzadeh et al. [16] analyzed the effect of shoulder diameter during the joining of dissimilar aluminum alloy to copper alloy when the applied Zn metal is an interlayer material. They found that due to plastic deformation and friction, the heat input increases as the shoulder diameter increases from 16 to 20 mm. Additionally, spot pass increasing from 1 to 3 plays a prominent role in the reduction of grain size of weld samples. It has been known that the presence of a pin leads to an aesthetically undesirable keyhole and groove leading to inadequate material flow, subsequently, affecting the micro and macrostructure evolution and mechanical performance of the joint sample [17]. Moreover, the end of the tool pin, particularly for joining materials with high strength, may be damaged. Consequently, a pinless tool has received more attention to obtaining high-strength joints [18]. Nanoparticles with a low coefficient of thermal expansion (CTE) have been used as the filler material to develop the microstructure and mechanical characteristics of the joint. Bagheri et al. [19] used the nanoparticle of SiC between the dissimilar joint of aluminum to copper alloys. It was concluded that the grain size of the joint sample decreased as SiC particles were used while the strength of joint samples improved. It should be mentioned that the quality of the joint sample is basically dependent on the heat input produced due to the frictional/contact conditions. That is, lower heat input leads to inadequate material flow for the brazed area and as a result, weak joining in the interface will be fabricated [20]. High heat input, on the other hand, leads to extra softening of filler metal, and consequently, a weak joining zone will be generated [21]. Regarding the nanoparticles, higher input heat causes super-dissolution conditions in the matrix due to the high frictional heat. The effect of rotational speed and vibration frequency on the quality of joint samples produced by the brazing process was investigated by Bagheri et al. [22]. It was indicated that the hardness and strength as well as the IMC thickness of joint samples decreased as rotational speed improved from 850 to 1150 rpm. In addition, the grain size of joint samples decreases and the heat input increases as vibration is applied during the brazing process.

3.3 Two Case studies

In this section, we analyze two studies of FSB processes with SiO_2 and SiC nanoparticles with the aid of Sn–Pb filler metal.

3.3.1 Joining of Low Carbon Steels by Application of Friction Stir Brazing and Sn–Pb Alloy as Filler

Rizi et al. [23] compared the effect of SiO_2 reinforcing particles and vibration during the FSB process. They brazed low carbon steel with %67Sn–34%Pb as filler metal with/without reinforcing particles and vibration. Figure 3.2 shows macrostructures of the cross-sectional areas of the joints made by FSB and FSVB. FS-brazed specimens show porosities and lack of fusion areas while these regions are not observed in macrostructure of FSV-brazed specimens. Additionally, a coherent and continuous interface is observed for FSV-brazed specimen, while the interface for FS-brazed specimen is not coherent. These observations can be justified based on temperature and pressure changes occurring during FSVB.

Temperature changes for regions around the joint area relating to FSB and FSVB processes are demonstrated in Figure 3.3. Although the maximum temperature for both processes are higher

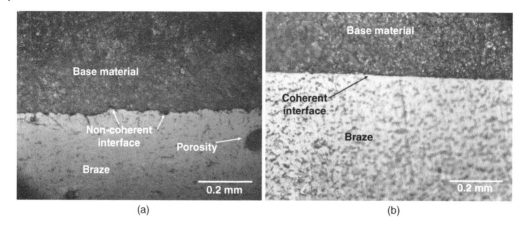

Figure 3.2   A comparison of interface generation during (a) FSB and (b) FSVB [23] Rizi et al., 2022 / Springer Nature.

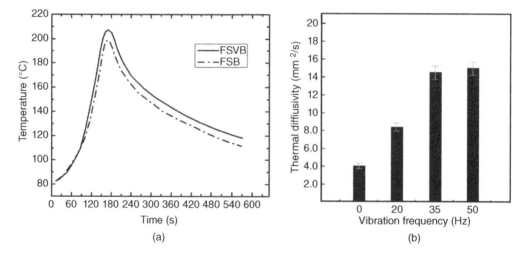

Figure 3.3   (a) Temperature variation of joint samples for FSB and FSVB processes. (b) Thermal diffusivity of the joint as a function of vibration frequency [23].

than the melting point of filler material (187 °C), the temperature for FSVB is higher than that for FSB. In fact, more extensive friction condition during FSVB results in higher heat and temperature. Higher temperature increases the mobility of molten braze. Furthermore, application of vibration in FSVB can break and destroy the oxide films on joining specimens' surfaces. For explanation, it can be noted that vibration changes the molten braze pressure alternately. When the molten in a small area near the surfaces is pulled off, a negative pressure develops, and a cavity is generated when the negative pressure is large. With the explosion of cavitation bubbles, strong shock waves are produced that break oxide films.

Lack of oxide films as well as higher mobility of molten braze brings the possibility of molten metal to contact the joining specimens directly and wet the joining specimens' surfaces. This results in the enhancement of the diffusion and dissolution processes between the braze metal and joining specimens [24]. Coherent interfacial boundaries make stronger bonds between base materials and braze metal. Therefore, mechanical vibration during the brazing process affected the heat input and affected the reaction time.

With the help of a laser flash method, the thermal property of the brazed samples was analyzed, as shown in Figure 3.3b. The thermal diffusivity for the multi-material structure of the sample has been considered the fundamental factor. According to Figure 3.3b, the highest thermal diffusivity belongs to the joints brazed by the highest vibration frequency, in contrast to the brazed samples by the conventional FSB method with lower amplitude. It may be attributed to the role of the homogeneous IMC formation in heat dissipation and the interface between base materials and brazing metal.

The microstructure of the joint zones for FSB and FSVB methods is presented in Figure 3.4. According to Figure 3.4a, larger grains are detected in the joining area fabricated by conventional FSB. The low size of grains in microstructure of joint made by FSVB can be related to the effect of vibration on solidification of molten filler. Joining specimens' surfaces are the initial sites for formation of nuclei. By introducing the mechanical vibration during the joining process, the stress field arises and the mobility of molten braze increases. Shakes and enhanced mobility of molten braze are two important factors that detach the initial nuclei produced on parent metal surfaces. Detached particles act as new nucleation sites, and consequently, the nucleation sites increase remarkably. In this regard, fine grains improve in the joint area after FSVB [25].

The SEM images of the joint and the EDS mapping images of SiO_2 nanoparticle distribution fabricated by FSVB at various vibration frequencies are shown in Figure 3.5. It is observed that by increasing the vibration frequency, the agglomeration decreases and particles get more homogenous distribution. Particles are small and they are at a very small distance from each other when vibration frequency is 50 Hz. As vibration frequency increases, the friction between tool and the joining specimen increases and higher temperature develops in the brazed zone. Higher mobility and movement of molten braze in different directions (while FSVB is carried out) enhance the homogenous distribution of SiO_2 particles. As a result, the distribution of SiO_2 particles in the brazing area becomes more uniform, and agglomeration of particles decreases as vibration frequency increases.

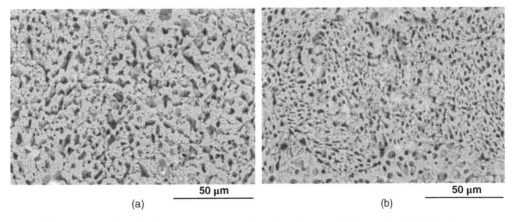

(a) 50 μm   (b) 50 μm

Figure 3.4 A comparison of microstructure evolution of brazing area fabricated by (a) FSB and (b) FSVB [23] Rizi et al., 2022 / Springer Nature.

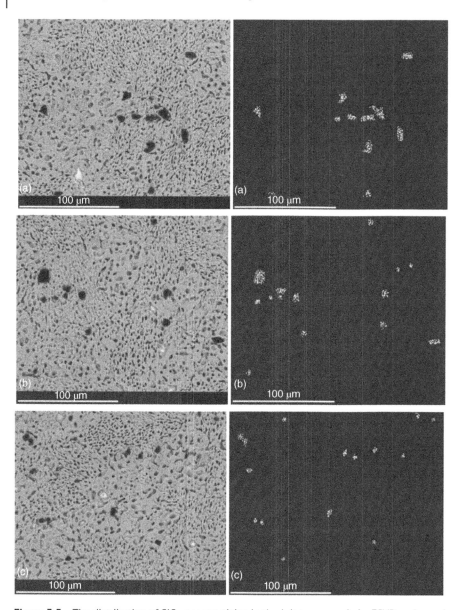

Figure 3.5 The distribution of SiO$_2$ nanoparticles in the joint area made by FSVB under various vibration frequencies. (a, b) 20 Hz, (c, d) 35 Hz, and (e, f) 50 Hz [23] Rizi et al., 2022 / Springer Nature.

Hardness distribution along paths normal to joint line for different joining conditions is observed in Figure 3.6. Based on Figure 3.6, hardness values near the joining specimens are higher than those in the braze metal. Additionally, hardness increases as vibration frequency increases, and hardness enhances as SiO$_2$ particles are added to braze zone. Higher hardness near the joining specimens can be related to partial diffusion of atoms relating to parent metals into the braze zone in these regions which decreases the mobility of dislocations. It is known that strength and hardness increase as dislocation mobility decreases.

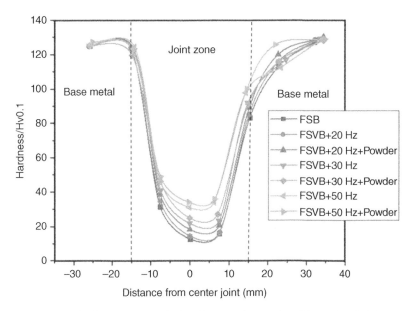

Figure 3.6 Hardness variation for brazing samples produced by various joining conditions [23].

It was observed in Figure 3.5 that agglomeration decreased, and the particles got more enhanced homogenous distribution as vibration frequency increased. According to Orowan and Ashby theory [24], strengthening by addition of strengthening particles (second-phase particles) depends on the size of particles as well as the distance among the particles. For a constant volume fraction of particles, small particles in small distance from each other lead to considerable strength increase. In fact, this distribution of particles prevents the movement of dislocations effectively. According to theory [24], the dislocations in opposition to particles act in two ways: cut the particles and pass through them or make a dislocation ring around the particle and then bypass them. For both conditions, the mobility of dislocation decreases, and more force is necessary to move the dislocations. When particles are small and they are in small distance from each other, the force to cut the particles or bypass the particles is large, and accordingly, the strength and hardness increase are high.

The fracture surface of the tensile samples for FSB and FSVB conditions is presented in Figure 3.7. It is evident that the tensile fracture surface, in the low magnification photograph, is rough and no noticeable necking is observed before fracture. Based on Figure 3.7a,b, there are many pores in the fracture surface, especially near the joining specimens' surfaces. This phenomenon indicates the fact that in the rapid solidification process, the time for gas scape is too short and consequently, the porosity defect is generated in the joint area. In addition, many quasi-cleavage splinter layers relating to brittle fracture, along with dimples relating to ductile fracture are detected. Worth noticing is that the nonequilibrium solidification process improves the brittle Pb–Sn phase in the conventional FSB process. Therefore, the production of the stress concentration phenomenon is intensified in this region.

Based on Figure 3.7c,d, the number of brazing pores is noticeably decreased during the FSVB. To be more specific, because of the smaller grain size, the cracks cannot expand easily, causing the higher mechanical characteristics of the brazing samples. Additionally, in the fracture area, the cleavage step reduces and the dimple number improves slightly, indicating the improvement in the plasticity of the joint area.

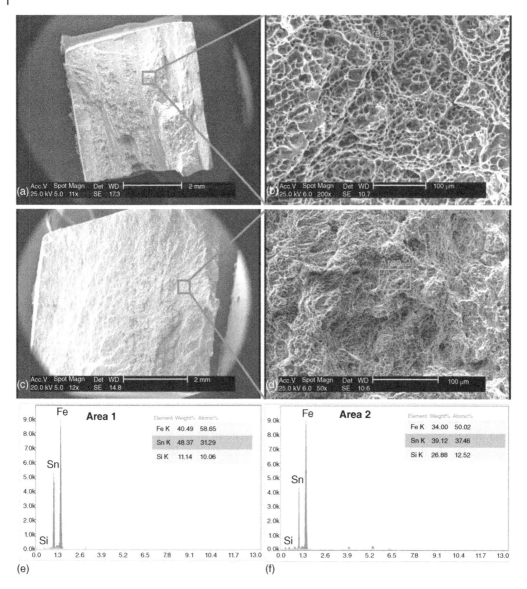

Figure 3.7 Fracture surface analysis of the brazing area under different magnifications. (a, b) FSB, (c, d) FSVB. EDS analysis of the noted areas for various joint methods. (e) FSB, (f) FSVB [23] Rizi et al., 2022 / Springer Nature.

3.3.2 Intermetallic Compound Formation and Mechanical Characteristics of Brazed Samples Made by Friction Stir Vibration Brazing with Sn–Pb Filler Material and SiC Reinforcing Particle

Abbasi et al. [26] used SiC reinforcing nanoparticles for the brazing of steel alloys. They compared the influence of reinforcing particles and mechanical vibration frequency on the formation of intermetallic layers and the mechanical characteristics of brazed samples made under conventional FSB and the FSVB process. According to LOM images presented in Figure 3.8, grains in the braze zone made by FSVB are smaller than grains developed by FSB. This was associated to nucleation procedure during molten braze solidification. During FSVB, turbulency due to vibrations

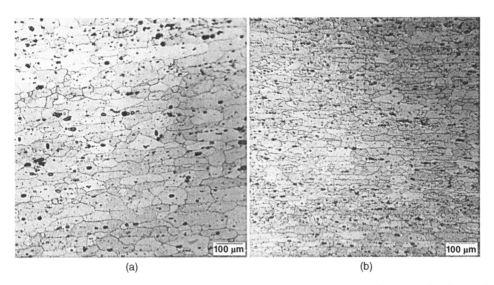

(a) (b)

Figure 3.8 A comparison of microstructure evolution in the brazing area under various brazing methods. (a) FSB and (b) FSVB [26] Abbasi et al., 2021 / Springer Nature.

makes small waves on the surfaces of the joining surfaces. These waves can fragment the aggregated clusters of nuclei. The detached clusters increase the nucleation sites for solidification, and as a result, fine grains develop in the solidified braze zone.

Figure 3.9 shows the SEM image of reinforcing distribution in brazing samples fabricated by FSB and FSVB processes. The distribution of reinforcing particles, based on Figure 3.9, improves as mechanical vibration is applied. Reinforcing particles prevent the growth of grains and decrease the growth of grains of the matrix by pinning the grain boundaries and impeding their movement. This phenomenon has been known as the Zener pinning effect [27]. It should be mentioned that the SiC particles improve the number of dislocations owing to difference in the thermal expansion coefficients between the parent metal matrix and the reinforcing particles. Furthermore, the SiC particles, by introducing a high discrepancy in the elastic behavior between the parent metal

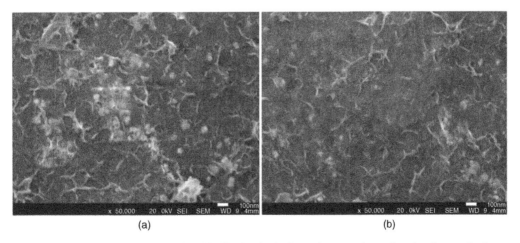

(a) (b)

Figure 3.9 A SEM image of reinforcing distribution in the brazed area under various brazing methods. (a) FSB and (b) FSVB [26] Abbasi et al., 2021 / Springer Nature.

matrix and the particle, hamper the micro-crack propagation. Consequently, higher strength for FSV-brazed samples with SiC particles compounded by smaller grain size and interparticle spacing is predicted. This point will be discussed in the following sentences.

The TEM analysis of reinforcing distribution in the brazing samples under different vibration frequencies during FSVB is shown in Figure 3.10. It is evident that the homogeneity and distribution of SiC particles increase as the mechanical vibration frequency increases. That is, the brazed region undergoes more plastic deformation as the vibration frequency levels up from 30 to 60 Hz, and consequently, the breakdown of reinforcing particles enhances.

Figure 3.11a presents the thickness of the IMC for the top and bottom interface of brazed samples made by FVB with various vibration frequencies. Worth noticing is that temperature and the

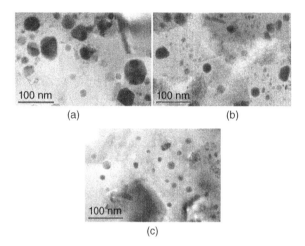

Figure 3.10 TEM image of SiC distribution in the brazed samples under various vibration frequencies. (a) 30 Hz, (b) 45 Hz, and (c) 60 Hz [26] Abbasi et al., 2021 / Springer Nature.

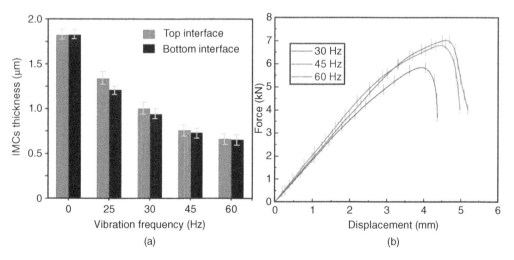

Figure 3.11 (a) Variation in IMC thickness for top and bottom layer of brazed samples and (b) shear strength of brazed samples under various brazing methods [26].

rate of severe plastic deformation have a fundamental influence on the formation of IMCs during the brazing process. According to the results in Figure 3.11a, the IMC thickness decreases from conventional FSB to FSB with the highest vibration frequency. It is shown the fact that the IMC layer breaks into smaller parts and distributes uniformly in the brazed zone as vibration is applied. Since the mechanical vibration is performed under the workpiece, the discrepancy in IMC thickness between the top and bottom layers is reasonable for lower vibration frequencies.

Figure 3.11b presents the result of the shear test for brazed samples made at various vibration frequencies with SiC reinforcing particles. It is observed from the data that the strength and ductility of brazed samples increase as vibration frequency during the brazing process enhances. It may be associated with higher particle distribution or less agglomeration, fine grains, or many grain boundaries, resulting in decreased crack propagation in the brazed zone.

3.4 Application of FSB and Its Variants in Industry

To improve the fuel efficiencies of various vehicles, multi-material structures have attracted much more attention by reducing their weight. Not only the selection of suitable materials for each structure is important, but also finding proper joining technique to make joints without defects is necessary. Quite problematic is to make sound joints between dissimilar metal alloys applying conventional fusion joining techniques due to their prominent dissimilarity in physical/chemical/mechanical characteristics as well as their high tendency to generate IMCs with brittle behavior at elevated temperatures. Consequently, various solid-state joining techniques have been improved to join various parts for various applications in different industries such as aircraft, shipping, automotive, electronics, and so on. FSB, due to low heat input makes it possible to join various similar and dissimilar alloys, such as Al–Cu, Al–Steel, Steel–Cu, and Al–Mg, by avoiding or control of joining defects, hook, distortion, residual stress, and formation of thick IMC.

3.5 Summary and Future Directions

This chapter briefly presented the latest improvements and rapidly growing knowledge about the FSB process, with special emphasis on the development of microstructure and mechanical behavior of brazed specimens. FS-brazed specimens do not contain defects normally observed in joints made by liquid-state welding methods and additionally, the constitution of keyhole which is characteristic of joints made by FSW does not occur in FSB. Recent investigations have presented important insight regarding the understanding of the progress on the relationship among the heat input, stress distribution, strain rate, and their distribution during FSB. The new version of the brazing process, titled FSVB was discussed in detail. The introduction of FSVB, its application, and various effective parameters during the joining of materials was presented. In addition, two case studies of FSVB with different reinforcing particles (SiO_2, SiC) were presented. Although extensive investigations have been done regarding the understanding of FSB over the past 20 years, the modifying and optimization of the process to improve suitable joints with proper specifications and relevant qualities are still demanding. Efforts to standardize the FSB process are also needed.

References

1 Jing, Y., Gao, X., Su, D. et al. (2018). The effects of Zr level in Ti-Zr-Cu-Ni brazing fillers for brazing Ti-6Al-4V. *Journal of Manufacturing Processes* **31**: 24–130.

2 Sharma, A., Lee, SJ., Choi, D. Y., Jung, J. P., Effect of brazing current and speed on the bead characteristics, microstructure and mechanical properties of the arc brazed galvanized steel sheets. *Journal of Materials Processing and Technology*, 2017. **249**: p. 212–220.

3 Zhang, G., Su, W., Zhang, J., and Wei, Z. (2011). Friction stir brazing: a novel process for fabricating Al/steel layered composite and for dissimilar joining of Al to steel. *Metals and Materials Transaction A* **42** (A): 2850–2861.

4 Mishra, R.S. and Ma, Z.Y. (2005). Friction stir welding and processing. *Materials Science Engineering R* **50**: 1–78.

5 Abdollahzadeh, A., Bagheri, B., and Shamsipur, A. (2023). Development of Al5083/Cu/SiC bimetallic nano-composite by friction stir spot welding. *Materials and Manufacturing Processes*. **38** (11): 1416–1425.

6 Vaneghi, A.H., Bagheri, B., Shamsipur, A. et al. (2022). Investigations into the formation of intermetallic compounds during pinless friction stir spot welding of AA2024-Zn-pure copper dissimilar joints. *Welding in the World* **66** (11): 2351–2369.

7 Wei, C., Shuqi, L., Jiuchun, Y., and Xing, Z.H. (2018). *Microstructure and mechanical performance of composite joints of sapphire by ultrasonic-assisted brazing. Journal of Materials Processing and Technology* **257**: 1–6.

8 Vianco, P., Hosking, F., and Rejent, J. (1996). *Ultrasonic soldering for structural and electronic applications. Welding Journal* **75**: 343–355.

9 Xu, G., Leng, X., Jiang, H. et al. (2020). *Microstructure and strength of ultrasonic-assisted brazed joints of Si3N4/6061Al composites. Journal of Manufacturing Processes* **54**: 89–98.

10 Zhang, C.H., Ji, H., Xu, H. et al. (2020). Interfacial microstructure and mechanical properties of ultrasonic assisted brazing joints between Ti–6Al–4V and ZrO$_2$. *Ceramic International* 46: 7733–7740.

11 Abbasi, M. and Bagheri, B. (2021). New attempt to improve friction stir brazing. *Materials Letters* **304**: 130688.

12 Cui, W., Li, SH., Yan, J., He, J., Liu, Y., Ultrasonic-assisted brazing of sapphire with high strength Al–4.5Cu–1.5Mg alloy *Ceramic International*, 2015. **41**: p. 8014–8022

13 Zhang, G., Yang, X., Zhu, D., and Zhang, L. (2020). Cladding thick Al plate onto strong steel substrate using a novel process of multilayer friction stir brazing (ML-FSB). *Materials and Design* **185**: 108232.

14 Yang, Z.W., Zhang, L.X., Chen, Y.C. et al. (2013). Interlayer design to control interfacial microstructure and improve mechanical properties of active brazed Invar/SiO$_2$–BN joint. *Materials Science and Engineering A* **575**: 199–205.

15 Chang, W., Rajesh, S.R., Chun, C., and Kim, H. (2011). Microstructure and mechanical properties of hybrid laser-friction stir welding between AA6061-T6 Al alloy and AZ31 Mg alloy. *Journal of Materials Science and Technology* **27** (3): 199–204.

16 Abdollahzadeh, A., Bagheri, B., Vaneghi, A.H. et al. (2022). Advances in simulation and experimental study on intermetallic formation and thermomechanical evolution of Al-Cu composite with Zn interlayer: effect of spot pass and shoulder diameter during the P-FSSW process. *Proceedings of the Institution of Mechanical Engineers Part L Journal of Materials Design and Applications*. **237** (6): 1–20.

17 Tozaki, Y., Uematsu, Y., and Tokaji, K. (2010). A newly developed tool without probe for friction stir spot welding and its performance. *Journal of Materials Processing and Technology* **210**: 844–851.

18 Bagheri, B., Abdollahzadeh, A., and Shamsipur, A. (2023). A different attempt to analysis friction stir spot welding of AA5083-copper alloys. *Materials Science and Technology* **39** (9): 1083–1089.

19 Bagheri, B., Shamsipur, A., Abdollahzadeh, A., and Mirsalehi, S.E. (2022). Investigation of SiC nanoparticle size and distribution effects on microstructure and mechanical properties of Al/SiC/Cu composite during the FSSW process: experimental and simulation. *Metals and Materials International* **29**: 1095–1112.

20 Gao, P., Zhang, Y., and Mehta, K.P. (2021). Metallurgical and mechanical properties of Al–Cu joint by friction stir spot welding and modified friction stir clinching. *Metals and Materials International* **27**: 3085–3094.

21 Chen, X., Xie, R., Lai, Z.H. et al. (2016). Interfacial structure and formation mechanism of ultrasonic-assisted brazed joint of SiC ceramics with Al–12Si filler metals in air. *Journal of Materials Science and Technology* **33** (5): 492–498.

22 Bagheri, B., Abbasi, M., Sharifi, F., and Abdollahzadeh, A. (2021). A different attempt to improve friction stir brazing: effect of mechanical vibration and rotational speed. *Metals and Materials International* **28** (9): 2239–2251.

23 Rizi, V.S., Abbasi, M., and Bagheri, B. (2022). Investigation on intermetallic compounds formation and effect of reinforcing particles during friction stir vibration brazing. *Journal of Material Engineering and Performance* **31** (4): 3369–3381.

24 Davies, S.H. (2001). *Theory of Solidification*. New York: Cambridge University Press.

25 Wu, K., Yuan, X., Li, T. et al. (2018). Effect of ultrasonic vibration on TIG welding–brazing joining of aluminum alloy to steel. *Journal of Materials Processing and Technology* **266**: 230–238.

26 Abbasi, M., Bagheri, B., Sharifi, F., and Abdollahzadeh, A. (2021). Friction stir vibration brazing (FSVB): an improved version of friction stir brazing. *Welding in the World* **65** (11): 2207–2220.

27 Zhao, Y.X., Wang, M.R., Cao, J. et al. (2015). Brazing TC$_4$ alloy to Si$_3$N$_4$ ceramic using nano-Si$_3$N$_4$ reinforced Ag-Cu composite filler. *Materials and Design* **76**: 40–46.

4

Fundamentals of Friction Stir Processing

Atul Kumar[1], Devasri Fuloria[1], Manu Srivastava[2], and Sandeep Rathee[3]

[1] *School of Mechanical Engineering, Vellore Institute of Technology, Vellore, India*
[2] *Department of Mechanical Engineering, Hybrid Additive Manufacturing Laboratory, PDPM Indian Institute of Information Technology, Design and Manufacturing, Jabalpur, India*
[3] *Department of Mechanical Engineering, National Institute of Technology Srinagar, Jammu & Kashmir, India*

4.1 Friction Stir Processing (FSP): Background

The FSP is an upgraded version of the friction stir welding (FSW) process. It was originally proposed by Mishra et al. [1, 2] and works on the fundamentals of FSW. FSW is a solid-state joining process, which was designed in 1991 at The Welding Institute (TWI) in United Kingdom. It was primarily employed to aluminum alloys [3]. The FSW technique finds several industrial applications, such as in fast-moving train fabrication, ship manufacturing, and in aviation along with the production of butt joints. This solid-state joining method involves mechanical mixing of the specimen under extreme deformation conditions [4–7]. However, in case of FSP process, in order to obtain desired properties in a particular region of a specimen, a revolving tool comprising a shoulder and a probe descends into the workpiece to modify the microstructure by stirring and plastically deforming the workpiece. In comparison to traditional metal processing processes, FSP holds many noteworthy advantages, such as it is a flexible technique with exhaustive performance for the production of materials, which is executed on simple milling machines [8]. FSP can be categorized as a green technology since the heat generated during the process as a result of plastic straining and friction does not lead to the production of poisonous gases and fumes. In FSP, the measurements of the product do not change after the processing; consequently, it is likely to attain the preferred properties by reiterating the process [9]. The additional benefits of the FSP are: it is possible to automate the FSP process, and to control the cooling rate via design of a simple fixture. Hence, FSP has a potential for applications at the commercial level. All these special advantages put FSP above various other SPD methods. Figure 4.1 shows the potentiality of the FSP as a diversified process [10].

4.2 Working Principle of FSP

The process illustration of FSP is represented in Figure 4.2. The FSP is acknowledged as one of the booming methods for processing metals and alloys for material processing and microstructural

Friction Stir Welding and Processing: Fundamentals to Advancements, First Edition.
Edited by Sandeep Rathee, Manu Srivastava, and J. Paulo Davim.
© 2024 John Wiley & Sons, Inc. Published 2024 by John Wiley & Sons, Inc.

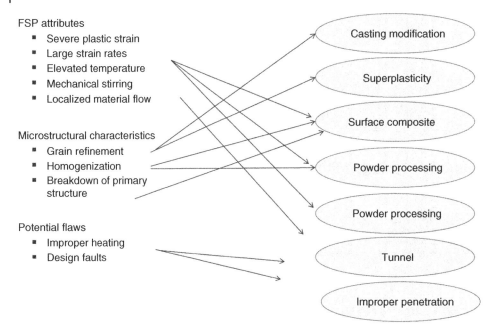

Figure 4.1 List of attributes and the potentiality of the FSP as a diversified process.

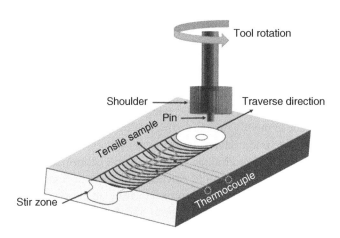

Figure 4.2 The process illustration of FSP.

modification. The working principle of FSP/FSW can be understood as: A specifically tailored revolving tool with a shoulder and a probe is dropped onto the material to be worked followed by traversing along the preferred path. The heat energy evolved as a result of friction between the work material and the tool, and plastic straining of work material is adequate to make the region of the metal surrounding the pin into a soft region. The mushy metal beneath the shoulder is exposed to the collective influence of tool rotation and translation, which leads to the traverse of the metal from front to back of the probe [11]. The combined influence of excessive plastic straining and heat evolution throughout the FSP leads to the microstructural refinement, thereby homogenizing the dissemination of reinforcement particles and removes the innate casting flaws, such as clustering of particles and porosity from the stir cast composites. The microstructural features evolved in the FSPed material considerably enhances the mechanical properties, such as tensile strength, hardness, ductility, and wear resistance [12–14].

The amount of heat evolution and plastic staining is different in different locations of the processed zone, due to which three distinct zones could be observed [11]; they are (i) heat-affected zone (HAZ), (ii) thermo-mechanically affected zone (TMAZ), and (iii) nugget zone. Different areas formed during FSP are shown in Figure 4.3.

The region of intense plastic deformation is called the nugget zone. Maximum heat is generated in this zone. Owing to excessive plastic straining and high temperature as a result of friction throughout the FSP process, the grains of NZ are fully recrystallized [16, 17]. The NZ is usually to pin diameter size or slightly greater than the pin size. Depending upon the variables (processing parameters, geometry of tool, and thermal conductivity of the workpiece), different nugget shapes have been observed. The NZ usually is classified into two types, (i) basin-shaped NZ which is wider at the top than at the bottom and (ii) elliptical NZ that has an onion ring structure (Figures 4.4a,b). Sato et al. [19] and Atul et al. [15] observed the basin-shaped NZ on FSP/FSW of Al6063-T5 plate and Al7075-T651 plate, respectively (Figures 4.3a,b). They suggested that the topmost region of the workpiece is subjected to the intense plastic strain and a high frictional heat by the shoulder of the cylindrical tool which leads to the generation of a NZ with a basin shape and having a wide top. While the elliptical-shaped NZ is reported to develop on FSW of Al7075-T651 by Rhodes et al. [20] and Mahoney et al. [21] (Figure 4.4b). Recently, Ma et al. [18] performed FSP on cast A356 aluminum and results showed that with same geometry of tool, various shapes of NZ were produced by altering processing parameters. Moreover, nugget shape changes from basin to elliptical with increase in rotation speed. NZ is bounded by TMAZ which comprises an extremely distorted grain structure as a result of stirring by the tool. Since the deformation strain generated in the TMAZ is insufficient to raise the temperature required for recrystallization, recrystallization has not been observed to occur in the TMAZ during FSP. Nevertheless, grain distortion/elongation could be observed in the TMAZ parallel to the direction of flow of material (Figure 4.3b). The TMAZ

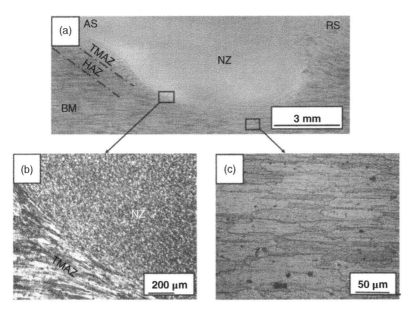

Figure 4.3 Optical micrograph of FSPed sample (a) at 720 rpm 85 mm/min, (b) TMAZ, and (c) base metal (Al7075 T651). Source: Kumar et al.[15]/Springer Nature.

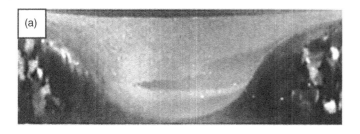

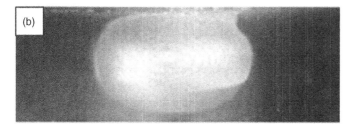

Figure 4.4 Effect of processing parameters on shapes of NZ in FSP of cast A356 aluminum at (a) an rpm of 300 and a feed rate of 51 mm/min and (b) an rpm of 900 and a feed rate of 203 mm/min. Source: MA et al. [18]/with permission from Elsevier.

is generally referred to as transition zone, which distinguishes the processed zone from the parent material as shown in Figure 4.3. Beyond the TMAZ, a different zone can be observed which is known as the HAZ. In this region, the sample experiences a thermal cycle but there is no mechanical deformation. The grain size in the HAZ is similar to base metal but thermal cycle affected the precipitate characteristics in a heat-treatable aluminum alloy [21].

4.3 Comparison with Other Severe Plastic Deformation (SPD) Techniques

SPD is possibly the best favorable route for fabricating ultrafine grained (UFG) metals and alloys straight away in bulk form among the various processes employed so far [22–25]. Grain refinement through SPD techniques required imposition of a large plastic strain on metals which causes dislocation accumulation and rearrangement and eventually grain fragmentation [26–28]. The common investigation method of SPD comprises high strain deformation by means of experimental methods, such as rolling, equal-channel angular pressing (ECAP), high-pressure torsion (HPT), multiaxial forging (MAF), equal-channel angular extrusion (ECAE), and FSP [29, 30]. These techniques have been successful in fabrication of UFG structure in metals and alloys which are ductile in nature and have shown initial strength not suitable for any structural applications [23]. Of the various techniques, FSP is comparatively somewhat dissimilar from other SPD methods. The high frictional and adiabatic heat originated during the process classifies it as a high-temperature SPD process. The materials which lacks sufficient ductility and are difficult to process by other SPD methods can be easily processed by FSP owing to the generation of local heating by friction between the workpiece and nonconsumable rotating tool which locally raises the temperature of the material to the point where it can be certainly plastically deformed, thereby leading to substantial amount of grain refinement. Additionally, it is very challenging to scale up the metals and alloys processed by any other SPD approach, and at the same time to fabricate them economically.

The challenges in achieving UFG structure in bulk samples has been the key barrier to their applicability for structural use which can be easily addressed by FSP process.

FSP has developed as a comprehensive microstructure tailoring tool in the last decade. All the primary endeavors of grain refinement by FSP were restricted to fine-grained microstructure in the range between 1 and 10 μm [11]. It has been observed from the various reported experimental investigations that a specially designed tool or an external cooling medium is indispensable to fabricate UFG microstructure in metals and alloys via FSP. Lately, some endeavors to obtain UFG microstructure by varying the FSP input parameters, such as constant tool traverse velocity (v), tool rotational rate (ω), or the ratio (ω/v), have been observed [29, 31, 32].

4.4 Process Variables

Processing parameters of the FSP have great influence on movement as well as flow of material and thermal cycle during the process. Optimizing these parameters plays an important role in achieving fine recrystallized and defect-free microstructure of the processed samples with enhanced mechanical properties. These parameters fall into four major categories: (i) tool geometry, (ii) machine variables, (iii) number of passes, and (iv) cooling condition/rate, which are shown in the flow diagram (Figure 4.5).

4.4.1 Tool Geometry

The FSP is a thermomechanical deformation process in which the tool's temperature generally approaches the workpiece's solidus temperature. Therefore, selection of the tool material as well

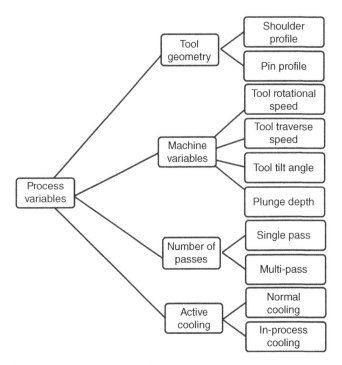

Figure 4.5 Classification of FSP process variables.

as design of it should be carried out carefully. The design of the tool greatly influences the final properties of the processed material. The correct tool material should have sufficient strength and must maintain dimensional stability at elevated temperatures. The geometry of the tool is a very crucial aspect of the process as it generates the required heat for processing and governs the material flow and thereby processed zone formation (NZ, TMAZ, and HAZ) [33]. A typical FSP tool comprises two most important parts, pin and shoulder, as shown in Figure 4.6. Tool pin is designed to produce frictional and deformational heating with the workpiece at the initial stage of tool plunge. Furthermore, it causes the material to flow from front to back of the tool, and also helps in uniform dispersion of second -phase particles, if any. The processing depth during the FSP mainly depends upon the dimension of the probe. For improvement of the processed zone, different types of pin profile, such as tapered, tapered with threads, cylindrical, cylindrical with thread, square, and triangle, have been used as shown in Figure 4.6. Tool shoulder is designed to serve as a reservoir for softened material underneath the shoulder so that it avoids splashing of the plasticized material from the processed region. The shoulder also produces the largest part of the heat needed to plasticize the material and provides the necessary downward forging action for consolidation. To improve the friction between shoulder and specimen, different styles of shoulder end surfaces have been used as shown in Figure 4.6. These include grooves, knurling, concentric circles, scrolls, and flat or featureless profile at the end surface of the shoulder. The diameter of the shoulder is directly related with the heat input, as increasing one will automatically increase the other and vice versa. The shoulder profile incudes flat, convex, and concave shape. So, to get a defect-free processed zone, a correct combination of the size of the shoulder and pin should be chosen. The influence of shoulder diameter to probe diameter (D/d) ratio on the mechanical behavior of FSPed LM25AA-5% SiCp MMCs was studied by Vijayavel et al. [34]. They concluded that the D/d ratio of 3 yielded better mechanical properties with defect-free processed zone.

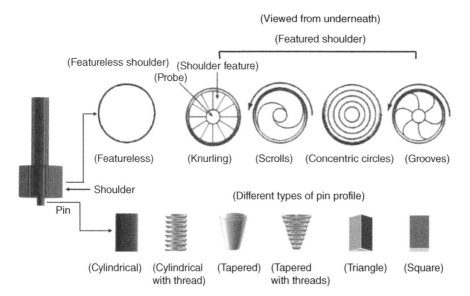

Figure 4.6 Basic tool geometry along with different types of pin profile and shoulder end surface features. Source: Adapted from Mishra et al. [11]/with permission from Elsevier.

4.4.2 Machine Variables

Tool traverse and rotational speed, tilt angle of the tool, and plunge depth are the machine variables. Among them, traverse speed and rotational speeds are the key parameters as the amount of heat input in the processed zone is dependent on these parameters. In general, the heat input to the work piece has direct relation to the rotational speed and inverse relation to the traverse speed. If the ratio of tool rotational speed to traverse speed is high, then the volume of heat input will be higher in the processed zone. Too high heat input can cause a turbulent flow of the material in the processed zone. On other hand, if the ratio of tool rotational speed to traverse speed is low, then the amount of heat input will be low. Insufficient heat input can cause nonuniform flow of processed material during the process and also results in the development of defects and voids within the NZ. In both the cases, defective processed zone could be obtained.

To simplify the effects of the rotational and traverse speed, many studies [15, 35–37] have been attempted on the FSP parameter optimization. For example, Zhang et al. [37] reported that the mechanical properties of FSW Al alloy joints improved with reducing tool rotational speed and increasing traverse speed. Azimzadegan et al. [16] carried out the FSW of Al7075 alloy with various parameters and reported that to achieve the highest strength, there was an optimum rotational and traverse speed. Based on the experimental work done on the FSW of Al alloys, an empirical relation between maximum NZ temperature (T_{max}) and the process parameters (rotational speed (ω), traverse speed (v)) has been established [38]:

$$\frac{T_{max}}{T_m} = K \left(\frac{\omega^2}{v \times 10^4} \right)^\alpha \tag{4.1}$$

where, K and α are the constants, ranges between 0.65–0.75 and 0.04–0.06, respectively, and T_m alloy melting temperature (in Kelvin). The ratio of ω^2 and v is considered as pseudo heat index. FSW of Al2024 plates [39] demonstrated that the relation between the maximum NZ temperature (T_{max}) and the heat index value is in accordance with the Eq. (4.1). The heat input in the processed zone controls the material flow and grain size, which consequently affects the properties of the processed material. To obtain desired microstructural changes and mechanical properties with defect-free processed zone, the combination of tool rotational and traverse speed should be effectively optimized. In addition to the tool traverse and rotational speed, other factors like tool tilt angle and plunge depth also affect the evolution of the processed zone; but normally these factors are kept constant [40]. A suitable tool tilt angle and plunge depth helps in proper filling of the key hole created by the moving tool and maintains necessary forging force for proper consolidation.

4.4.3 Number of Passes

The dimensions of NZ depend upon the tool pin dimensions. The width of the FSP processed material is not large from real-world application perspective. Hence, the width of the processing zone can be increased by two ways [41]. One way is to enlarge the pin dimensions which would use more power of the machine and apply high torque on the tool. The other way is by applying more number of passes along with some overlapping that could lead to the development of necessary width with enhanced properties in the overall zone. Figure 4.7 denotes the representation of various travel configurations of the FSP tool.

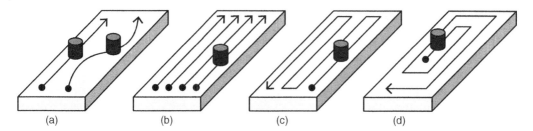

Figure 4.7 Tool travel patterns for FSP: (a) linear and curved single passes, (b) number of parallel passes at equal space, (c) raster configuration, (d) spiral configuration [41, 42].

A linear or curved single-pass configuration is shown in Figure 4.7a. The various configurations of multi-pass FSP are shown in Figures 4.7b,d. The multi parallel passes configuration can be performed by altering advancing side to retreating side or retreating side to advancing side in every consecutive pass with necessary overlapping (Figure 4.7b). While the raster pattern as presented in Figure 4.7c is formed by overlapping of advancing/retreating side of one pass with the advancing/retreating side of the consecutive pass. Therefore, this configuration may lead to gradient structure in the processed region [41]. The usual spiral configuration is denoted in Figure 4.7d. In this, the advancing side is swapped by the retreating side or the other way round for every succeeding pass that aids in minimizing the microstructure gradients inside the stir zone. The overlap-ratio (OR) during multi-pass FSP for processing the bulk samples is given as $OR = 1 - \left(\dfrac{l}{d_{pin}}\right)$ [43], where l is the space among two successive passes and d_{pin} is the pin dia. Usually, the space between two consecutive passes, l, is considered as half of the diameter of the pin that alleviates the irregularity in the microstructure of stir zone [42]. Figure 4.8 represents the cross-sectional view of the multi-pass FSPed Al5083 and Al7022 and with distinct values of OR. The samples processed with an overlap ratio, OR = 1/2 results in continuous improved surface with no defects. However, for the overlap value OR = 0 and −1, the samples displayed unrefined areas between the stir zones. Therefore, an overlap ratio, OR = 1/2 is considered as most effective for processing bulk samples during multi-pass FSP. Furthermore, it is observed that the way of overlapping, whether the retreating side or the advancing side overlaps also impacts the subsequent microstructure in stir zone owing to asymmetry in material flow. It was further observed that overlapping by retreating side had led to smooth surfaces, while overlapping by advancing side had led to the development of layers of uniform thickness [44].

4.4.4 Cooling-Assisted FSP

FSP technique is near to complete its 20 years journey and many types of variations have been implemented to enhance the efficiency of the method. It is well reported that the grain size reduction during FSP leads to improvement in physical and mechanical properties of the processed materials [32, 45, 46]. Though, in case FSP of heat-treatable Al alloys, a competition between softening due to metastable precipitates overaging and strengthening due to reduction in grain size are observed [15, 47]. Many researchers [15, 21, 48, 49] have investigated the effect of the FSP

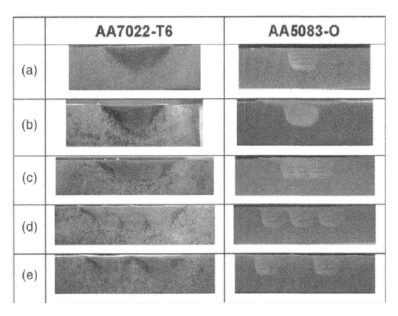

Figure 4.8 Cross-sectional view of the specimens with various OR: (a) single pass with OR of 1, (b) 4 passes with OR of 1, (c) 3 passes with OR of 1/2, (d) 3 passes with OR of 0, and (e) 2 passes with OR of 1. Source: Nascimento et al. [43]/with permission from Elsevier.

thermal cycle on precipitate kinetics of heat-treatable Al alloys and concluded that coarsening or overaging during the processing would be the prime cause for degradation in mechanical properties of the material, despite the evolution of fine grain structure. Moreover, Xu et al. [50] reported that fine grains evolved during processing coarsened very rapidly owing to sluggish rate of cooling after FSP. Abovementioned problems associated with the FSP hamper the potential of the process. Maximum work of the literature was focused on the optimization of the pin design and speed parameters during FSP. Coarsening of the precipitates and grain size especially in case of the heat-treatable Al alloys was the main concern in FSP.

Lately, researchers have investigated the effect of external cooling the FSP and reported that FSP with external cooling is an efficient way to improve the mechanical and metallurgical characteristics of the processed samples. For example, Sharma et al. [51] did the external cooling during FSW through flowing the water, liquid nitrogen, and compressed air on the surface of the sample and reported that the water cooling was more efficient in enhancing the mechanical properties of the 7039AA FSW weld. Upadhyay et al. [52] performed the underwater FSW on Al7050-T7451 alloys, and it was found that the UTS and hardness of the nugget zone were increased for the submerged/sub-ambient welding conditions. Mofid et al. [53] welded AA5083 and AZ31 with similar parameters but in different cooling mediums. They reported that formation of intermetallic compounds for submerged welds is much lower than that of the weld made in the air. Other researchers also stated the advancement in the mechanical properties of the processed materials through water cooling [54–57]. Schematic illustration of cooling-assisted FSP is depicted in Figure 4.9.

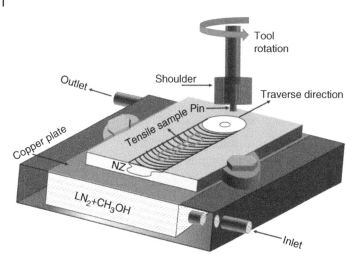

Figure 4.9 Schematic illustration of cooling-assisted FSP.

4.5 Mechanisms of Microstructural Evolution During FSP

Microstructural evolution in the NZ is governed by dynamic recrystallization (DRX), which occurred due to high frictional heating and intense plastic deformation [58–60]. Several mechanisms have been proposed for DRX, including continuous dynamic recrystallization (CDRX), discontinuous dynamic recrystallization (DDRX), and geometric dynamic recrystallization (GDRX) [11, 49, 61–65]. All these mechanisms lead to grain refinement. However, the exact mechanism of DRX operating during FSP is still debated [66]. Generally, the mechanism of grain refinement mainly depends on the type of metals or alloys. Based on the spatial locations and transient conditions, combination of these models (CDRX, GDRX, and DDRX) may all be possible mechanisms during the FSP [11, 61, 62]. Various stages of microstructural evolution in these mechanisms (CDRX, GDRX, and DDRX) of DRX is schematically shown in Figure 4.10 [67, 68]. The DDRX is characterized by generation of new strain-free grain structure at high-angle boundaries (Figure 4.10a) [65].

In Al and its alloys, dynamic recovery is observed to precede DDRX because of their high stacking fault energy [62, 69]. Jata et al. [70] was the first one who proposed that CDRX is the primary mechanism in refining the grain size during FSW. The CDRX is generally reported in the materials in which motion of dislocation is hindered either by solute drag or due to few number of active slip systems [71]. The dislocations rearrange themselves to achieve lower energy configurations, which leads to formation of low-angle subgrain structure. As deformation progress, gradual relative rotation of adjacent subgrains leads to the formation of recrystallized grains with high-angle grain boundaries (Figure 4.10b) [71]. In the GDRX mechanism, a large reduction in cross section during deformation, flattening, or elongating the original grains due to collective effect of compression, tension, and torsion lead to the formation of serrations at initial grain boundaries (Figure 4.10c). With further deformation, the original grain thickness is reduced to the thickness of the boundary serrations and causes the serrated boundaries to touch each other and forms newly isolated grains [63, 72].

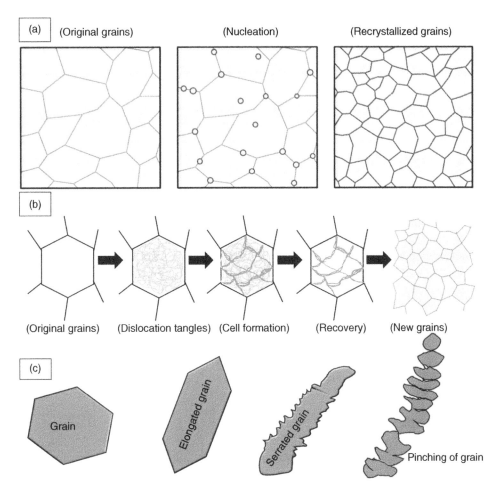

Figure 4.10 Schematic of various stages of dynamic recrystallization mechanisms (a) DDRX, (b) CDRX, and (c) GDRX.

4.6 Critical Issues in FSP

FSP has aptly developed as the present-day inclination for the fabrication of functionally graded systems and composites. Because of its intrinsic potential for the fabrication of materials with enhanced mechanical, functional, structural, and high-quality surface characteristics, it is being widely used in automobile, aerospace, and marine industries. Regardless having numerous advantages over traditional manufacturing processes, FSP has showed many critical issues related with the composite fabrication. One such is extreme tool wear owing to the interaction with the hard reinforcement ceramic particles. Though the tool wear has been observed to become sluggish once the tool approaches a steady worn profile [73], this does not provide a universal solution. The FSP of high melting point materials, for example, steel, titanium, and superalloys [74], is also difficult because of excessive tool wear. One most important issue that is frequently ignored in the literature is the enormously laborious manufacturing of composites, which usually consist of holes drilling or a groove cutting followed by stuffing the voids with powders, and applying multi-pass FSP. Even though a substantial amount of refinement in the processes has developed, there has been an inadequate accomplishment in attaining an even distribution of particles during a single-pass composite fabrication. Even though there are a number of studies on different reinforced particle composites, a perfect correspondence between the particle properties with the mechanical properties of composites remains uncertain. Moreover, agglomeration of the reinforcement particle is another concern. This is a principal drawback since the distribution of reinforcement particles eventually describes the enhancement in several properties [74]. The insufficient heat production and asymmetric material flow throughout FSP may lead to the agglomeration of reinforcement particles. Nevertheless, this concern can be effectively resolved by improving the dispersion of reinforcement particle by means of numerous steps. Out of many parameters, the most important ones are the use of optimum rotation and traverse speed, multi-pass FSP, optimum ratio of groove width to probe diameter, plunge depth, shoulder diameter, and optimum tilt angle of tool. Further, change of tool rotation between consecutive passes [75] and change of probe shape between consecutive passes [76] lead to sufficient heat evolution and defect-free surface composites. Therefore, the appropriate choice of reinforcement approaches and process parameters lead to good-quality and homogeneously distributed surface composites.

The FSP facilities are accessible at a commercial scale and it has showed surface treatment for the components up to more than 1 m in magnitude. In spite of this, the fabrication of bulk composites through FSP processing is quite problematical and incompetent when matched with other fabrication processes, such as casting [77].

4.7 Future Scope

Although the FSP of polymers has showed encouraging results, still more research is required to confirm their successful fabrication with enhanced properties owing to their chain configuration and low melting temperature. The FSP of hard materials, that is all those materials having high melting temperature, i.e. steel, titanium, or ceramic particle reinforced composites, leads to excessive tool wear. For this, tools fabricated with tungsten-based alloys/carbides and cubic boron nitride are suggested. On the other hand, the low fracture toughness and the high cost of these tools restrict their employment. As a result, the fabrication of surface composites of hard materials via FSP still anticipates the evolution of robust and less costly tools. Further, the uniform dispersion of reinforcement particles during FSP in a single pass only is still a key area of research. Also, there is a need of successful tooling system in order to avoid the step of covering the holes and grooves during FSP. Though FSP is highly established in the field of metal matrix composite, there is a higher need of optimization of parameters for fabricating polymer matrix composites with various base metals in order to popularize the process. Apart from composite and functionally graded system development, FSP has proven to have great potential for the fabrication of magnetostrictive composites and piezoelectric ceramic composites [78]. FSP in combination with additive manufacturing has emerged as an advance manufacturing technique known as friction stir additive manufacturing or additive friction stir manufacturing process which has led to the revolutionization of the field of manufacturing. FSP in conjunction with ultrasonic vibrations can increase the process speed without influencing the microstructure and properties of the surface composites [79]. But the ultrasonic vibrations can have constructive effects on the properties of FSPed materials [80]. Further, modeling aspects of FSP can also be explored. Since FSP does not lead to the evolution of any dust, fumes, or toxic gases, it is considered as an environment friendly, energy efficient, and versatile technology and has a potential to be considered as a standard metalworking process that can modify and control the microstructures of surface composites. With more study and research, a better understanding of the FSP route will be developed that will lead to the identification of various applications of FSP in the processing and fabrication of metallic materials. Even though several issues are present, still FSP provides highly competitive prospects for commercial glory.

References

1 Mishra, R.S., Mahoney, M.W., McFadden, S.X. et al. (1999). High strain rate superplasticity in a friction stir processed 7075 Al alloy. *Scr. Mater.* 42 (2): 163–168.
2 Mishra, R.S. and Mahoney, M.W. (2001). Friction stir processing: a new grain refinement technique to achieve high strain rate superplasticity in commercial alloys. In: *Materials Science Forum*, vol. 357, 507–514. Trans Tech Publications Ltd.
3 Thomas, W.M., Nicholas, E.D., Needham, J.C., et al. (1991). GB Patent application no. 9125978.8. International patent application no. PCT/GB92/02203.
4 Çam, G. and İpekoğlu, G. (2017). Recent developments in joining of aluminum alloys. *Int. J. Adv. Manuf. Technol.* 91 (5): 1851–1866.
5 Cam, G. and Mistikoglu, S. (2014). Recent developments in friction stir welding of Al-alloys. *J. Mater. Eng. Perform.* 23: 1936–1953.
6 Çam, G. (2011). Friction stir welded structural materials: beyond Al-alloys. *Int. Mater. Rev.* 56 (1): 1–48.
7 Çam, G., İpekoğlu, G., Küçükömeroğlu, T., and Aktarer, S.M. (2017). Applicability of friction stir welding to steels. *J. Achiev. Mater. Manuf. Eng.* 80 (2): 65–85.
8 Galvao, I., Loureiro, A., and Rodrigues, D.M. (2012). Influence of process parameters on the mechanical enhancement of copper-DHP by FSP. In: *Advanced Materials Research*, vol. 445, 631–636. Trans Tech Publications Ltd.
9 Shaeri, M.H., Salehi, M.T., Seyyedein, S.H. et al. (2014). Microstructure and mechanical properties of Al-7075 alloy processed by equal channel angular pressing combined with aging treatment. *Mater. Des.* 57: 250–257.
10 Mishra, R.S. and Mahoney, M.W. (2007). *Friction Stir Welding and Processing*, ASM International. Material Park, Ohio, The Materials Information Society.
11 Mishra, R.S. and Ma, Z.Y. (2005). Friction stir welding and processing. *Mater. Sci. Eng.: R: Rep.* 50 (1-2): 1–78.
12 Cavaliere, P. (2005). Mechanical properties of friction stir processed 2618/Al_2O_3/20p metal matrix composite. *Composites, Part A* 36 (12): 1657–1665.
13 Rahsepar, M. and Jarahimoghadam, H. (2016). The influence of multipass friction stir processing on the corrosion behavior and mechanical properties of zircon-reinforced Al metal matrix composites. *Mater. Sci. Eng., A* 671: 214–220.
14 Kumar, A., Pal, K., and Mula, S. (2017). Simultaneous improvement of mechanical strength, ductility and corrosion resistance of stir cast Al 7075-2% SiC micro-and nanocomposites by friction stir processing. *J. Manuf. Processes* 30: 1–13.
15 Kumar, A., Sharma, S.K., Pal, K., and Mula, S. (2017). Effect of process parameters on microstructural evolution, mechanical properties and corrosion behavior of friction stir processed Al 7075 Alloy. *J. Mater. Eng. Perform.* 26 (3): 1122–1134.
16 Azimzadegan, T. and Serajzadeh, S. (2010). An investigation into microstructures and mechanical properties of AA7075-T6 during friction stir welding at relatively high rotational speeds. *J. Mater. Eng. Perform.* 19 (9): 1256–1263.

17 Su, J.Q., Nelson, T.W., Mishra, R., and Mahoney, M. (2003). Microstructural investigation of friction stir welded 7050-T651 aluminium. *Acta Mater.* 51 (3): 713–729.

18 Ma, Z.Y., Sharma, S.R., and Mishra, R.S. (2006). Effect of friction stir processing on the microstructure of cast A356 aluminum. *Mater. Sci. Eng., A* 433 (1-2): 269–278.

19 Sato, Y.S., Kokawa, H., Enomoto, M., and Jogan, S. (1999). Microstructural evolution of 6063 aluminum during friction stir welding. *Metall. Mater. Trans. A* 30 (9): 2429–2437.

20 Rhodes, C.G., Mahoney, M.W., Bingel, W.H. et al. (1997). Effects of friction stir welding on microstructure of 7075 aluminum. *Scr. Mater.* 36 (1).

21 Mahoney, M.W., Rhodes, C.G., Flintoff, J.G. et al. (1998). Properties of friction stir welded 7075 T651 aluminum. *Metall. Mater. Trans. A* 29 (7): 1955–1964.

22 Valiev, R.Z., Korznikov, A.V., and Mulyukov, R.R. (1992). The structure and properties of metallic materials with a submicron-grained structure. *Phys. Met. Metall.* 73 (4): 373–384.

23 Valiev, R.Z., Islamgaliev, R.K., and Alexandrov, I.V. (2000). Bulk nanostructured materials from severe plastic deformation. *Prog. Mater Sci.* 45 (2): 103–189.

24 Tao, N.R., Wang, Z.B., Tong, W.P. et al. (2002). An investigation of surface nanocrystallization mechanism in Fe induced by surface mechanical attrition treatment. *Acta Mater.* 50 (18): 4603–4616.

25 Wu, X.L., Tao, N.R., Hong, Y.S. et al. (2002). Microstructure and evolution of mechanically-induced ultrafine grain in surface layer of AL-alloy subjected to USSP. *Acta Mater.* 50 (8): 2075–2084.

26 Bay, B., Hansen, N., Hughes, D.A., and Kuhlmann-Wilsdorf, D. (1992). Overview no. 96 evolution of FCC deformation structures in polyslip. *Acta Metall. Mater.* 40 (2): 205–219.

27 Bay, B., Hansen, N., and Kuhlmann-Wilsdorf, D. (1989). Deformation structures in lightly rolled pure aluminium. *Mater. Sci. Eng., A* 113: 385–397.

28 Huang, J.Y., Zhu, Y.T., Jiang, H., and Lowe, T.C. (2001). Microstructures and dislocation configurations in nanostructured Cu processed by repetitive corrugation and straightening. *Acta Mater.* 49 (9): 1497–1505.

29 Kwon, Y.J., Saito, N., and Shigematsu, I. (2002). Friction stir process as a new manufacturing technique of ultrafine grained aluminum alloy. *J. Mater. Sci. Lett.* 21 (19): 1473–1476.

30 Ma, Z.Y. and Mishra, R.S. (2005). Development of ultrafine-grained microstructure and low temperature (0.48 Tm) superplasticity in friction stir processed Al–Mg–Zr. *Scr. Mater.* 53 (1): 75–80.

31 Sato, Y.S., Urata, M., and Kokawa, H. (2002). Parameters controlling microstructure and hardness during friction-stir welding of precipitation-hardenable aluminum alloy 6063. *Metall. Mater. Trans. A* 33 (3): 625–635.

32 Ma, Z.Y., Mishra, R.S., and Mahoney, M.W. (2002). Superplastic deformation behaviour of friction stir processed 7075Al alloy. *Acta Mater.* 50 (17): 4419–4430.

33 Kumar, K., Kailas, S.V., and Srivatsan, T.S. (2011). The role of tool design in influencing the mechanism for the formation of friction stir welds in aluminum alloy 7020. *Mater. Manuf. Processes* 26 (7): 915–921.

34 Vijayavel, P., Balasubramanian, V., and Sundaram, S. (2014). Effect of shoulder diameter to pin diameter (D/d) ratio on tensile strength and ductility of friction stir processed LM25AA-5% SiCp metal matrix composites. *Mater. Des.* 57: 1–9.

35 Balasubramanian, V. (2008). Relationship between base metal properties and friction stir welding process parameters. *Mater. Sci. Eng., A* 480 (1-2): 397–403.

36 Liu, H.J., Zhang, H.J., and Yu, L. (2011). Effect of welding speed on microstructures and mechanical properties of underwater friction stir welded 2219 aluminum alloy. *Mater. Des.* 32 (3): 1548–1553.

37 Zhang, F., Su, X., Chen, Z., and Nie, Z. (2015). Effect of welding parameters on microstructure and mechanical properties of friction stir welded joints of a super high strength Al-Zn-Mg-Cu aluminum alloy. *Mater. Des.* 67: 483–491.

38 Arbegast, W.J. and Hartley, P.J. (1999). Friction stir weld technology development at Lockheed Martin Michoud Space System-an overview. *ASM Int. Trends Weld. Res. (USA)* 541–546.

39 Chen, Z., Li, J., Borbely, A. et al. (2015). The effects of nanosized particles on microstructural evolution of an in-situ TiB2/6063Al composite produced by friction stir processing. *Mater. Des.* 88: 999–1007.

40 Reddy, P.J., Kailas, S.V., and Srivatsan, T.S. (2011). Effect of tool angle on friction stir welding of aluminum alloy 5052: Role of sheet thickness. *Adv. Mater. Res.* 410: 196–205.

41 Węglowski, M.S. (2018). Friction stir processing–State of the art. *Arch. Civ. Mech. Eng.* 18 (1): 114–129.

42 McNelley, T.R. (2010). Friction stir processing (FSP): refining microstructures and improving properties. *Rev. Metal.* 46: 149–156.

43 Nascimento, F., Santos, T., Vilaça, P. et al. (2009). Microstructural modification and ductility enhancement of surfaces modified by FSP in aluminium alloys. *Mater. Sci. Eng., A* 506 (1-2): 16–22.

44 Gandra, J., Miranda, R.M., and Vilaça, P. (2011). Effect of overlapping direction in multipass friction stir processing. *Mater. Sci. Eng., A* 528 (16-17): 5592–5599.

45 Jana, S., Mishra, R.S., Baumann, J.B., and Grant, G. (2010). Effect of friction stir processing on fatigue behavior of an investment cast Al-7Si-0.6Mg alloy. *Acta Mater.* 58 (3): 989–1003.

46 Kwon, Y.J., Shigematsu, I., and Saito, N. (2003). Mechanical properties of fine-grained aluminum alloy produced by friction stir process. *Scr. Mater.* 49 (8): 785–789.

47 Feng, X., Liu, H., and Lippold, J.C. (2013). Microstructure characterization of the stir zone of submerged friction stir processed aluminum alloy 2219. *Mater. Charact.* 82: 97–102.

48 Chen, Y.C., Feng, J.C., and Liu, H.J. (2009). Precipitate evolution in friction stir welding of 2219-T6 aluminum alloys. *Mater. Charact.* 60 (6): 476–481.

49 Rhodes, C.G., Mahoney, M.W., Bingel, W.H., and Calabrese, M. (2003). Fine-grain evolution in friction stir processed 7050 aluminum. *Scr. Mater.* 48 (10): 1451–1455.

50 Xu, N., Ueji, R., and Fujii, H. (2014). Enhanced mechanical properties of 70/30 brass joint by rapid cooling friction stir welding. *Mater. Sci. Eng., A* 610: 132–138.

51 Sharma, C., Dwivedi, D.K., and Kumar, P. (2012a). Influence of in-process cooling on tensile behaviour of friction stir welded joints of AA7039. *Mater. Sci. Eng., A* 556: 479–487.

52 Upadhyay, P. and Reynolds, A.P. (2010). Effects of thermal boundary conditions in friction stir welded AA7050-T7 sheets. *Mater. Sci. Eng., A* 527 (6): 1537–1543.

53 Mofid, M.A., Abdollah-Zadeh, A., Ghaini, F.M., and Gür, C.H. (2012). Submerged friction-stir welding (SFSW) underwater and under liquid nitrogen: an improved method to join Al alloys to Mg alloys. *Metall. Mater. Trans. A* 43: 5106–5114.

54 Liu, H.J., Zhang, H.J., and Yu, L. (2011). Effect of traverse speed on microstructures and mechanical properties of underwater friction stir welded 2219 aluminum alloy. *Mater. Des.* 32: 1548–1553.

55 Xu, W.F., Liu, J.H., Chen, D.L. et al. (2012). Improvements of strength and ductility in aluminum alloy joints via rapid cooling during friction stir welding. *Mater. Sci. Eng., A* 548: 89–98.

56 Zhao, Y., Jiang, S., Yang, S. et al. (2015). Influence of cooling conditions on joint properties and microstructures of aluminum and magnesium dissimilar alloys by friction stir welding. *Int. J. Adv. Manuf. Technol.* 83: 673–679.

57 Sree Sabari, S., Malarvizhi, S., and Balasubramanian, V. (2016). Characteristics of FSW and UWFSW joints of AA2519-T87 aluminium alloy: effect of tool rotational speed. *J. Manuf. Processes* 22: 278–289.

58 Rajan, H.M., Dinaharan, I., Ramabalan, S., and Akinlabi, E.T. (2016). Influence of friction stir processing on microstructure and properties of AA7075/TiB$_2$ in situ composite. *J. Alloys Compd.* 657: 250–260.

59 Su, J.Q., Nelson, T.W., and Sterling, C.J. (2005a). Microstructure evolution during FSW/FSP of high strength aluminum alloys. *Mater. Sci. Eng., A* 405 (1-2): 277–286.

60 Yadav, D. and Bauri, R. (2015). Friction stir processing of Al-TiB$_2$ in situ composite: effect on particle distribution, microstructure and properties. *J. Mater. Eng. Perform.* 24 (3): 1116–1124.

61 Feng, X., Liu, H., and Babu, S.S. (2011). Effect of grain size refinement and precipitation reactions on strengthening in friction stir processed Al-Cu alloys. *Scr. Mater.* 65 (12): 1057–1060.

62 McNelley, T.R., Swaminathan, S., and Su, J.Q. (2008). Recrystallization mechanisms during friction stir welding/processing of aluminum alloys. *Scr. Mater.* 58 (5): 349–354.

63 McQueen, H.J., Evangelista, E., and Kassner, M.E. (1991). The classification and determination of restoration mechanisms in the hot working of Al alloys. *Z. Metallkd.* 82 (5): 336–345.

64 Gourdet, S. and Montheillet, F. (2003). A model of continuous dynamic recrystallization. *Acta Mater.* 51 (9): 2685–2699.

65 Humphreys, F.J. and Hatherly, M. (2012). *Recrystallization and Related Annealing Phenomena*. Elsevier.

66 Charit, I. and Mishra, R.S. (2005). Low temperature superplasticity in a friction stir processed ultrafine grained Al-Zn-Mg-Sc alloy. *Acta Mater.* 53 (15): 4211–4223.

67 Giribaskar, S., Prasad, R., and Ramkumar, J. (2008). TEM studies on recovery and recrystallisation in equal channel angular extrusion processed Al-3%Mg alloy. *Trans. Indian Inst. Met.* 61 (2-3): 173–176.

68 Robson, J.D. and Campbell, L. (2010). Model for grain evolution during friction stir welding of aluminium alloys. *Sci. Technol. Weld. Joining* 15 (2): 171–176.

69 Raghavan, R.S., Tiwari, S.M., Mishra, S.K., and Carsley, J.E. (2014). Recovery quantification and onset of recrystallization in aluminium alloys. *Philos. Mag. Lett.* 94 (12): 755–763.

70 Jata, K. and Semiatin, S. (2000). Continuous dynamic recrystallization during friction stir welding of high strength aluminum alloys (No. AFRL-ML-WP-TP-2003-441). Air Force Research Lab Wright-Patterson AFB OH Materials and Manufacturing Directorate.

71 Besharati-Givi, M.K. and Asadi, P. (2014). *Advances in Friction Stir Welding and Processing*. Elsevier.

72 Prangnell, P.B. and Heason, C.P. (2005). Grain structure formation during friction stir welding observed by the 'stop action technique'. *Acta Mater.* 53 (11): 3179–3192.

73 Prado, R.A., Murr, L.E., Soto, K.F., and McClure, J.C. (2003). Self-optimization in tool wear for friction-stir welding of Al 6061+ 20% Al$_2$O$_3$ MMC. *Mater. Sci. Eng., A* 349 (1-2): 156–165.

74 Legendre, F., Poissonnet, S., Bonnaillie, P. et al. (2009). Some microstructural characterisations in a friction stir welded oxide dispersion strengthened ferritic steel alloy. *J. Nucl. Mater.* 386: 537–539.

75 Heydarian, A., Dehghani, K., and Slamkish, T. (2014). Optimizing powder distribution in production of surface nano-composite via friction stir processing. *Metall. Mater. Trans. B* 45: 821–826.

76 Bahrami, M., Nikoo, M.F., and Givi, M.K.B. (2015). Microstructural and mechanical behaviors of nano-SiC-reinforced AA7075-O FSW joints prepared through two passes. *Mater. Sci. Eng., A* 626: 220–228.

77 Bhadeshia, H.K.D.H. (1997). Recrystallisation of practical mechanically alloyed iron-base and nickel-base superalloys. *Mater. Sci. Eng., A* 223 (1-2): 64–77.

78 Das, S., Martinez, N.Y., Das, S. et al. (2016). Magnetic properties of friction stir processed composite. *JOM* 68: 1925–1931.

79 Farshbaf Zinati, R. (2015). Experimental evaluation of ultrasonic-assisted friction stir process effect on in situ dispersion of multi-walled carbon nanotubes throughout polyamide 6. *Int. J. Adv. Manuf. Technol.* 81: 2087–2098.

80 Ahmadnia, M., Seidanloo, A., Teimouri, R. et al. (2015). Determining influence of ultrasonic-assisted friction stir welding parameters on mechanical and tribological properties of AA6061 joints. *Int. J. Adv. Manuf. Technol.* 78: 2009–2024.

5

Role of FSP in Surface Engineering

Setu Suman and Kazi Sabiruddin

Department of Mechanical Engineering, Indian Institute of Technology Indore, Indore, India

5.1 Introduction

During recent decades, the need to protect various types of engineering components due to degradation while in service by surface-related problems occurs such as corrosion, erosion, wear, fatigue, and many more. This led to evolving the concept of "surface engineering," which is an interdisciplinary branch of science and technology. Surface engineering basically provides information related to the surface phenomenon occurring at the material, which includes wear resistance, surface roughness, topography, scratch and texture, etc. The primary function of surface engineering is to improve specific properties of a component's surface independent of the underlying substrate material. It also helps to increase the service life of engineering components exposed to an open environment during operation. Surface engineering is the process of modifying the properties of a material's surface or near-surface to enable modifications accomplished through various surface modification processes, such as mechanical alloying, thin film coating, plasma spray coating, and solid-state-based coating.

One of the newest and most efficient solid-state surface modification techniques is friction stir processing (FSP). This technique produces a homogeneous grain refinement structure, leads to lower residual stress, and absorbs more energy, due to the solid-state nature of this process [1]. The chemical makeup of the base material was not altered when the FSP technique was applied. Wayne Thomas at The Welding Institute in the United Kingdom developed FSW for the first time in 1991 [2]. FSP is an expanded version of FSW. The same fundamental idea underlies FSW and FSP [2, 3]. Initially, FSP was used to produce superplastic aluminum alloys with fine-grain boundary misorientations and high-grain boundary misorientations [1]. The process of changing the heterogeneous to homogeneous microstructure, microstructural refinement, and mechanical property enhancement is gaining popularity as a surface modification technique. FSP has shown promise in improving the mechanical, corrosion, wear, and degradation properties of materials. The FSP technique was primarily beneficial for aluminum metal matrix composites (AMMCs) containing nanoparticles [4].

This chapter gives a fundamental understanding of FSP and its role in surface engineering, considering plasma spray, and solid-state coating-FSP. Further discussed are the parameters of FSP affecting wear and corrosion behavior of the surface. In the end, improper parameters of FSP leading to various defects occurring during surface modification technique have been assessed.

5.2 Role of Surface Modification Techniques

In everyday life, whenever we see an object or purchase a product, the surface appearance of those things is the first thing we notice. We reject the item even if it has cut marks or scratches on the surface. As a result, the appearance of the surface has a significant impact on whether or not it affects functionality. Before knowing the surface modification techniques, understand the importance of surface modification. Surface modification technique is a sub-domain of surface engineering (refer to Figure 5.1).

Surface engineering is a generic idea for improving the life cycle performance of the engineering component. Some points are the following:

- Improve **mechanical properties**: tensile strength, hardness, ductility, fracture toughness, and residual stress.
- Improve **chemical properties**: corrosion, wear and scratch resistance, oxidation, fatigue, and erosion.
- Improve **physical properties**: thermal insulation, thermal expansion of coefficient, and thermal conductivity.
- Improve **dimensional properties**: flatness and surface roughness.
- Improve **visual appearance** characteristics: color and texture.

The main components of the surface modification process are (a) changes to the substrate's surface or subsurface zone and (b) the addition of a second layer of a suitable material to the surface to achieve the desired properties for extending component life [5].

Surface modification at the substrate's surface or subsurface zone is classified into three types [6, 7]:

(a) A modification technique, such as a change in the base material's metallurgy or surface texture, functions without altering the substrate's chemical composition.

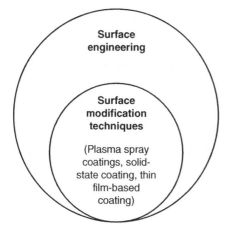

Figure 5.1 Surface modification technique is a sub-domain of surface engineering.

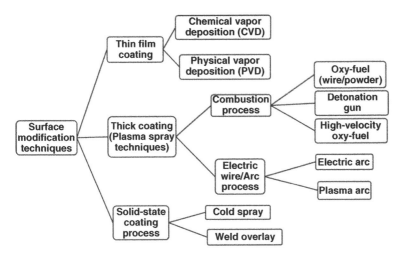

Figure 5.2 Surface modification techniques are categorized [8].

(b) A modification technology that alters the chemical composition of the substrate by diffusing a known element in the base material's surface layer.
(c) A coating is a technique that involves depositing a new material on top of the base material. Extra material is sprayed onto the substrate in the form of different thicknesses during this process.

FSP, cold spray, and weld overlay technique are surface modification processes that do not change the chemical composition of the substrate material. This method modifies the surface material's microstructure without changing its chemical composition.

Ion implantation and electrochemical processes are processes where the component's surface can be modified by changing its chemical composition or surface chemistry. In these procedures, the reactive surface is bombarded with atoms or ions, which causes a layer to form with the desired properties [9]. By adding a layer of fresh material on top of the base material, surface modification can be achieved (i.e. coating or cladding). Figure 5.2 depicts a general classification of surface modification techniques.

In this chapter, we shall discuss the general classes of surface engineering process outline and coating processes which play an essential role in surface modification.

5.3 Thermal Spray Technique

At the time, this was due to the increasing development and demand for technologies such as thermal spray, plasma, and laser. Specifically, laser surface technology has advanced significantly in hardening, alloying, and cladding applications and is now widely used in industry. The thermal spray process is a material processing technology that employs consumable powder and wire. A very high-velocity compressed air jet is used during this procedure to propel the particles or molten droplets produced at the nozzle's tip onto the substrate surface, where they quickly cool to form a solid splat. This process is repeated until the coating has the desired thickness [10]. There are various methods for melting the consumables, and they can be introduced into the heat source as wire or powder, as depicted in Figure 5.3.

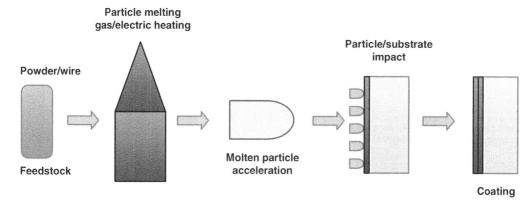

Figure 5.3 Basic mechanism of the thermal spraying coating process.

Thermal spray technology can melt and deposit any type of material as a coating on a wide range of substrate materials. A lot of research and development opportunities are needed in this field. Furthermore, many researches have been made to improve the properties of the deposited coating such as microhardness, fracture toughness, and porosity content by spraying a mixture of different powders in different proportions. The following seven types of thermal spray coating techniques are currently used in industrial applications [11]:

(a) Plasma spray
(b) Electric arc spray
(c) Flame spray
(d) Warm spray
(e) Detonation gun (D-gun)
(f) High-velocity oxy-fuel spraying (HVOF)
(g) High-velocity air-fuel spraying (HAVF)

Coating technology encompasses the thermal spray process. This is an efficient technology that allows for the depositing of coatings ranging from very thin to very thick for the enhancement of component surface properties and the modification of component surface functionality. Thermal spraying is used in various sectors, including automotive, chemical process equipment, power generation, boiler components, orthopedics, ships, marine turbines, and so on [12].

5.4 FSP – Solid-State Coating Process

To meet the requirements, there is an increasing demand to modify the surface properties of components, sub-layers of the surface, or specific areas. As a result, surface engineering has emerged as a vital field for mechanical and material engineers. Surface engineering is concerned with processes that alter the structural and chemical properties of the surface layer of a substrate and modify the surface or subsurface to a particular depth.

FSP is a new method in the area of surface engineering, and it has shown to be a qualified and promising candidate for surface modification technology. FSP is used in specific areas for the modification and microstructural change of surface-processed metallic components to enhance specific substrate properties [1]. This method is very helpful for improving surface modification so that the processed zone can achieve significant microstructural refinement, densification, and homogeneity [1, 13, 14]. FSP was initially used to improve surface properties through microstructure refinement and to remove porosity and shrinkage defects in the casting process [15]. In line with current trends, FSP can be used to modify the surface characteristics or texture of the base metal by incorporating reinforcement particles. Various reinforcement particles include as a ceramic, i.e. SiC, TiC, ZrC, TiN, TiO_2, ZrO_2, SiO_2, Tib_2, Al_2O_3, and B_4C [16–18]. The different types of reinforcement particles can be categorized as single reinforcement, multi-reinforcement, and non-reinforcement particles [19].

Thermal spray technology is classified broadly and it has various applications in surface engineering. The coating combines fully melted or partially melted particles with intense heat. One significant disadvantage of this technique is that it undergoes a phase change when exposed to high temperatures. This has an adverse effect on the coating properties. It is also interconnected to other inborn irregularities like porosity, splat boundaries, and uneven microstructure. To overcome all of the associated coating problems, any material produced without the melting of a substrate is preferred. As a result, the most popular method for producing a coating on a variety of substrate materials with the least amount of dilution is FSP and its variations. FSP largely eliminates the problems associated with thermal spray techniques due to its ability to impart large strains and strain rates [8]. The solid-state coating process, i.e. FSP, has been given more attention in this chapter.

5.4.1 Friction Stir Processing Technique

FSP is mainly intended for joining of materials. Furthermore, in addition to joining samples, the process modifies the local microstructure of the workpiece and achieves the desired surface microstructure modification [20]. FSP is the simplest process and consists of a nonconsumable rotating tool that is initially plunged into the workpiece and then moved at a specific speed in the direction of interest. The pictorial view of FSP along with pin-less tool are illustrated in Figure 5.4. Tool is the heart of this process; it serves basically two functions: (i) deformation of workpiece material and (ii) producing heat due to friction generated between rotating tool shoulder and workpiece. The material undergoes intense plastic deformation due to the shoulder rotating under the influence of an applied axial force. The processed area also experiences localized frictional heating, which typically results from the refined grain structure of the plasticized metal [21, 22]. The temperature produced during the operation is approximately 0.5–0.9 of the melting point (T_m), and no

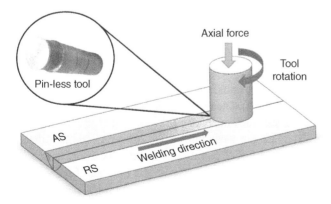

Figure 5.4 Schematic diagram of FSP along with the pin-less tool.

melting is observed. As a result, this technique is known as the solid-state joining technique [1, 23]. One study found that Mg–Al–Zn alloy contained fine-grain microstructures with an average size of particles 100–300 nm [14]. These micro or nanostructures are the primary cause of the hardness and wear behavior increases discovered by various researchers. Multi-pass FSPs are essential for grain refinement and particle distribution.

Suman et al. [24] studied the effect of multi-pass FSP on microstructure and mechanical characteristics revealing that at the fourth pass, homogeneous distribution of particles and fine grains was seen.

Depending on the application, FSP can be applied as a post- or pre-surface modification method. Prior to fusion welding, pretreatment processing of the base metal, such as laser melting, can be used to avoid grain boundary liquation cracking [25, 26]. While introducing a revolutionary technique for post-processing known as stationary friction processing (SFP) [27]. The thermal sprayed coatings were tested after SFP to improve their erosion resistance, hardness, and refined microstructure [27].

The most popular technique used by scientists who study surface modification is FSP. There are numerous variables that can be manipulated during this process to refine the surface properties. These variables are broadly classified into six categories: (i) tool design, (ii) machine variable, (iii) material properties, (iv) number of passes, (v) particle reinforcement, and (vi) active cooling technique [22, 28, 29]. Additionally, certain factors including tool geometries, number of passes, and cooling method play a significant role, which may allow a higher strain rate due to super-plasticity [30].

5.5 Process Parameters of FSP: Surface Engineering

The surface integrity is influenced by three primary parameters: tool rotational speed (TRS), tool traverse speed (TTS), and number of passes. Higher tool rotational and traverse speed produce more temperature due to high frictional heating. It encourages uniform material mixing and reinforcement particle distribution. The rotational and traverse speed of the tool are critical parameters that influence the microstructure and, consequently, the mechanical properties of the processed material [31]. In addition, it discusses the impact of FSP parameters on wear and corrosion behavior, as well as the inappropriate properties that affect surface integrity as a result of some defects.

5.5.1 Influence of FSP Parameters on Wear Behavior

Wear is a surface phenomenon that occurs when two solid surfaces slide each other. The wear resistance can be calculated using Eq. (5.1) weight loss method (5.2) and wear rate.

$$\text{Net weight} = \text{Weight before wear} - \text{Weight after wear} \tag{5.1}$$

$$\text{Wear rate} = \text{Net weight/time (g/s)} \tag{5.2}$$

The wear rate of SiC/380 Al alloy surface composites was investigated, and it was discovered that increasing the plunge depth reduces the wear rate [32]. The wear rate is significantly influenced by multi-pass FSP. As shown in Figure 5.5, the wear track appears on the surface of the friction stir processed (FSPed) AA6061-T6. We can calculate the wear rate in g/s using Eq. (5.2). Waheed S. AbuShanab et al. [33] investigate the effect of FSP parameters on the wear resistance of AA2024 wrought aluminum alloy. They noticed that with increases in the number of passes, the wear rate tended to decrease as compared to rates obtained using base alloy and single pass, respectively. Maschera et al. [34] have studied the tribological properties of AZ31 alloy after FSP. The incorporation of ZrO_2 nanoparticles reinforced in AZ31 alloy to obtain surface nanocomposite material.

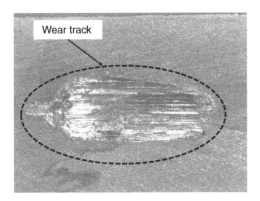

Figure 5.5 Microscopic image of wear track.

After the fourth pass, it was seen that the wear rate of AZ31-ZrO$_2$ nanocomposites had decreased by about 40%. The wear rate of FSP is significantly influenced by its rotational speed. Similar to the trend seen with traverse speed, increasing rotational speed tends to increase wear rate [33].

5.5.2 Influence of FSP Parameters on Corrosion Behavior

The phenomenon of corrosion is one that happens on a material's surface. Corrosion is a serious problem that deteriorates the surface as well as shortens the life of the material. Essentially, we want to prevent or control corrosion-related damage. In this section, we looked at how FSP can be used to improve corrosion resistance. There are primarily three factors that can improve corrosion resistance when using FSP [35, 36]. Improvements in corrosion resistance can be attributed to (i) FSP's effects on grain refining (which improve corrosion resistance), (ii) FSP's surface-strengthening effects (which slow down corrosion in processed samples), and (iii) the substrate's densification (which removes defects that increase corrosion resistance) [36].

Up to now, most researchers used FSP to improve the mechanical properties of the material and corrosion resistance [37–40]. The most significant FSP parameters affecting corrosion behavior have been determined to be rotational speed, TTS, and the number of passes. Hamed Seifiyan et al. [41] investigate the effect of FSP condition on the corrosion behavior of AZ31B magnesium alloy. They reported that maximum corrosion resistance occurred at a TTS of 50 mm/min and a tool rotation speed of 1000 rpm. It is also observed that enhanced corrosion resistance is supported by refined microstructures after FSP.

5.6 Inappropriate Characteristics of Surface Modification

The modification of the surface or subsurface zone requires the use of surface engineering. Poor surface modification of the substrate can result in a variety of issues, including (i) inappropriate surface topography; (ii) cracks, pores, and inclusions in the forms of oxide formed during surface modification, which involves melting and solidification; (iii) bonding strength between the upper layer of the coating material and substrate; (iv) physical, mechanical, and metallurgical incompatibility between the surface and substrate may result in residual stress, thermal stress, deterioration of surface properties, flaking or peeling off of the coating from the substrate [5]. It can also be seen in Figure 5.6a during cyclic oxidation test after some cycle coating is peeling off from substrate

Figure 5.6 (a) Coating peel-off due to thermal stress during cyclic oxidation and (b) FESEM image of cracks and pores formation at the surface of alumina-chromium coating.

material (AISI 1020 low alloy steel). The surface morphology of alumina-based coated material shows cracks formation after the oxidation test was observed by scanning electron microscopy (FESEM) equipment at higher magnification (refer to Figure 5.6b).

Inappropriate parameters of FSP can lead to various defects (refer to Table 5.1) occurring during surface modification like

1) Inadequate stirring action of the tool causes poor mixing of material, incomplete root penetration, and groove formation on the surface.
2) Tunnel defect, porosity, and excessive flash produced on the surface due to inadequate stirring action of the tool during FSP.
3) Wormhole formation due to tool's significant advance per revolution. This is basically an extreme case of surface braking.
4) Insufficient plunging of the tool attributed to kissing bond defects. There is a possibility that a sufficient breakup of the oxide layer during FSP.
5) Insufficient heat input and incorrect tool orientation cause a lack of penetration [42].

Table 5.1 Macroscopic view of all defects produces during FSP.

Sl. no	Macrostructure	Defect name
1		Incomplete root penetration
2		Worm hole

Sl. no	Macrostructure	Defect name
3		Kissing bond
4		Tunnel defect
5		Pin hole
6		Root gap

5.7 Summary

The field of surface engineering encompasses thermal spray coating. It has been widely used to enhance the surface properties such as adhesion strength, wear, and corrosion resistance and also increase the service life of the components. One major limitation of the thermal spray process is the change of phase at higher temperatures interacting with intense heat, thereby deteriorating the coating properties. FSP is emerging as a suitable method to eliminate the drawback associated with thermal spraying. In FSP, friction plays a vital role in generating appropriate heat at the tool–workpiece interface, causing the material to soften and the material to join due to severe plastic deformation. FSP and its variations are hence solid-state by nature. FSP modifies the surface integrity such as toughness, ductility, wear, and corrosion resistance without altering the qualities of the underlying material. Surface modification is significantly influenced by FSP process parameters such as tool rotational speed (rpm) and tool traverse speed (mm/s) are the two major parameters to consider. Recent advancements encompass the fabrication of surface alloys and metal matrix composites by inserting the reinforcement particles in order to change the material's surface properties by employing the FSP technique. As a result, FSP plays a significant role in surface engineering as a modification technique.

References

1 Mishra, R.S. and Ma, Z.Y. (2005). Friction stir welding and processing. *Mater. Sci. Eng.: R: Rep. 50* (1-2): 1–78.

2 Thomas, W.M., Nicholas, E.D., Needham, J.C., et al. (1991). Friction stir welding. International Patent Application No. PCT/GB92/02203 and GB Patent Application No. 9125978.8. *US Patent*, (5), pp.460–317.

3 Dawes, C.J. (1995). Friction stir joining of aluminum alloys. *TWI Bull. 36*: 124.

4 Sharifitabar, M., Sarani, A., Khorshahian, S., and Afarani, M.S. (2011). Fabrication of 5052Al/Al$_2$O$_3$ nanoceramic particle reinforced composite via friction stir processing route. *Mater. Des. 32* (8-9): 4164–4172.

5 Kumar, D. (2018). *Surface Engineering: Enhancing Life of Tribological Components*. India: Springer.

6 Jambagi, S.C. (2017). *Property Improvement of Thermally Sprayed Coatings Using Carbon Nanotube Reinforcement* (Doctoral dissertation, IIT, Kharagpur).

7 Jambagi, S.C. (2017). Scratch adhesion strength of plasma sprayed carbon nanotube reinforced ceramic coatings. *J. Alloys Compd. 728*: 126–137.

8 Bajakke, P.A., Jambagi, S.C., Malik, V.R., and Deshpande, A.S. (2020). Friction stir processing: an emerging surface engineering technique. *Surf. Eng. Mod. Mater.* 1–31.

9 Gupta, K. (ed.) (2020). *Surface Engineering of Modern Materials*. Berlin: Springer.

10 Davis, J.R. and Handbook, A.S.M. (2013). *Thermal Spray Technology*. ASM International.

11 Oksa, M., Tuurna, S., and Varis, T. (2013). Increased lifetime for biomass and waste to energy power plant boilers with HVOF coatings: high temperature corrosion testing under chlorine-containing molten salt. *J. Therm. Spray Technol.* 22: 783–796.

12 Tucker, R.C. (2002). Thermal spray coatings: broad and growing applications. *Int. J. Powder Metall. 38* (7): 45–53.

13 Nascimento, F., Santos, T., Vilaça, P. et al. (2009). Microstructural modification and ductility enhancement of surfaces modified by FSP in aluminium alloys. *Mater. Sci. Eng., A 506* (1-2): 16–22.

14 Chang, C.I., Du, X.H., and Huang, J.C. (2007). Achieving ultrafine grain size in Mg–Al–Zn alloy by friction stir processing. *Scr. Mater. 57* (3): 209–212.

15 Ma, Z.Y., Sharma, S.R., and Mishra, R.S. (2006). Effect of multiple-pass friction stir processing on microstructure and tensile properties of a cast aluminum–silicon alloy. *Scr. Mater.* 54 (9): 1623–1626.

16 Kumar, A., Mahapatra, M.M., Jha, P.K. et al. (2014). Influence of tool geometries and process variables on friction stir butt welding of Al–4.5% Cu/TiC in situ metal matrix composites. *Mater. Des.* 59: 406–414.

17 Rathee, S., Maheshwari, S., Siddiquee, A.N., and Srivastava, M. (2018). Distribution of reinforcement particles in surface composite fabrication via friction stir processing: suitable strategy. *Mater. Manuf. Processes 33* (3): 262–269.

18 Yuvanarasimman, P. and Malayalamurthi, R. (2018). Studies on fractures of friction stir welded Al matrix SiC-B 4 C reinforced metal composites. *Silicon 10*: 1375–1383.

19 Wu, B., Ibrahim, M.Z., Raja, S. et al. (2022). The influence of reinforcement particles friction stir processing on microstructure, mechanical properties, tribological and corrosion behaviors: a review. *J. Mater. Res. Technol.* .

20 Węglowski, M.S. (2018). Friction stir processing—state of the art. *Arch. Civ. Mech. Eng. 18*: 114–129.

21 McNelley, T.R. (2010). Procesado por fricción batida (FSP): afino de la microestructura y mejora de propiedades. *Rev. Metall. 46* (Extra): 149–156.

22 Sharma, V., Prakash, U., and Kumar, B.V.M. (**2015**). Surface composites by friction stir processing: a review. *J. Mater. Process. Technol.* 224: 117–134.

23 Ma, Z.Y. (2008). Friction stir processing technology: a review. *Metall. Mater. Trans. A 39*: 642–658.

24 Suman, S., Sethi, D., Bhargava, M., and Roy, B.S. (2022). Investigating the effect of multi-pass friction stir processing of SiC particles on temperature distribution, microstructure and mechanical properties of AA6061-T6 plate. *Silicon* 1–13.

25 Mousavizade, S.M., Ghaini, F.M., Torkamany, M.J. et al. (2009). Effect of severe plastic deformation on grain boundary liquation of a nickel–base superalloy. *Scr. Mater. 60* (4): 244–247.

26 Rule, J.R., Rodelas, J.M., and Lippold, J.C. (2013). Application of friction stir processing as a pretreatment to fusion welding. *Welding J. 92* (10).

27 Arora, H.S., Rani, M., Perumal, G. et al. (2020). Structural rejuvenation of thermal spray coating through stationary friction processing. *Surf. Coat. Technol. 389*: 125631.

28 Rathee, S., Maheshwari, S., and Siddiquee, A.N. (2018). Issues and strategies in composite fabrication via friction stir processing: a review. *Mater. Manuf. Processes 33* (3): 239–261.

29 Patel, V.V., Badheka, V., and Kumar, A. (2016). Metallography microstructure. *Analysis 5*: 278.

30 Mishra, R.S., Mahoney, M.W., McFadden, S.X. et al. (1999). High strain rate superplasticity in a friction stir processed 7075 Al alloy. *Scr. Mater. 42* (2): 163–168.

31 Mishra, R.S. (2000). Chapter 2 Introduction to friction stir processing (FSP). *J. Mater. Process. Technol. 2*: 1–28.

32 Mohammed, M.H. and Subhi, A.D. (2021). Exploring the influence of process parameters on the properties of SiC/A380 Al alloy surface composite fabricated by friction stir processing. *Eng. Sci. Technol. Int. J. 24* (5): 1272–1280.

33 AbuShanab, W.S. and Moustafa, E.B. (2020). Effects of friction stir processing parameters on the wear resistance and mechanical properties of fabricated metal matrix nanocomposites (MMNCs) surface. *J. Mater. Res. Technol. 9* (4): 7460–7471.

34 Mazaheri, Y., Jalilvand, M.M., Heidarpour, A., and Jahani, A.R. (2020). Tribological behavior of AZ31/ZrO$_2$ surface nanocomposites developed by friction stir processing. *Tribol. Int. 143*: 106062.

35 Selvam, K., Ayyagari, A., Grewal, H.S. et al. (2017). Enhancing the erosion-corrosion resistance of steel through friction stir processing. *Wear 386*: 129–138.

36 Lotfollahi, M., Shamanian, M., and Saatchi, A. (2014). Effect of friction stir processing on erosion-corrosion behavior of nickel–aluminum bronze. *Mater. Des. 1980-2015* (62): 282–287.

37 Ahmadkhaniha, D., Sohi, M.H., Zarei-Hanzaki, A. et al. (2015). Taguchi optimization of process parameters in friction stir processing of pure Mg. *J. Magnesium Alloys 3* (2): 168–172.

38 Zeng, R.C., Chen, J., Dietzel, W. et al. (2009). Corrosion of friction stir welded magnesium alloy AM50. *Corros. Sci. 51* (8): 1738–1746.

39 Ralls, A.M., Kasar, A.K., and Menezes, P.L. (2021). Friction stir processing on the tribological, corrosion, and erosion properties of steel: a review. *J. Manuf. Mater. Process.* *5* (3): 97.

40 Argade, G.R., Kandasamy, K., Panigrahi, S.K., and Mishra, R.S. (2012). Corrosion behavior of a friction stir processed rare-earth added magnesium alloy. *Corros. Sci.* *58*: 321–326.

41 Seifiyan, H., Sohi, M.H., Ansari, M. et al. (2019). Influence of friction stir processing conditions on corrosion behavior of AZ31B magnesium alloy. *J. Magnesium Alloys* *7* (4): 605–616.

42 Fowler, S., Toumpis, A., and Galloway, A. (2016). Fatigue and bending behaviour of friction stir welded DH36 steel. *Int. J. Adv. Manuf. Technol.* *84*: 2659–2669.

6

Surface Composite Fabrication Using FSP

Baidehish Sahoo[1], Jinu Paul[2], and Abhishek Sharma[3]

[1] School of Mechanical Engineering, MIT World Peace University, Pune, India
[2] Department of Mechanical Engineering, National Institute of Technology Calicut, Calicut, India
[3] Research Division of Materials Joining Mechanism, Joining & Welding Research Institute, Osaka University, Osaka, Japan

6.1 Introduction

Surface composites refer to materials that exhibit composite properties on surface unharming the bulk properties of the matrix material [1]. With these properties, improved surface characteristics, such as wear resistance, surface hardness, and corrosion resistance, in addition to superior mechanical and thermal properties, these composites are appropriate for various automotive applications. Friction stir processing (FSP) became a popular and efficient tool for manufacturing surface composites and altering microstructural features in the past two decades. This technique was developed by Mishra et al. [2] as a modification of the friction stir welding (FSW) process devised by The Welding Institute (TWI) in the United Kingdom in 1991. Initially, FSP was used to achieve superplasticity in aluminum alloys through microstructural grain refinement and promoting grain boundary migration [2]. Subsequently, the technique was further developed to incorporate various reinforcements on the surface of aluminum alloys [3].

FSP is a solid-state manufacturing method that uses severe plastic deformation (SPD) to improve material mixing and modify microstructures. It is a relatively new process that presents a unique challenge to established techniques like accumulative roll-bonding (ARB), equal-channel angular pressing (ECAP), high-pressure torsion (HPT), and multidirectional forging [4]. FSP is especially effective for microstructural refinement because it employs a dynamic recrystallization approach that creates a significant quantity of high-angle grain boundaries (HAGBs) [5, 6]. Unlike other SPD processes that modify the bulk properties, FSP focuses only on surface modification retaining the bulk properties of the base material.

The basic form of FSP involves plunging of a nonconsumable rotating tool into the material up to a definite depth and then moving in a specific direction, as depicted in schematic Figure 6.1. However, the manufacturing of surface composites through FSP requires some modifications to the process. Surface composite production through FSP comprises of two steps. The starting step involves the manufacturing of groove(s) or array of holes on surface of the base material and filling the manufactured groove or array of holes with reinforcement. The second stage involves carrying out FSP utilizing a nonconsumable rotating tool with a linear motion in the filled groove or array

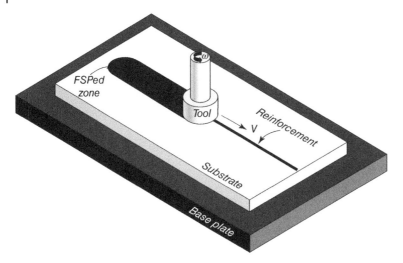

Figure 6.1 Schematic illustration of FSP.

of holes direction (Figure 6.1). The tool used in FSP assists in (i) generating heat by utilizing friction amid the tool shoulder and the workpiece [7] and (ii) deforming the base material. Intense plastic deformation occurs inside the treated region, stemming to a dynamically recrystallized fine-grain composition [8].

The past two decades have been noticing substantial evolution in the advancement of surface composites with aluminum, magnesium, copper, and titanium [9–14]. This chapter focuses specifically on recent advancements in the surface composites manufacturing using FSP. It delivers a complete analysis of the influence of various process parameters, tool geometry, microstructural evolution, and strengthening mechanisms.

6.2 Reinforcement Incorporation Approaches

The first step toward the processing of surface composites through FSP encompasses the placement of reinforcement on the substrate surface. Since the inception of this process in 2003, researchers have developed numerous efficient methods for the incorporation of reinforcing particles on the base material surface, shown in Figure 6.2.

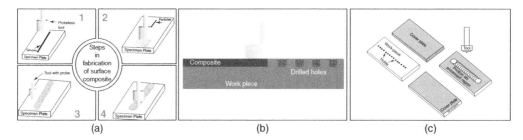

Figure 6.2 Conventional reinforcement filling methods for the surface composites fabrication by (a) groove, (b) drilled holes, and (c) using cover plate [14].

The most common approach involves the fabrication of groove(s) on top of the substrate which is then stuffed with reinforcing particles and FSP is carried out along the filled groove. However, in this approach, reinforcement removal by the linear motion of the tool pin was frequently observed. Moreover, the airborne effect at high rotational speed also contributed significantly to the reinforcement removal from the groove during the FSP. Thus, as a solution to this problem, a capping pass at a lower tool rotational speed was introduced. A capping pass is used to seal the reinforcement-filled groove. A tool without a pin or probe is used like the FSP tool for the capping pass. Moreover, the shoulder plunge depth (PD) of 0.02–0.05 mm is used during the capping pass. The schematic representation of capping pass approach used for FSP is depicted in Figure 6.2a. The capping pass arrangement helped in preventing particle ejection from the groove during the FSP. But due to this arrangement, a thin softened aluminum layer is formed at the capping pass which partially fills the groove and the particles were pushed toward the end of the groove. So, even after using the capping pass, the initial and final volume fractions of the particles will not be the same. To overcome this situation, many researchers started using a cover plate instead of packing filled grooves with a pin-less tool. For example, Lim et al. [15] employed a cover plate having thickness of 1.1 mm for packing MWCNT in the fabricated grooves. This thin cover sheet avoids the ejection of reinforcement particles [16]. A similar procedure was also applied by Avettand-Fènoël et al. [17] where a 0.2 mm thick plate was used for reinforcing Y_2O_3 in the Cu matrix.

The methods involving packing of filled grooves (i.e. a cover sheet or pin-less tool passing) consumes a considerable amount of time and resources. Thus, an alternative solution was proposed by Li et al. [12] where blind holes were created on the surface of matrix to serve as reinforcement reservoirs before starting the FSP procedure as shown in Figure. 6.3a. The decline in particle loss volume with this strategy is attributed to the closing of blind holes by the half shoulder ahead of the pin as exhibited in Figure 6.3b. This mechanism prevents the squeezing out of the particle from the holes and avoids the reinforcement loss. Also, the fabrication of grooves creates a discontinuity in the matrix material along its length. This material discontinuity is believed to be harmful for the formability of the surface composites fabricated via FSP and can be well avoided by incorporating reinforcements through holes instead of grooves. Akramifard et al. [18] also incorporated SiC by drilling holes on the Cu matrix surface and uniform distribution of reinforcements was detected within the stir zone as a consequence of using holes.

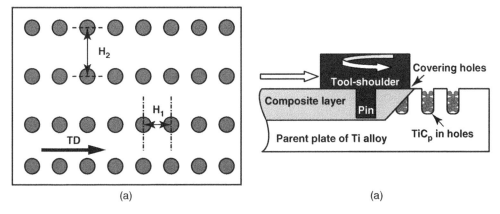

(a)   (a)

Figure 6.3 Schematic representation of (a) reinforcement filling through drilling holes and (b) material flow during FSP [12].

The filling of reinforcements on the matrix surface either by making a groove or hole drilling method is a time-consuming effort. Moreover, pre-machining of reservoirs is a laborious process and may put further design restrictions in commercial applications. This problem can be solved by the incorporation of reinforcement directly on the matrix surface before FSP. In this league, Miranda et al. [19] examined three different reinforcement strategies for incorporating SiC and Al_2O_3 in the AA5083 matrix by the use of FSP. The authors observed that the direct deposition of reinforcements is the quickest method since it eliminates any extra material preparation, such as groves or holes. Similarly, Jeon et al. [20] fabricated graphene oxide (GO)/AA5052 matrix composite by applying GO directly onto the AA5052 surface in the form of GO/water colloid before FSP (Figure 6.4).

Besides direct pasting methodology, some advanced spray and coating techniques can also be used for the better dispersion of reinforcements before FSP. Mazaheri et al. [21] deposited the Al_2O_3 reinforcement on the A356 T6 alloy substrate by high-velocity oxyfuel (HVOF) spraying technique before FSP. The results disclosed that the Al_2O_3 particles are well disseminated within the matrix with excellent bonding amid the particles and substrate. In another study, Hodder et al. [22] used a cold spraying strategy for applying Al_2O_3 reinforcement on top of AA6061 alloy before FSP. The results discovered a prominent rise in surface hardness of the composite. Similarly, Huang et al. [23] combined cold spraying and FSP to fabricate Al5056/SiC surface composite. The refined SiC particles are homogeneously distributed inside FSPed coating. Also, a sweeping rise in microhardness was observed as a result of the redeployment and refinement of reinforcing particles inside the composite.

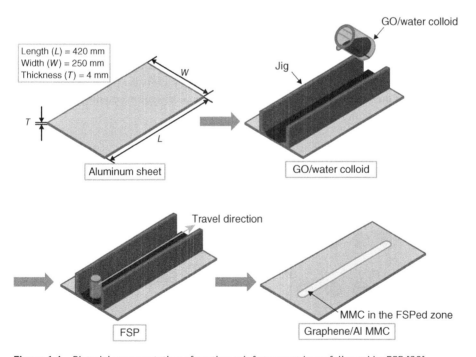

Figure 6.4 Pictorial representation of pasting reinforcement layer followed by FSP [20].

6.3   Effect of Process Parameters

The important process parameters that regulate the consolidation and properties of the surface composites are rotational and linear traverse speed of the tool [24], Khayyamin et al. [25]. Additionally, the PD and tool tilt angle (with respect to the vertical axis) also contribute to governing the effective material flow during FSP [26, 27].

In FSP, the plastic deformation and friction amid the interacting surfaces of rotating tool and base material result in severe heat generation. This heat input during FSP serves as a stimulus for the material flow and microstructural development within the stir region which in turn unambiguously impacts the mechanical and tribological behavior of the composite. The best material flow and properties can be obtained only with the perfect amalgamation of tool rotational and traverse speed. With higher rotational velocity and lower traverse velocity of the tool excessive, heat is generated which can cause growth of the grain and excessive softening in the processing zone. On contrary, a combination of lower rotational velocity and higher traverse velocity of the tool comes up with deficient heat production which in turn leads to various processing defects, such as tunnel defects, cold shut, and poor material flow [28]. Thus, an optimum balance between the rotational and traverse velocity of the tool is necessary for obtaining a defect-free composite. The heat input or the maximum temperature (T_{max}) achieved during FSP can be calculated through an empirical relationship between T_{max} in the SZ and the processing conditions like tool rotational velocity (ω) and traverse velocity (v) [29]:

$$\frac{T_{max}}{T_m} = K \left(\frac{\omega^2}{v \times 10^4} \right)^\alpha \tag{6.1}$$

where K is a constant (0.65–0.75), α is the exponent (0.04–0.06), and T_m (°C) represents the alloy's melting point temperature. Here, the (ω^2/v) ratio is counted as the pseudo-heat index. Fu et al. [30] confirmed the relationship in Eq. (6.1) through experimental studies and concluded an inverse relationship between T_{max} and heat index. Frigaard et al. [31] suggested another model for calculating average heat input per unit length which is expressed mathematically as:

$$Q = \frac{4}{3}\pi^2 \frac{\mu PNR^3}{v} \tag{6.2}$$

where μ represents friction coefficient, P denotes the pressure (Pa), N represents RPS (rotations per second), and R represents tool radius (m). Thus, from Eq. (6.2), the heat input during FSW/P is directly proportional to the rotational speed of the tool while it is inversely proportional to the tool traverse speed. For a constant μ, P, and R, Eq. (6.2) can be simplified to:

$$Q \propto \frac{\omega}{v} \tag{6.3}$$

Chen and Kovacevic [32] proposed a model by taking into consideration the radius of both the shoulder and the tool pin. The expression can be represented mathematically as:

$$Q = \frac{2\pi\omega\mu P \left(R_0^3 - r_0^3 \right)}{3} \tag{6.4}$$

where μ represents friction coefficient, ω denotes angular speed, P represents axial pressure, R_0 denotes tool shoulder radius, and r_0 denotes tool pin radius.

6.3.1 Effect of Tool Rotational and Traverse Speed

As mentioned earlier, the heat input in the SZ depends upon rotational plus traverse speed of the tool which in turn governs the microstructure as well as the mechanical properties of the composite. In FSP, the distribution and fragmentation of reinforcements also depend upon the rotational or angular velocity of the tool. With increased rotational speed, the particles are further uniformly distributed, breaking up of reinforcement clusters takes place, and the individual particles also get fragmented into the small ones under the severe stirring action generated by the tool. However, at higher rotational speed, augmentation in grain size occurs due to the higher heat input in the SZ [24]. Thus, an optimum combination of tool rotational and traverse speed is required for the homogeneous distribution of particles in the SZ with reduced grain size.

Kurt et al. [33] manufactured Al-SiC surface composite by FSP. In this study, the authors reported that higher rotational with low traverse speed of the tool resulted in higher heat contribution which impacts the final thickness of the surface composite layer, the size of the grains, and precipitates as well as reinforcing particles distribution. With higher heat input, bonding amid the surface composite layer and aluminum substrate also gets enhanced. The higher heat input during FSP also serves the purpose of annealing of the material in SZ, and consequently, the dislocation density gets reduced. With the reduction in dislocation density, the contribution of the dislocation strengthening mechanism is also minimized, and thus the best results are obtained at lower rotational speed in case of Al alloys reinforced with copper powder [34]. Shahraki et al. [35] also observed that the low rotational or high traverse velocity alone is not effective enough for achieving the homogenous dispersion of ZrO_2 particles in AA5083 alloy. Additionally, selecting a low rotational speed in combination with high traverse speed causes poor plastic flow of material, clustering of particles, and porosity formation.

Khayyamin et al. [25] carried out the investigation with three sets of traverse speeds for dispersing SiO_2 in the AZ91 matrix. The higher traverse speed revealed the augmented grain refinement by means of short exposure of material to the higher temperature and thereby avoiding any grain growth. Morisada et al. [36] manufactured a surface composite of AZ91/SiC by dispersing SiC particles by varying the traverse velocity from 25 to 200 mm/min with a fixed rotational velocity of the tool as 1500 rpm. It was detected that the FSPed sample's grain size with SiC particles at varying travel speeds of 25, 50, 100, and 200 mm/min are ~8, ~6, ~5, and ~2 μm, respectively. On the contrary, FSPed sample's grain size deprived of SiC particles at varying travel speeds of 25, 50, 100, and 200 mm/min are ~16, ~13, ~10, and ~5 μm, respectively, as depicted in Figure 6.5. This phenomenon is attributed to the augmentation of generated strain and the pinning effect provided by the SiC particles. In another study, Morisada et al. [37] found that 25 mm/min (high heat input) is the perfect traverse speed for removing entangled carbon nanotubes (CNTs) and dispersing it uniformly in AZ91 matrix instead of 100 mm/min (low heat input). This fact is ascribed to more appropriate viscosity of CNT within AZ31 matrix at reduced traverse velocity. Salehi et al. [38] during the fabrication of AA6061/SiC surface composite, observed that the most influential process parameter is the rotational speed with a 43.70% contribution followed by the traverse speed.

6.3.2 Effect of Tool Tilt Angle

Tool tilt angle also performs a major role due to the plasticized material flow about the tool pin during stirring. The variation in tool tilt angle comes up with a significant change in material flow behavior while carrying out FSP. Tool shoulder is judged as the primary cause of heat generation during FSP. Hence, the heat production at the tool shoulder/workpiece interface (~1585 W) is

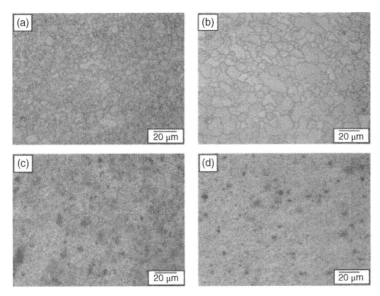

Figure 6.5 OM micrographs from the SZ of the FSPed AZ31 (a and b) and the FSPed AZ31 with the SiC particles (c and d). Travel speed of the rotating tool was 200 mm/min for (a) and (c), and 25 mm/min for (b) and (d) [36] Morisada et al., 2006 / with permission from ELSEVIER.

greater than that at the bottom surface measured along the thickness direction (~19 W) [39]. This higher heat generation at the top surface results in a significant decrease in material yield stress on the upper surface when equated to the decreased yield stress of the material near the bottom surface. As the pin tool progresses, the material ahead of it is extruded and due to the difference in yield stress, the upper region material is first compelled to move toward the trailing edge of the pin tool. Then, material at the bottom region (higher yield stress) naturally flows in an upward direction under the extrusion effect produced by the proceeding pin. In the core, the combined effect of heterogenous heat distribution (83% from the shoulder, 16% from the probe sides, and 1% from the probe tip) and the extrusion effect of the tool pin lead toward the vertically upward flow of material in the processing zone. When the material flows toward pin tool's trailing edge, the force insisted on is fueled radically owing to the pin tool tilting. This phenomenon enables the material in rotation driven by the shoulder to move downward along the thickness direction [40]. Thus, the tool tilt angle portrays a major role in the dissemination of reinforcement and material flow inside the processing zone.

The tilting of the tool from the vertical axis results in the upward lifting of the front end while lowering the rear end of the tool. The lifted front end of the tool behaves as the reservoir for the material flow and in this way, material loss is reduced. Meanwhile, at the rear end, the lowered edges exert a tremendous compressive force over the materials. Decreasing the tilt angle causes a decline in the optimum penetration depth. Thus, the lower tilt angle (≤1°) results in poor forging, the pressure which in turn causes insufficient heat generation and poor material flow. Consequently, defects like surface cracks, particle agglomeration, and tunneling defects were created inside the processing zone. The higher tool tilt angle (>2°) leads to excessive heating, and consequently, grain size also increases. Moreover, with higher tool tilt angle, the plasticized material can outflow simply from the bottom of the tool shoulder which in turn creates a break in the weld leading to the creation of tiny holes on the surface. With an excess and inappropriate increase of tilt angle, the extruded materials from the tool front collide with the tool body on the surface of the workpiece and will be unable to fill the SZ. The Taguchi technique implemented for the best distribution and microhardness of AA6061/SiC composite revealed the 2.5° tool tilt angle as the perfect one with the amalgamation of 1400 rpm tool rotational and 50 mm/min tool travel velocity [26].

6.3.3 Effect of Tool Plunge Depth

The magnitude of defects and homogenous dissemination of reinforcing particles inside surface composite fabricated by FSP also relies upon appropriate tool PD. Similar to the tool tilt angle, the PD remains constant throughout the process and cannot be altered during the process after the process has started. The PD significantly affects frictional heat generation amid the tool shoulder and base material. In case of any variation in PD, the contact region amid the tool shoulder and base material varies which in turn governs the frictional heat generation. For shallow PD, the contact region amid the tool shoulder and base material becomes less and vice versa [26]. The PD also governs the vertical pressure applied by the tool onto the base material. Increased PD indicates the increased pressure which results in improved forging, material flow, and particle dispersion [41]. The lower PD comes up with cavity formation in the SZ and is the most observed processing defect.

The increase in PD beyond a threshold limit causes a surplus heat production which helps in the extra softening of material adjacent to the tool/matrix interface [42]. This additional softening results in the generation of a large amount of flash. The increased amount of flash generation comes with the expulsion of reinforcement from the channels and consequently less dispersion of particles in the composite. Moreover, increasing PD beyond the threshold limit changes the mode of friction amid tool and shoulder. Specifically, with enlarged PD, the sliding friction condition amid tool and substrate changes to sticking friction mode [43]. In the case of sliding friction condition, the real contact area between the tool and substrate is beneath the apparent contact region. The increased PD, in turn, increases the normal load amid the tool and workpiece, and this results in further enhancement of the actual region of contact and the friction force. With increasing normal force, as the actual contact region starts approximating the apparent contact region, the friction mode changes from sliding to sticking. With domination of sticking friction, frictional force becomes equal to the shear force of the softer material which does not depend on the normal force. Further increase in the PD after this situation drastically surges the pressure on the contact surfaces which ultimately helps in sticking of tool and workpiece. Thus, an optimum tool PD is essential for the effective manufacturing of surface composite.

6.3.4 Effect of Number of FSP Passes

The most crucial issue in manufacturing surface composites through FSP is the agglomeration of reinforcements in the matrix since it adversely affects its strength [44]. The effect of agglomeration becomes more noticeable for nano-sized reinforcements when equated to micro-sized reinforcements. The magnitude of the particle clusters can be effectively reduced by escalating the count of FSP passes. The multiple FSP passes are also helpful in uniformly dispersing reinforcements and refining the matrix grain size as shown from the EBSD maps in Figure 6.6 obtained while fabricating Al/TiC surface composite [45]. A decline in grain size from 9 μm in a single FSP to 4 μm in a double pass is observed which is endorsed by the availability of higher grain boundary region after every individual FSP pass as grain boundaries and particle/matrix interfaces act as dislocation sources.

The distortion in succeeding FSP passes instigates the generation of supplementary dislocations from these sources. The high stacking fault energy (SFE) of aluminum drives dynamic recovery and the creation of low-angle sub-grain boundaries. Zohoor et al. [34] while fabricating Al–Mg/Cu

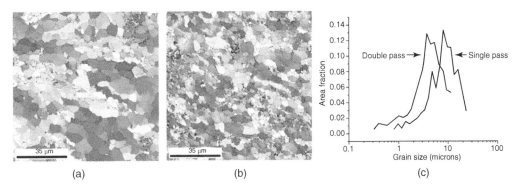

Figure 6.6 EBSD images (IPF + grain boundary map) of the Al–TiC composite subjected to (a) single and (b) double pass FSP. (c) Grain size distribution [45] Bauri et al., 2011 / with permission from ELSEVIER.

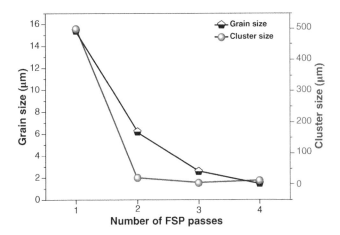

Figure 6.7 Grain size and cluster size variation with the number of FSP passes [47].

composite via FSP found that the particle distribution is strongly affected by the number of FSP passes. Similar observations were made by Lee et al. [46] while fabricating SiO_2/AZ61 nanocomposite via FSP. Here, SiO_2 clusters reduced from 0.1–3 μm with 1-pass FSP to 150 nm with 4-pass FSP. As a consequence, a substantial amendment in the yield strength of the composite was noted as compared to the as-received AZ61 alloy. Khodabakhshi et al. [47] have also observed an extreme drop in the grain and cluster size during the fabrication of AA5052/TiB_2 nanocomposite through reactive FSP. Here, the grain size decreased from four passes as presented in Figure 6.7.

6.4 Microstructural Evolution and Mechanical Properties

FSP produces a significant amount of frictional heat and SPD at the SZ. This heat and plastic deformation at the SZ lead to dynamic recrystallization, because of which refined and equiaxed recrystallized grains are observed in the SZ as depicted in Figure 6.8. Though the grain refinement at SZ is controversial, it is usually assumed that dynamic recrystallization refines the grain size [49]. The crucial factors which affect the nucleation along with the growth during dynamic recrystallization will control the final grain size at SZ. During processing of surface composites via FSP, intense plastic deformation and material amalgamation are observed inside the processed region. The true

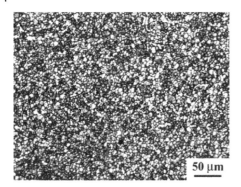

Figure 6.8 Optical micrograph displaying fine and equiaxed grains in FSP of 7075Al-T651 [48] Ma et al., 2008 / Springer Nature.

strain was predicted to be as high as 40 in the case of FSP [50]. This demonstrates the possibility of ceramic reinforcements incorporation into metallic substrates to fabricate surface composites.

Ma et al. [51] observed the homogenous allocation of reinforcements (SiC_P particles) in the Al matrix along with worthy bonding amid the surface composite and substrate. A surface composite of ~100 μm layer was fabricated on top of A356 and 5083 Al substrates as shown in Figure 6.9a,b. The authors have reported uniform dissemination of reinforcing particles within the matrix and good bonding amid the surface composite and substrate. The microhardness of SiC/5083 Al surface composite (see Figure 6.10) is also increased drastically with increased volume fraction of the SiC.

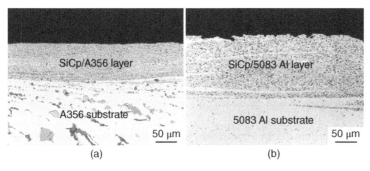

(a)   (b)

Figure 6.9 Optical micrograph of FS-processed surface composites (a) SiC/A356 and (b) SiC/ 5083 Al [51] Ma et al., 2003 / The Minerals, Metals & Materials Society.

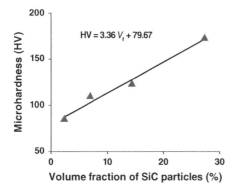

Figure 6.10 Hardness variation with varying volume fraction of SiC in SiC/5083 Al surface composite [48].

In the last decade, several methods were adopted by researchers for the uniform allocation of reinforcing particles in the surface composites manufactured through FSP. One of the techniques is to introduce one or more grooves along the path of FSP in which the reinforcements can be pre-deposited [37]. The suggested approach resulted in the effective dispersion of reinforcement, refined grains, and improved hardness of the fabricated composite. Additionally, an interesting conclusion in this context was made when the consequence of groove depth on the microstructure was calculated. With a groove direction perpendicular to the processing direction, particle squeezing out can be avoided [37]. With increased groove depth, the distribution of particles in base material becomes more homogenous with reduced agglomeration as represented in Figure 6.11 [52].

6.4.1 Microstructural Evolution

Traverse speed and rotational speed (v and ω) are two important aspects which influence the microstructure generation at SZ as these parameters determine the heat input. With lower heat input, more refinement in grain structure is observed, and vice versa. But the heat origination at the processed zone must be necessary enough to plasticize the material. For manufacturing of surface composites, higher tool rotational speed is expected to break and distribute the reinforcements. But increased rotational speed affects grain refinement by generating a higher amount of heat.

Particle clustering is considered a leading concern in the processing of composites as it impacts the strength in a negative way [7]. To break this clustering and to achieve homogenous dissemination of particles, multi-pass FSP is used which results in producing hindrance to the grain size [53]. The reason for this phenomenon is the availability of more grain boundary regions in each subsequent FSP pass which acts as a source for the dislocation generation. More and more dislocations are generated with an increased number of FSP passes. With materials having high SFE (like aluminum), increased dislocation generation steers to dynamic recovery and forms low-angle sub-grain boundaries [54].

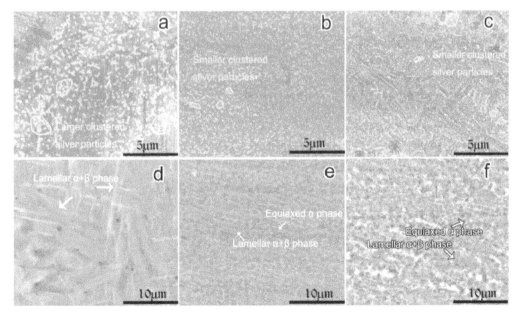

Figure 6.11 SEM image of the stir zone with groove depth (a,d) 1 mm, (b,e) 1.5 mm, and (c,f) 2 mm [52] Yag et al., 2018 / American Chemical Society.

Microstructural inhomogeneity in surface composites processed through FSP is also scrutinized as particle segregation with banded configuration inside the SZ region. Tool geometry performs a vital function in resolving the microstructure as it affects heat generation and material flow. Normally, a concave tool shoulder is preferred for processing as it offers an escape section or reservoir to the plasticized material displaced by the probe. Here, the angle of tilting is also an important factor in maintaining the material pool below the tool as well as facilitating the following edge for processed material extrusion. Again, a tool shoulder with a large diameter helps in generating a high amount of heat and improved material flow. With small diameter tool shoulder, defects such as poor material flow and tunnel defects are generated in the fabricated composites [55].

Adding reinforcement to base material through FSP will change base material structure and composition and the surface composite thus fabricated will exhibit properties influenced by both grain size reduction and reinforcing particles. The occurrence of second-phase particle hinders the grain boundary movement due to recrystallization or grain growth [56]. In this case, the grain diameter (d_z) is expressed as:

$$d_z = \frac{4r}{3V_f} \tag{6.5}$$

where r represents the radius of the particles while V_f represents the volume fraction of the particles.

Analyzing Eq. (6.5), it is noted that grain alteration is inspired by the reduced size and compounded volume fraction of the particles. A graphic illustration of grain growth restrained by the reinforcing particles was presented by Shamsipur et al. [57] as depicted in Figure. 6.12. The authors reported an almost 10 times drop in the size of the titanium grains with the addition of SiC particles. Similar observations like a 2.5 times reduction in AZ31 grain size at SZ with the addition of SiC particles were reported by Morisada et al. [36].

Elucidating the impact of diverse ceramic reinforcements on the same matrix would be interesting as different ceramic particles possess different properties. A similar category study was executed by Abushanab et al. [58] where the authors observed that adding ceramic particles like VC, Al_2O_3, BN, and SiC to AA5250 alloy produces refined and equiaxed grains at the SZ as an outcome of dynamic recrystallization. Homogenous distribution of particles within the matrix is expected for better performance which is persuaded by the category of particles and process parameters which was observed in this study. Not only ceramic reinforcement, but carbon-based reinforcements also assist significantly in reducing the grain size. When C_{60} reinforcements were added to

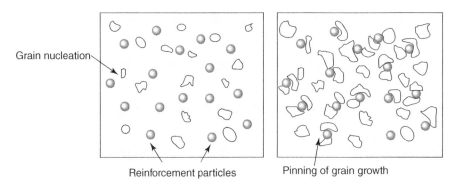

Figure 6.12 Schematic grain growth restraining by reinforcement particles [14].

AZ31, because of the pinning effect exhibited by the C_{60} particles grain size of AZ31 was reduced to a size of ~100 nm [59]. The grain boundary pinning effect can be boosted by augmenting the volume fraction and count of FSP passes. However, the clustering of reinforcement harms the pinning effect which in turn results in generating a bigger grain size than the limiting grain size (dz) [60]. Other allotropes of carbon like CNT, graphene, and graphite are also used for fabricating surface composites because of their enormously high strength and elastic modulus along with superior thermal and electrical properties [49]. The carbon structures establish a superior bonding with the matrix material as portrayed in Figure 6.13a,b. But with increased number of passes, survivability of CNT becomes a critical issue [4]. A strong bond amid reinforcement and the matrix is observed without any intermetallic compound or nano-pores. With pinning effect provided by MWCNT, the grain size at SZ was observed to be 50–100 nm [61]. When powder forging route was adopted to ensure stability of CNTs during FSP, clustering of CNTs was observed. This clustering can be minimized by increasing the number of passes in FSP, but it decreases the CNT length in the composite. The decreased CNT length has a negative impact on the strength of the composite.

Other reinforcement like nanohydroxyapatite (nHA) when used for fabricating nanocomposites with titanium using multi-pass overlapping FSP also reduced the grain size of titanium at SZ without any porosity [62]. Similarly, when Ag is used as reinforcement for fabricating Ti-6Al-4V/Ag nanocomposite, it was observed that the Ag nanoparticles were distributed in a streamlined and stripped manner. Particle aggregation inside the matrix which led to increased dislocation density and associated recrystallization reduced the grain size of the SZ to 10 μm [63].

Addition of second-phase particles helps in reducing the SZ area. This effect was observed by Satishkumar et al. [64] when they fabricated Cu/B_4C surface composite. In this study, the SZ area with addition of B_4C was observed to be ~24 mm² whereas without addition it is observed as ~44 mm². The reason for this behavior can be explained as the hinderance provided by B_4C to the flow of plasticized copper.

In situ composite comes with many advantages like defect-free bonding, improved thermodynamic stability of reinforcements, enhanced compatibility, and improved bonding strength amid reinforcements and matrix [65]. The major challenge in fabricating in situ composite by conventional technique is the segregation of reinforced particles. The SPD during FSP offers an excellent

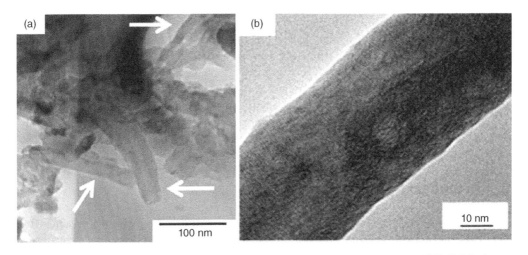

Figure 6.13 TEM images of (a) MWCNT reinforced in AA5059 alloy after two passes of FSP [26] Rathee et al., 2017 / with permission from ELSEVIER and (b) interface showing good bonding between MWCNT and AA1016 alloy [61] Liu et al., 2013 / with permission from ELSEVIER.

material mixing and high heat generation which is beneficial for the in situ reactions and consolidation to form a fully dense composite material. The reinforcements which are produced in this method are usually characterized under nanometric size.

FSP can be also be used for the production of hybrid surface composites where more than one type of reinforcement can be used to achieve the desired functionality [66]. In case of hybrid composite, one softer phase and one harder phase (generally ceramic) particles are added to induce strength and ductility into the matrix. Optimum ratio of the softer and harder phase must be chosen for the maximum enhancement in properties of the matrix. For example, Patil et al. [67] fabricated AA7075-T6/TiC/Graphite (Gr) hybrid surface composite via FSP by varying the TiC:Gr ratios as 90 : 10, 75 : 25, and 60 : 40. The microstructural analysis of the composites showed three different zones in the SZ, i.e. (i) no particle zone, (ii) uniform distribution area, and (iii) dense distribution zone. Refined grains were observed in SZ as a result of secondary particle addition through dynamic recrystallization.

6.4.2 Mechanical Properties

Incorporation of reinforcing particles to the matrix improves the mechanical properties of the matrix when compared to the unreinforced one. The increased mechanical properties depend on the type and size of the reinforcement, uniform dispersion of the reinforcement in the matrix, and its effect on the grain size generation.

Mehdi et al. [68] observed that reinforcing SiC nanoparticles to AA6061 matrix improved the microhardness and tensile strength with increased number of FSP passes. After six FSP passes, the microhardness was found to be 134HV at the centerline of SZ whereas the tensile strength was increased by 26%. SiC particles with varying volume fraction strategies were used by El-Mahallawy et al. [69] for fabricating surface composite. In their study, the surface hardness and tensile strength of the fabricated composite was improved by 112% and 9.7%, respectively. Jain et al. [70] introduced $Fe_{60}Al_{40}$ nanoparticles into the AA5083 matrix via FSP. Here, the introduction of $Fe_{60}Al_{40}$ nanoparticles started the in situ reaction between the reinforcement and the matrix in the presence of high temperature and plastic strain generated during FSP. The matrix structure decomposed, and fine grains were observed at SZ which led to improved microhardness. In the context of hybrid surface composites, Rahman et al. [71] used Ti and SiC particles to improve the mechanical behavior of AA5083 alloy. After FSP, microhardness of the SZ was increased by 38%. Eskandari et al. [72] have also successfully impregnated TiB_2/Al_2O_3 in 8026 aluminum alloy where the authors observed 105% increase in hardness and 86% increase in tensile strength. Nazari et al. [73] also carried out a similar type of study where they used graphene and TiB_2 to fabricate composite with AA6061. In this study, the microhardness across the SZ (as shown in Figure 6.14) was increased by 67% as a consequence of grain refinement. Other mechanical property like yield strength was also increased by 300%. Though the mechanical properties were substantially increased, it registered a 9% loss in elongation.

6.5 Strengthening Mechanisms

Generally, the mechanisms which are operative in bulk composites, also perform a key character in improving the properties of surface composite. There are four major strengthening mechanisms pointed out in case of a particle-reinforced composite, i.e. (i) grain refinement strengthening, (ii) strengthening ascribed by increase in dislocation density arising from variance in coefficient

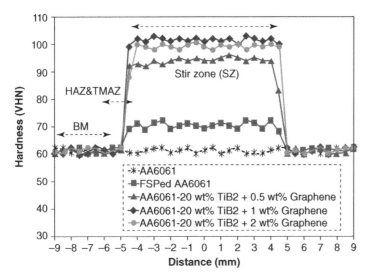

Figure 6.14 Microhardness profiles of AA6061-Gr-TiB2 nanocomposites [73].

of thermal expansion (CTE) amid matrix and reinforcement, (iii) Orowan strengthening, and (iv) Shear lag model [74]. The above-stated mechanisms contribute to the observed overall strength of composites fabricated via FSP.

6.5.1 Grain Refinement Strengthening

Grain refinement during FSP as a consequence of dynamic recrystallization is a widely accepted and reported mechanism [75, 76]. In the presence of reinforcement, the grains are refined to a higher degree due to the pinning effect provided by the reinforcement by sitting on the grain boundaries of the recrystallized grains [77]. This effect is also known as Hall–Petch effect. The augmented strength due to grain refinement can be stated as [78]:

$$\Delta\sigma_y = \frac{k}{\sqrt{d}} \tag{6.6}$$

where $\Delta\sigma y$ represents increased yield strength of matrix after inclusion of reinforcement, d represents average grain size, and k denotes material constant. Thus, Eq. (6.6) indicates that the strength of the composite increases with the decrease in the grain size.

6.5.2 Dislocation Strengthening Due to CTE Mismatch

When large variance in thermal expansion coefficient (CTE) amid matrix and reinforcement exists, it generates prismatic punching of dislocations which causes augmented strength of composite. Strength of the composite is highly influenced by increased dislocation density. The dislocation density depends on the particle's surface area. The relationship between the dislocation density and particle's surface are can be described as [79]:

$$\rho = \frac{10V_f\,\varepsilon}{bt\left(1-V_f\right)} \tag{6.7}$$

where ρ denotes dislocation density, V_f refers to the volume fraction of reinforcement, ε denotes strain caused thermally, b denotes Burgers vector, and t stands for thickness of particle.

The thermal strain (ε) generated owing to disparity in CTE amid the matrix and reinforcement is stated as:

$$\varepsilon = \int_{T_R}^{T_P} \left(\alpha_m - \alpha_p \right) dT = \Delta\alpha\Delta T \tag{6.8}$$

where α_m and α_p represents CTE of matrix and particle, respectively, T_R and T_P represents room temperature and processing temperature, respectively.

The dislocation density (ρ) is also related to the disparity in coefficient of thermal expansion (ΔCTE), difference in processing and room condition temperature (ΔT), volume fraction of the particles (V_f), Burger vector (b), and diameter of the particle (d) as [80]:

$$\rho = \frac{12 \times \Delta CTE \times \Delta T \times V_f}{bd} \tag{6.9}$$

From Eq. (6.9), it can be inferred that with increased disparity amid CTE, dislocation density increases. Due to this phenomenon, increased strength of composite can be written as [80]:

$$\Delta\sigma_y = \alpha Gb \sqrt{\frac{12 \times \Delta CTE \times \Delta T \times V_f}{bd\left(1-V_f\right)}} = \alpha Gb \sqrt{\frac{\rho}{\left(1-V_f\right)}} \tag{6.10}$$

where $\Delta\sigma y$ is the increased yield strength, α denotes coefficient of thermal expansion, G denotes matrix shear modulus, b represents Burgers vector, d denotes particle diameter, V_f denotes volume fraction of the particles, ΔCTE represents disparity in CTE, ΔT represents difference in processing and room condition temperature, and ρ represents increased dislocation density.

6.5.3 Orowan Strengthening

Eq. (6.10) confirms that the increased dislocation density induces strength to the composite. This mechanism is activated when dislocations find restrictions to their movement caused by the inclusion of nano-reinforcements. When dislocations interact with nano-reinforcements, their movement is hindered and due to this blockade, the dislocations start bending between two reinforcements. The bending of the dislocations produces back stress on the further movement of dislocations which in turn contributes to strength of the composite. The increased strength of composite due to Orowan strengthening can be expressed as [81]:

$$\Delta\sigma_{\text{Orowan}} = \frac{0.8 \times MGbln\left(\frac{d}{b}\right)}{2\pi \times \sqrt{(1-v)}(\lambda - d)} \tag{6.11}$$

where $\Delta\sigma_{\text{Orowan}}$ represents the increased strength due to Orowan mechanism, M represents a constant, G denotes the shear modulus of the matrix, b represents Burgers vector, d represents particle diameter, v represents Poisson's ratio, and λ represents interparticle distance.

6.5.4 Strengthening by Load Sharing (Shear Lag Model)

Interfacial shear stress helps in transferring load from matrix to reinforcement and is explained by shear lag model. This model counts for particles having high aspect ratio and thus adds to the strength of the composite. The effectiveness of load transference from matrix to reinforcement relies on the effective linking amid the matrix and reinforcement. The increase in shear strength can be represented as:

$$\sigma_{SL} = \sigma_m \left(1 + 0.5 S V_f\right) \tag{6.12}$$

where σ_{SL} represents yield strength prediction by shear lag model, S is the aspect ratio of particle, and V_f represents volume fraction of reinforcement.

The overall strength when all the mechanisms are acting combinedly can be written as [82]:

$$\sigma_{CY} = \sigma_{mY} \left(1 + 0.5 S V_f\right) \tag{6.13}$$

$$\sigma_{mY} = \sigma_O + \sigma_g + \sigma_{CTE} + \sigma_{OR} \tag{6.14}$$

where σ_{mY} and σ_{CY} represents matrix and composite yield strength, respectively, S represents particle aspect ratio ($S = 1$ for equiaxed one), V_f represents volume fraction of reinforcement, σ_O represents the initial matrix strength, σ_g, σ_{CTE}, and σ_{OR} represents the strength enhancement due to grain refinement, difference in CTE and Orowan strengthening, respectively.

6.6 Defects

When composites are manufactured using FSP, defects may develop at FSPed zone because of inappropriate selection of process parameters similar to the defects in FSW. Obstruction in material flow and heat generation quantity during FSP can cause defect generation in composites. Defects like oxide entrapment, surface galling, lack of penetration, wormholes, scalloping, and lacking in surface fill can occur in FSP [83].

In FSP, heat input quantity plays a key role in invoking different types of defects [84]. For example, while high heat input generates flash, low heat input produces cavity in the processed zone. Similarly, improper selection of tilt angle and tool speed can lead to cavity generation due to abnormal stirring action [55]. Improper or reduced contact time amid material and tool arises due to the decreased value of the ratio between rotational and traverse speed. Decrease in contact time helps in producing tunnel defect along the seam because of restricted material flow and plastic deformation [57]. Improper tool angle can also invoke tunnel defect and large voids [85]. Conversely, when the ratio between rotational speed and traverse speed increases, it produces more heat which can create cavity toward the advancing side, especially in case of low-melting point materials as more material movement happens from forward-moving side to the retreating side. Excess heating can also produce thin flash in materials. Particles clumping also helps in defect generation, i.e. particle clumping produces empty and filled cavity on the advancing and retreating side, respectively. A detailed description of defects along with reasons are listed in Table 6.1.

Table 6.1 Defects observed in the surface composites fabricated via FSP.

Defect type	Defect description	Reasons	Pictorial representation	References
Oxide entrapment	Materials having high oxygen affinity forms oxide which is entrapped in the processing zone.	Inadequate surface preparation before processing.		[83] Zettler et al., 2010 / with permission from ELSEVIER
Dearth of penetration	Insufficient mechanical and chemical bonding with workpiece.	Inaccurate pin height or tool position.		[83] Zettler et al., 2010 / with permission from ELSEVIER
Voids or wormhole	Cavity generation in the direction of advancing side.	(i) Insufficient heat input and material flow. (ii) High value of ratio of rotational to traverse speed.		[86] Zettler et al., 2010 / with permission from ELSEVIER
Surface galling	Rip up of material from top of the weld surface.	Bonding amid workpieces and tool pin while processing.		[1] Zettler et al., 2010 / with permission from ELSEVIER
Improper filling of surface	Void or cavities formation inside processing region.	Insufficient plunge force.		[83] Zettler et al., 2010 / with permission from ELSEVIER

Defect type	Defect description	Reasons	Pictorial representation	References
Improper fusion	Inadequate fusion laps	Occurrence of impurities on edge and surface of material.		[83] Zettler et al., 2010 / with permission from ELSEVIER
Root flow defect	Tendency of sticking to the backing plate by the workpiece due to softening.	(i) Over penetration by the tool due to excessive pin length. (ii) Excessive heat generation.		[29 Arbegast et al., 2008 / with permission from ELSEVIER, 87 Fakih et al., 2018 / with permission from ELSEVIER]
Excessive flash	Extreme exclusion of material from workpiece surface in the course of processing.	(i) Extreme heat generation. (ii) High axial load.		[86] Fakih et al., 2018 / with permission from ELSEVIER
Nugget failure	Inadequate growth of nugget shape.	(i) High heat flow toward stir zone. (ii) Excess material movement toward stir zone. (iii) High traverse velocity.		[83] Zettler et al., 2010 / with permission from ELSEVIER
Scalloping	Occurrence of number of small voids toward forward moving side of the processing zone in the weld area.	Improper pin and shoulder dimensions.		[88] Kwon. et al., 2003 / with permission from ELSEVIER

6.7 Summary and Future Directions

The present chapter demonstrated that the solid-state FSP is a novel and versatile technique for the fabrication of surface composites. The outcome of this chapter can be summarized as follows:

(i) The grain refinement through dynamic recrystallization, homogeneous dispersion of reinforcements, and reinforcement particle fragmentations are some of the fascinating key features associated with the FSP. Additionally, the FSP permits a wide variety of particles ranging from ceramic to metallic, carbonaceous reinforcements (such as graphene, CNTs, and fullerenes), natural fibers, etc., to be incorporated into the matrix material so as to form an efficient surface composite.

(ii) The process parameters such as tool rotational speed, travel speed, tilt angle, and PD significantly affect the particle dispersion characteristics in the composite fabricated via FSP. Thus, the suitable parameter selection is required for the efficient composite fabrication.

(iii) The tool design such as shoulder diameter and probe length provide an easy access to control the composite microstructure such as depth of particle impregnation.

(iv) FSP offers an advantage of rapid removal of reaction products from the interface during in situ composite fabrication which helps in further proceeding of the reaction. Additionally, the particles formed during in situ reaction are of nano-meter size and thus the bimodal distribution of particles can be easily obtained by the FSP.

(v) The Hall–Petch strengthening, dislocation strengthening, Orowan strengthening, and load transfer from matrix to reinforcement are the primary strengthening mechanisms in the surface composites fabricated via FSP.

Future directions: Most of the reports dealing with the fabrication of surface composites via FSP are primarily focused on the flat plates. However, the advancements should be made for the surface modification of complex geometries with the FSP. Moreover, the surface alloying through FSP should also be focused for the improved surface properties of various alloys.

References

1 Gandra, J., Miranda, R., Vilaa, P. et al. (2011). Functionally graded materials produced by friction stir processing. *J. Mater. Process. Technol.* 211: 1659–1668.

2 Mishra, R.S.S. and Ma, Z.Y.Y. (2005). Friction stir welding and processing. *Mater. Sci. Eng.: R: Rep.* 50: 1–78.

3 Su, J.-Q., Nelson, T.W., and Sterling, C.J. (2005). Microstructure evolution during FSW/FSP of high strength aluminum alloys. *Mater. Sci. Eng., A* 405: 277–286.

4 Sharma, A., Gupta, G., and Paul, J. (2021). A comprehensive review on the dispersion and survivability issues of carbon nanotubes in Al/CNT nanocomposites fabricated via friction stir processing. *Carbon Lett.* 31: 339–370.

5 Sharma, A., Morisada, Y., and Fujii, H. (2021). Influence of aluminium-rich intermetallics on microstructure evolution and mechanical properties of friction stir alloyed Al Fe alloy system. *J. Manuf. Processes* 68: 668–682.

6 Sharma, A., Morisada, Y., and Fujii, H. (2022). Bending induced mechanical exfoliation of graphene interlayers in a through thickness Al-GNP functionally graded composite fabricated via novel single-step FSP approach. *Carbon* 186: 475–491.

7 Sharma, A., Sharma, V.M., and Paul, J. (2019). A comparative study on microstructural evolution and surface properties of graphene/CNT reinforced Al6061−SiC hybrid surface composite fabricated via friction stir processing. *Trans. Nonferrous Met. Soc.* 29: 2005–2026.

8 Pradeep, S., Jain, V.K.S., Muthukumaran, S. et al. (2021). Microstructure and texture evolution during multi-pass friction stir processed AA 5083. *Mater. Lett.* 288: 129382.

9 Raja, R., Jannet, S., and Thankachan, T. (2021). Investigation of hybrid copper surface composite synthesized via FSP. *Mater. Manuf. Processes* 36: 1377–1383.

10 Sunil, B.R., Reddy, G.P.K., Patle, H. et al. (2016). Magnesium based surface metal matrix composites by friction stir processing. *J. Magnesium Alloys* 4: 52–61.

11 Navazani, M. and Dehghani, K. (2016). Fabrication of Mg-ZrO_2 surface layer composites by friction stir processing. *J. Mater. Process. Technol.* 229: 439–449.

12 Li, B., Shen, Y., Luo, L. et al. (2013). Fabrication of TiCp/Ti–6Al–4V surface composite via friction stir processing (FSP): process optimization, particle dispersion-refinement behavior and hardening mechanism. *Mater. Sci. Eng., A* 574: 75–85.

13 Sarmadi, H., Kokabi, A.H.H.H., and Seyed Reihani, S.M.M.M. (2013). Friction and wear performance of copper–graphite surface composites fabricated by friction stir processing (FSP). *Wear* 304: 1–12.

14 Sharma, V., Prakash, U., and Kumar, B.V.V.M. (2015). Surface composites by friction stir processing: a review. *J. Mater. Process. Technol.* 224: 117–134.

15 Lim, D.K.K.K., Shibayanagi, T., and Gerlich, A.P.P.P. (2009). Synthesis of multi-walled CNT reinforced aluminium alloy composite via friction stir processing. *Mater. Sci. Eng., A* 507: 194–199.

16 Choi, D.-H., Kim, Y.-H., Ahn, B.-W. et al. (2013). Microstructure and mechanical property of A356 based composite by friction stir processing. *Trans. Nonferrous Met. Soc. China* 23: 335–340.

17 Avettand-Fènoël, M.-N., Netto, N., Simar, A. et al. (2022). Design of a metallic glass dispersion in pure copper by friction stir processing. *J. Alloys Compd.* 907: 164522.

18 Akramifard, H.R., Shamanian, M., Sabbaghian, M. et al. (2014). Microstructure and mechanical properties of Cu/SiC metal matrix composite fabricated via friction stir processing. *Mater. Des.* (1980-2015) 54: 838–844.

19 Miranda, R.M.M., Santos, T.G., Gandra, J. et al. (2013). Reinforcement strategies for producing functionally graded materials by friction stir processing in aluminium alloys. *J. Mater. Process. Technol.* 213: 1609–1615.

20 Jeon, C.-H., Jeong, Y.-H., Seo, J.-J. et al. (2014). Material properties of graphene/aluminum metal matrix composites fabricated by friction stir processing. *Int. J. Precis. Eng. Manuf.* 15: 1235–1239.

21 Mazaheri, Y., Karimzadeh, F., and Enayati, M.H. (2011). A novel technique for development of A356/Al_2O_3 surface nanocomposite by friction stir processing. *J. Mater. Process. Technol.* 211: 1614–1619.

22 Hodder, K.J., Izadi, H., McDonald, A.G. et al. (2012). Fabrication of aluminum−alumina metal matrix composites via cold gas dynamic spraying at low pressure followed by friction stir processing. *Mater. Sci. Eng., A* 556: 114–121.

23 Huang, Y., Li, J., Wan, L. et al. (2018). Strengthening and toughening mechanisms of CNTs/ Mg-6Zn composites via friction stir processing. *Mater. Sci. Eng., A* 732: 205–211.

24 Azizieh, M., Kokabi, A.H., and Abachi, P. (2011). Effect of rotational speed and probe profile on microstructure and hardness of AZ31/Al_2O_3 nanocomposites fabricated by friction stir processing. *Mater. Des.* 32: 2034–2041.

25 Khayyamin, D., Mostafapour, A., and Keshmiri, R. (2013). The effect of process parameters on microstructural characteristics of AZ91/SiO_2 composite fabricated by FSP. *Mater. Sci. Eng., A* 559: 217–221.

26 Rathee, S., Maheshwari, S., Siddiquee, A.N. et al. (2017). Effect of tool plunge depth on reinforcement particles distribution in surface composite fabrication via friction stir processing. *Def. Technol.* 13: 86–91.

27 Vigneshkumar, M., Padmanaban, G., and Balasubramanian, V. (2019). Influence of tool tilt angle on the formation of friction stir processing zone in cast magnesium alloy ZK60/SiCp surface composites. *Metall. Microstruct. Anal.* 8: 58–66.

28 Lombard, H., Hattingh, D.G., Steuwer, A. et al. (2008). Optimising FSW process parameters to minimise defects and maximise fatigue life in 5083-H321 aluminium alloy. *Eng. Fract. Mech.* 75: 341–354.

29 Arbegast, W.J. (2008). A flow-partitioned deformation zone model for defect formation during friction stir welding. *Scr. Mater.* 58: 372–376.

30 Fu, R., Zhang, J., Li, Y. et al. (2013). Effect of welding heat input and post-welding natural aging on hardness of stir zone for friction stir-welded 2024-T3 aluminum alloy thin-sheet. *Mater. Sci. Eng., A* 559: 319–324.

31 Frigaard, Ø., Grong, Ø., and Midling, O.T. (2001). A process model for friction stir welding of age hardening aluminum alloys. *Metall. Mater. Trans. A* 32: 1189–1200.

32 Chen, C.M. and Kovacevic, R. (2003). Finite element modeling of friction stir welding—thermal and thermomechanical analysis. *Int. J. Mach. Tools Manuf.* 43: 1319–1326.

33 Kurt, H.I. (2016). Influence of hybrid ratio and friction stir processing parameters on ultimate tensile strength of 5083 aluminum matrix hybrid composites. *Composites, Part B* 93: 26–34.

34 Zohoor, M., Besharati Givi, M.K., and Salami, P. (2012). Effect of processing parameters on fabrication of Al–Mg/Cu composites via friction stir processing. *Mater. Des.* 39: 358–365.

35 Shahraki, S., Khorasani, S., Abdi Behnagh, R. et al. (2013). Producing of AA5083/ZrO$_2$ nanocomposite by friction stir processing (FSP). *Metall. Mater. Trans. B* 44: 1546–1553.

36 Morisada, Y., Fujii, H., Nagaoka, T. et al. (2006). Effect of friction stir processing with SiC particles on microstructure and hardness of AZ31. *Mater. Sci. Eng., A* 433: 50–54.

37 Morisada, Y., Fujii, H., Nagaoka, T. et al. (2006). MWCNTs/AZ31 surface composites fabricated by friction stir processing. *Mater. Sci. Eng., A* 419: 344–348.

38 Salehi, M., Farnoush, H., and Mohandesi, J.A. (2014). Fabrication and characterization of functionally graded Al-SiC nanocomposite by using a novel multistep friction stir processing. *Mater. Des.* 63: 419–426.

39 Schmidt, H., Hattel, J., and Wert, J. (2004). An analytical model for the heat generation in friction stir welding. *Modell. Simul. Mater. Sci. Eng.* 12: 143–157.

40 Liu, H., Hu, Y., Peng, Y. et al. (2016). The effect of interface defect on mechanical properties and its formation mechanism in friction stir lap welded joints of aluminum alloys. *J. Mater. Process. Technol.* 238: 244–254.

41 Mahto, R.P., Bhoje, R., Pal, S.K. et al. (2016). A study on mechanical properties in friction stir lap welding of AA 6061-T6 and AISI 304. *Mater. Sci. Eng., A* 652: 136–144.

42 Rathee, S., Maheshwari, S., and Siddiquee, A.N. (2018). Issues and strategies in composite fabrication via friction stir processing: a review. *Mater. Manuf. Processes* 33: 239–261.

43 Asadi, P., Faraji, G., and Besharati, M.K. (2010). Producing of AZ91/SiC composite by friction stir processing (FSP). *Int. J. Adv. Manuf. Technol.* 51: 247–260.

44 Sharma, A., Morisada, Y., and Fujii, H. (2022). Through-thickness localized strain distribution and microstructural characterization of functionally graded Al/GNP composite fabricated by friction stir processing BT. *Light Metals* 2022: 274–282.

45 Bauri, R., Yadav, D., and Suhas, G. (2011). Effect of friction stir processing (FSP) on microstructure and properties of Al–TiC in situ composite. *Mater. Sci. Eng., A* 528: 4732–4739.

46 Lee, C., Huang, J., and Hsieh, P. (2006). Mg based nano-composites fabricated by friction stir processing. *Scr. Mater.* 54: 1415–1420.

47 Khodabakhshi, F., Simchi, A., and Kokabi, A.H. (2017). Surface modifications of an aluminum-magnesium alloy through reactive stir friction processing with titanium oxide nanoparticles for enhanced sliding wear resistance. *Surf. Coat. Technol.* 309: 114–123.

48 Ma, Z.Y.Y. (2008). Friction stir processing technology: a review. *Metall. Mater. Trans. A* 39: 642–658.

49 Sharma, A., Narsimhachary, D., Sharma, V.M. et al. (2019). Surface modification of Al6061-SiC surface composite through impregnation of graphene, graphite & carbon nanotubes via FSP: a tribological study. *Surf. Coat. Technol.* 368: 175–191.

50 Heurtier, P., Desrayaud, C., and Montheillet, F. (2002). A thermomechanical analysis of the friction stir welding process. *Mater. Sci. Forum* 396–402: 1537–1542.

51 Ma, Z.Y. and Mishra, R.S. (2003). Surface Engineering. *Materials Science II : Proceedings of a Symposia Sponsored by the Surface Engineering Committee of the Materials Processing and Manufacturing Division (MPMD) of TMS (The Minerals, Metals & Materials Society)*: 2003 TMS Annual Meet. 243–250.

52 Yang, Z., Gu, H., Sha, G. et al. (2018). TC4/Ag metal matrix nanocomposites modified by friction stir processing: surface characterization, antibacterial property, and cytotoxicity in vitro. *ACS Appl. Mater. Interfaces* 10: 41155–41166.

53 Sharma, A., Morisada, Y., and Fujii, H. (2023). Interfacial microstructure and strengthening mechanisms of SPSed Al/GNP nanocomposite subjected to multi-pass friction stir processing. *Mater. Charact.* 197: 112652.

54 Yadav, D. and Bauri, R. (2012). Effect of friction stir processing on microstructure and mechanical properties of aluminium. *Mater. Sci. Eng., A* 539: 85–92.

55 Elangovan, K., Balasubramanian, V., and Valliappan, M. (2008). Influences of tool pin profile and axial force on the formation of friction stir processing zone in AA6061 aluminium alloy. *Int. J. Adv. Manuf. Technol.* 38: 285–295.

56 Rohrer, G.S. (1948). Introduction to grains, phases, and interfaces—an interpretation of microstructure. *Trans. AIME* 175: 15–51, by C.S. Smith. (2010). *Metall. Mater. Trans. A* 41:1063–1100.

57 Shamsipur, A., Kashani-Bozorg, S.F., and Zarei-Hanzaki, A. (2011). The effects of friction-stir process parameters on the fabrication of Ti/SiC nano-composite surface layer. *Surf. Coat. Technol.* 206: 1372–1381.

58 Abushanab, W.S., Moustafa, E.B., Melaibari, A.A. et al. (2021). A novel comparative study based on the economic feasibility of the ceramic nanoparticles role's in improving the properties of the AA5250 nanocomposites. *Coatings* 11: 977.

59 Morisada, Y., Fujii, H., NAGAOKA, T. et al. (2006). Nanocrystallized magnesium alloy – uniform dispersion of C60 molecules. *Scr. Mater.* 55: 1067–1070.

60 Shafiei-Zarghani, A., Kashani-Bozorg, S.F., and Zarei-Hanzaki, A. (2009). Microstructures and mechanical properties of Al/Al$_2$O$_3$ surface nano-composite layer produced by friction stir processing. *Mater. Sci. Eng., A* 500: 84–91.

61 Liu, Q., Ke, L., Liu, F. et al. (2013). Microstructure and mechanical property of multi-walled carbon nanotubes reinforced aluminum matrix composites fabricated by friction stir processing. *Mater. Des.* 45: 343–348.

62 Rahmati, R. and Khodabakhshi, F. (2018). Microstructural evolution and mechanical properties of a friction-stir processed Ti-hydroxyapatite (HA) nanocomposite. *J. Mech. Behav. Biomed. Mater.* 88: 127–139.

63 Wang, L., Xie, L., Shen, P. et al. (2019). Surface microstructure and mechanical properties of Ti-6Al-4V/Ag nanocomposite prepared by FSP. *Mater. Charact.* 153: 175–183.

64 Sathiskumar, R., Murugan, N., Dinaharan, I. et al. (2013). Characterization of boron carbide particulate reinforced in situ copper surface composites synthesized using friction stir processing. *Mater. Charact.* 84: 16–27.

65 Sharma, A. and Paul, J. (2020). a review on the fabrication of *in situ* metal matrix composite during friction stir welding. *Mater. Sci. Forum* 978: 191–201.

66 Sharma, A., Sharma, V.M., Mewar, S. et al. (2018). Friction stir processing of Al6061-SiC-graphite hybrid surface composites. *Mater. Manuf. Processes* 33: 795–804.

67 Patil, N.A., Pedapati, S.R., Mamat, O. et al. (2021). Morphological characterization, statistical modeling and wear behavior of AA7075-titanium carbide-graphite surface composites via friction stir processing. *J. Mater. Res. Technol.* 11: 2160–2180.

68 Mehdi, H. and Mishra, R.S. (2022). Consequence of reinforced SiC particles on microstructural and mechanical properties of AA6061 surface composites by multi-pass FSP. *J. Adhes. Sci. Technol.* 36: 1279–1298.

69 El-Mahallawy, N.A., Zoalfakar, S.H., and A. Abdel Ghaffar Abdel Maboud. (2019). Microstructure investigation, mechanical properties and wear behavior of Al 1050/SiC composites fabricated by friction stir processing (FSP). *Mater. Res. Express* 6: 96522.

70 Jain, V.K., Yadav, M.K., Siddiquee, A.N. et al. (2021). Optimization of friction stir processing parameters for enhanced microhardness of AA5083/Al-Fe in-situ composites via Taguchi technique. *Mater. Sci. Eng. Appl.* 1: 55–61.

71 Rahman, Z., Siddiquee, A.N., and Khan, Z.A. (2021). Effect of Ti/SiC reinforcement on AA5083 surface composites prepared by friction stir processing. *IOP Conf. Ser.: Mater. Sci. Eng.* 1149: 12001.

72 Eskandari, H. and Taheri, R. (2015). A novel technique for development of aluminum alloy matrix/ TiB_2/Al_2O_3 hybrid surface nanocomposite by friction stir processing. *Procedia Mater. Sci.* 11: 503–508.

73 Nazari, M., Eskandari, H., and Khodabakhshi, F. (2019). Production and characterization of an advanced AA6061-Graphene-TiB_2 hybrid surface nanocomposite by multi-pass friction stir processing. *Surf. Coat. Technol.* 377: 124914.

74 Lloyd, D.J. (1994). Particle reinforced aluminium and magnesium matrix composites. *Int. Mater. Rev.* 39: 1–23.

75 Sharma, A., Sharma, V.M., Sahoo, B. et al. (2019). Effect of multiple micro channel reinforcement filling strategy on Al6061-graphene nanocomposite fabricated through friction stir processing. *J. Manuf. Processes* 37: 53–70.

76 Sharma, A., Sharma, V.M., Sahoo, B. et al. (2019). Effect of exfoliated few-layered graphene on corrosion and mechanical behaviour of the graphitized Al–SiC surface composite fabricated by FSP. *Bull. Mater. Sci.* 42: 204.

77 Liu, Z.Y.Y.Y., Xiao, B.L.L.L., Wang, W.G.G.G. et al. (2012). Singly dispersed carbon nanotube/ aluminum composites fabricated by powder metallurgy combined with friction stir processing. *Carbon* 50: 1843–1852.

78 Choi, H.J.J.J., Kwon, G.B.B., Lee, G.Y.Y. et al. (2008). Reinforcement with carbon nanotubes in aluminum matrix composites. *Scr. Mater.* 59: 360–363.

79 Sahoo, B. and Paul, J. (2018). Solid state processed Al-1100 alloy/MWCNT surface nanocomposites. *Materialia* 2: 196–207.

80 Luster, J.W., Thumann, M., and Baumann, R. (1993). Mechanical properties of aluminium alloy 6061–Al_2O_3 composites. *Mater. Sci. Technol.* 9: 853–862.

81 Yoo, S.J.J., Han, S.H.H., and Kim, W.J.J. (2013). Strength and strain hardening of aluminum matrix composites with randomly dispersed nanometer-length fragmented carbon nanotubes. *Scr. Mater.* 68: 711–714.

82 Ferguson, J.B., Lopez, H., Kongshaug, D. et al. (2012). Revised orowan strengthening: effective interparticle spacing and strain field considerations. *Metall. Mater. Trans. A* 43: 2110–2115.

83 Zettler, R., Vugrin, T., and Schmücker, M. (2010). Effects and defects of friction stir welds. *Frict. Stir Weld.* 245–276.

84 Kim, Y.G.G., Fujii, H., Tsumura, T. et al. (2006). Three defect types in friction stir welding of aluminum die casting alloy. *Mater. Sci. Eng., A* 415: 250–254.

85 Sharifitabar, M., Sarani, A., Khorshahian, S. et al. (2011). Fabrication of 5052Al/Al$_2$O$_3$ nanoceramic particle reinforced composite via friction stir processing route. *Mater. Des.* 32: 4164–4172.

86 Fakih, M.A., Mustapha, S., Tarraf, J. et al. (2018). Detection and assessment of flaws in friction stir welded joints using ultrasonic guided waves: experimental and finite element analysis. *Mech. Syst. Signal Process.* 101: 516–534.

87 Khodabakhshi, F., Simchi, A., Kokabi, A. et al. (2014). Strain rate sensitivity, work hardening, and fracture behavior of an Al-Mg TiO$_2$ nanocomposite prepared by friction stir processing. *Metall. Mater. Trans. A* 45: 4073–4088.

88 Kwon, Y. (2003). Mechanical properties of fine-grained aluminum alloy produced by friction stir process. *Scr. Mater.* 49: 785–789.

7

Friction Stir Welding of Dissimilar Metals

Narayan Sahadu Khemnar[1,2], Yogesh Ramrao Gunjal[1,3], Vijay Shivaji Gadakh[1,2], and Amrut Shrikant Mulay[4]

[1] *Department of Mechanical Engineering, Dr. Vithalrao Vikhe Patil College of Engineering, Savitribai Phule Pune University, Ahmednagar, Maharashtra, India*
[2] *Department of Automation and Robotics Engineering, Amrutvahini College of Engineering, Savitribai Phule Pune University, Ahmednagar, Maharashtra, India*
[3] *Department of Mechanical Engineering, Amrutvahini College of Engineering, Savitribai Phule Pune University, Ahmednagar, Maharashtra, India*
[4] *Department of Mechanical Engineering, Sardar Vallabhbhai National Institute of Technology, Surat, Gujarat, India*

7.1 Introduction

The dissimilar material welding attracts the attention of many industries due to its technical and economic benefits, including the automobile. Researchers worldwide now acknowledge the recognized benefits of employing the solid-state welding technique as a promising solution for difficult-to-weld materials and additive manufacturing [1]. FSW is a superior joining method in contrast to traditional joining methods due to its energy efficiency, versatility, solid-state nature, and environment friendliness [2]. FSW, usually recognized as the most notable development in this subject, has revolutionized the joining of similar and dissimilar materials. The dissimilar materials using the FSW process can offer several benefits, including improved performance, weight reduction, cost reduction, and increased material availability. Joining two dissimilar metals is very much crucial in industrial applications.

Due to new environmental rules, the transport industries must strictly limit greenhouse gas emissions. The vehicle's structure can be made lighter as one method of reducing the emissions. Research-minded automotive industries are now aiming for lightweight materials to enhance fuel efficiencies. Al alloys with other alloys offer dominant strength assurance without losing a lightweight nature. In dissimilar metals, particularly lightweight metals, Al alloy in combination with steel is an advantage in the automotive and aerospace industries. This chapter begins with an Introduction, followed by applications of FSW in dissimilar material joining, as depicted in Section 7.2. Section 7.3 describes the issues related to dissimilar material joining. Section 4 elucidates the FSW of dissimilar material joining of Aluminum (Al) alloys, Al–Copper (Cu), Al–Titanium (Ti), and Al–Steel alloys. The recent developments in tool design and tool materials are explained in Section 7.5. The parameter optimization and common defects that occur are depicted in Sections 7.6 and 7.7, respectively. Finally, future recommendations for dissimilar metal joining are described in Section 7.8.

7.2 Application Areas of Dissimilar Material Joining

The dissimilar material joining can help create more functional, cost-effective, and readily available products while ensuring that different materials can work together effectively. The need for dissimilar material joining becomes necessary in order to achieve a specific function or performance. For example, a metal component may need to be attached to a plastic component to create a more lightweight and durable product. Similarly, using different materials or multi-materials (Figure 7.1) for a product may be more cost-effective than a single material. On the other hand, tailor-welded blanks (Figure 7.2) offer great potential in the automobile sector, where industries aim for weight reduction, like side frames, doors, pillars, and rails [3].

7.3 Issues for Dissimilar Material Joining

Traditionally, fusion and solid-state joining methods have been employed for dissimilar material joining. The choice of joining techniques depends on various factors, such as the types of materials being joined, the desired joint strength, and the required production speed. Most issues in dissimilar material joining are common in fusion and solid-state joining methods, but these are less severe in solid-state joints. The process of joining dissimilar materials using various techniques is complex, mainly due to the IMC formation, thermal expansion differences, and significant residual stress [4]. The variations in melting point, thermal expansion, and thermal conductivity are common issues encountered in both joining techniques. However, large temperature gradients across the joint create high residual stresses that can lead to distortion, cracking, or other forms of damage; porosity, brittle phase formation, shape, size, and distribution are commonly found in fusion welding techniques. The brittle phase formation is also observed in solid-state joining

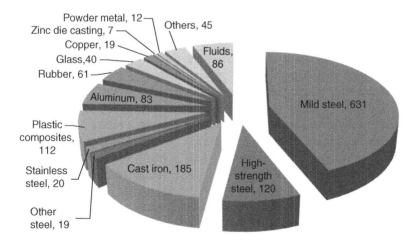

Figure 7.1 Distribution of material of total vehicle curb-weight in kilogram. Source: Kumar et al. [3]/with permission of Elsevier.

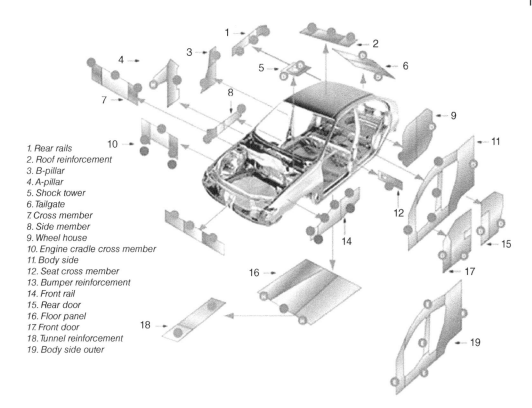

1. Rear rails
2. Roof reinforcement
3. B-pillar
4. A-pillar
5. Shock tower
6. Tailgate
7. Cross member
8. Side member
9. Wheel house
10. Engine cradle cross member
11. Body side
12. Seat cross member
13. Bumper reinforcement
14. Front rail
15. Rear door
16. Floor panel
17. Front door
18. Tunnel reinforcement
19. Body side outer

Figure 7.2 Use of tailor-welded blanks in automobile. Source: Kumar et al. [3]/with permission of Elsevier.

techniques, but they are less severe than in fusion welding techniques. Solid-state welding methods can minimize the above problem. Different solid-state welding techniques employed for dissimilar material joining include FSW, ultrasonic welding, explosive welding or cladding, pressure welding, accumulative roll bonding, electromagnetic or magnetic pulse welding, diffusion welding, brazing and brazing, and soldering [5].

In the FSW process, following are the critical issues related to dissimilar material joining, such as the coefficient of friction, thermal diffusivity, coefficients of thermal expansion, and softening characteristics. The heat generation and material flow decide the dissimilar welds' resulting structure properties. In addition to this, the process variables like tool rotational speed (TRS), weld speed (WS), tool tilt angle, position of materials, and tool pin offset (TPO) play a vital role in determining weld performance.

Addressing these challenges requires a deep understanding of the welding process and carefully selecting materials and process parameters. New interlayer or transition materials, tool designs, and optimization techniques can also be developed to enhance the welding joint's quality. Proper pre-weld and post-weld treatments can also help in minimizing the formation of IMCs and reduce the risk of corrosion.

7.4 FSW of Dissimilar Materials

Figure 7.3 shows the year-wise progression of FSW of dissimilar metal joining. FSW of dissimilar Al alloy joints was first reported in 1998 [6]. Later reported literature shows FSW of Al to Cu, Al to

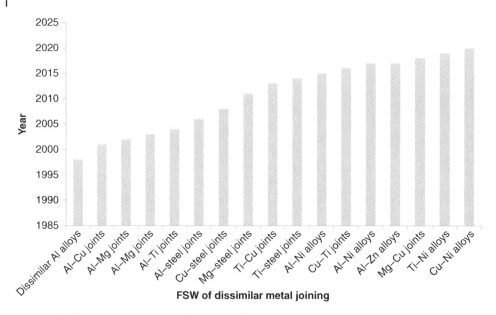

Figure 7.3   Year-wise progression of FSW of dissimilar metal joining.

Mg, Al to Ti, Al to Steel, steel to Cu joining, etc. Most of the works reported are process parameter optimization, tool design, joint geometry, structure–property correlation, and interlayer studies. The following sections cover FSW of dissimilar Al alloys, Al–Cu, Al–Ti, and Al–Steel joining with a discussion of microstructure, mechanical properties studies, different IMCs formed, etc.

7.4.1   FSW of Dissimilar Aluminum Alloys

Figure 7.4 shows some critical developments in FSW of dissimilar Al alloys joining. The first study of FSW of dissimilar Al alloys joining was reported in 1998, later reported works show different

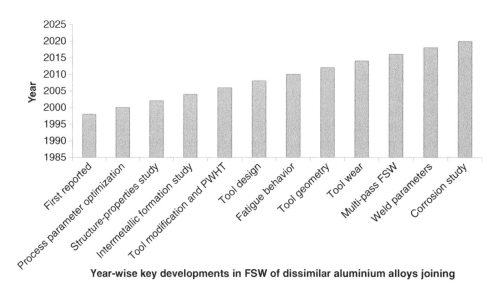

Figure 7.4   Year-wise progression of FSW of dissimilar aluminum alloy joining.

studies on the optimization of process parameters, microstructure and mechanical properties, IMC formation, joint strength improvement with tool modification and post-weld heat treatment (PWHT), tool design, tool wear, multiple passes, and corrosion studies.

Ghosh et al. [7] joined dissimilar A356 and 6061 Al alloys using FSW at constant TRS with different WSs from 70 to 240 mm/min. It is claimed that increasing the WS, strain rate, and Zener Holloman Parameter increases gradually, resulting in a fine grain structure of 6061 within the weld zone (WZ). Moreover, a reduction in heat input and a rise in temperature at the WZ are reported when TRS is increased.

The material flow is complex and has an intercalated flow pattern in the WZ using FSW of dissimilar material joining. These complex flow patterns are shown in Figure 7.5. The onion ring structures in the WZ are observed for dissimilar FSW of AA2024-AA 7039 alloy, as shown in Figure 7.5a; similarly, a complex intercalated structure is reported for joining AA2024-AA6061 alloy. Fadi et al. [9] developed a thermomechanical model for FSW of AA6061-T6 and AA5083-O Al alloys with different tool geometry. It is reported that placing AA6061-T6 toward the advancing side (AS) minimizes the process's maximum temperature and strain rate; however, there is an increase in tool reaction load. The feature tool pin profiles produced better mixing of the material, resulting in less volumetric defects and improved joint quality.

The starting temper and PWHT condition's impact on the structural characteristics of FSW butt joints between AA7075-O and AA6061-O, as well as AA7075-T6 and AA6061-T6, were investigated by İpekoğlu and Çam [10]. It is reported that the joint area exhibited increased hardness values in

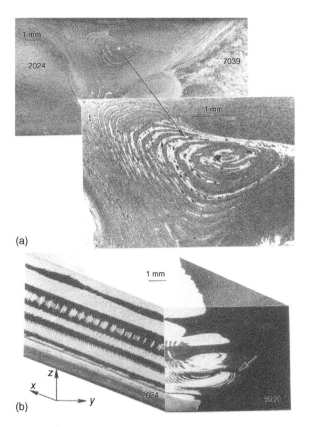

(a)

(b)

Figure 7.5 Complex intercalated material flow pattern in the WZ for FSW of dissimilar metals. (a) microstructure in AA 2024/AA7039 WZ and (b) intercalated material flow patterns for AA 2024/Al 6061. Source: Murr [8]/Springer Nature.

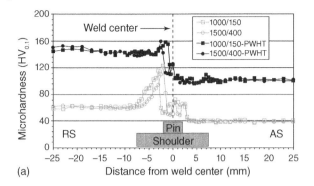

(a)

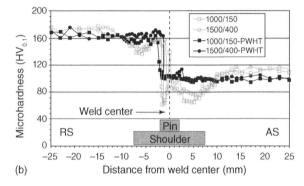

(b)

Figure 7.6 Hardness profiles in as-welded and PWHT condition: (a) O joints and (b) T6 joints. Source: İpekoğlu and Çam [10]/Springer Nature.

O-temper condition whereas reduced hardness was found in T6-temper condition, as depicted in Figure 7.6. Additionally, both temper conditions increased tensile strength values (greater than 85%) following PWHT.

7.4.2 FSW of Aluminum-to-Copper

Due to their excellent electrical and thermal conductivity, Cu and Al metals are utilized in electrical and thermal applications [11]. Fusion welding methods are undesirable to join Al–Cu materials due to the formation of large IMCs, which are very hard and brittle and cause several defects. The year-wise progression of FSW of Al–Cu material joining is depicted in Figure 7.7.

Recently, there has been a notable surge in research focused on Al–Cu joining due to its promising applications across various industries, including nuclear, aerospace, chemical, electrical, electronics, and power generation [12]. However, Al–Cu joining by traditional fusion welding is difficult due to poor weldability [13]. The following parameters are crucial for obtaining sound joints for FSW of Al–Cu alloy joints: tool design, TPO, process parameters, material position, reducing welding defects, and IMC formation.

Xue et al. [14] joined AA 1060 Al alloy with commercially pure (CP) Cu using friction stir lap joining. It is reported that good interface or metallurgical bonding is due to the IMC formation. Akbari et al. [15] studied the influence of materials' position in the lap FSW of Al–Cu alloy. It is reported that maximum temperature has a major impact in the increase of the joint's fracture load. Higher maximum temperatures are desirable for an increase in fracture load.

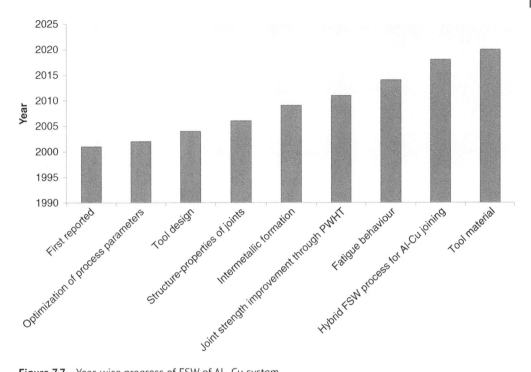

Figure 7.7 Year-wise progress of FSW of Al–Cu system.

Similarly, top-placed low thermal conductivity material (Al) and bottom-placed high thermal conductivity material (Cu) generate greater heat in the weld region, thereby defect-free fined grain structure and high joint's fracture load are obtained. On the contrary, if the material position is reversed, then poor joint strength is obtained. Hence, it is suggested that for butt joint configuration, Cu and Al are placed on AS and retreating side (RS), respectively, while Al is placed on top of Cu alloy for lap joint configuration. When a modest pin offset is applied on the Al side, a considerable amount of IMCs produced in the weld region reduces weld joint strength. A pin offset of 1.5–2 mm gives joints of high quality. Figure 7.8 illustrates the distribution and mixing of various irregular-sized Cu particles within the Al matrix in the WZ. The following IMCs CuAl, $CuAl_2$, Cu_3Al, and Cu_9Al_4 are formed when these Al–Cu particles are mixed. It is reported that the IMC layer in the range of 0.5 to 4 µm produces sound, defect-free joints with better joint strength [11].

In the realm of dissimilar Al–Cu system, recent advancements in FSW technology have introduced hybrid approaches, such as assisted heating and cooling, friction stir (FS) brazing, FS diffusion bonding, hybrid FSW, friction stir butt (FSb) barrier welding, friction stir spot welding (FSSW), FS lap welding with interlayer, micro-FSW, and underwater FSW for dissimilar Al–Cu system. These developments have emerged subsequent to the initial implementation of FSW [11].

7.4.3 FSW of Aluminum-to-Titanium

Al and Ti alloys are frequently employed in aerospace and automobile industries owing to their lightweight nature. Due to their superior mechanical properties, titanium and its alloys are utilized in numerous applications, including aircraft and biomedical equipment. Additionally, the joining of Al–Ti alloy is extensively utilized in aerospace, automobile, marine, and defense industries due to its ability to offer a favorable strength-to-weight ratio.

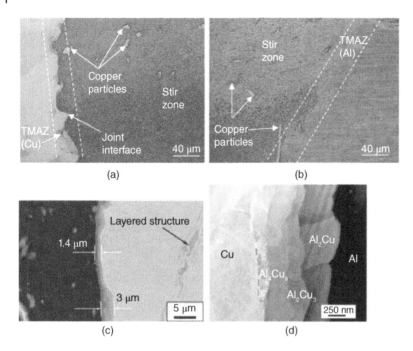

Figure 7.8 Microstructure of Al–Cu FSW system (a) TMAZ and WZ region, (b) TMAZ and WZ at Al, (c) and (d) Cu–Al interface layer and IMCs. Source: Mehta and Badheka [11]/Taylor & Francis.

The utilization of fusion welding processes for joining Al–Ti often results in joints with undesirable strength due to inferior metallurgical properties, high residual stresses, the formation of cracks, and brittle IMC formation. Therefore, these fusion welding methods are not recommended for achieving the desired strength in Al–Ti joints. Solid-state welding overcomes the abovementioned challenges pertaining to fusion welding.

Various solid-state welding processes have been employed for joining Al–Ti, including friction welding, FS lap welding, FS butt welding (FSbW), FSSW, ultrasonic vibration-assisted FSbW, GTAW FSW, ultrasonic-assisted FSW, heating-assisted FSW, FS extrusion, brazing, ultrasonic welding, explosive welding or cladding, pressure welding, electromagnetic or magnetic pulse welding, and accumulative roll bonding [5]. The year-wise progression of FSW of Al–Ti material joining is depicted in Figure 7.9.

The following IMCs reported in the literature, Ti_3Al, $TiAl$, Al_3Ti, $TiAl_2$, Ti_5Al_{11}, and Ti_9Al_{23}, are formed in the Ti–Al joints. The thickness of IMC layer decides the Ti–Al joint strength. The critical thickness of IMC layer of ~2 to 5 μm is recommended to produce high-quality joints [5].

It is essential to carefully select the welding parameters, including tool design, TRS, and WS, to address these issues. Using interlayers or transition materials can also help reduce the IMC formation and improve the mechanical properties of the weld. Proper pre-weld and post-weld treatments, such as cleaning and coating, can also help minimize the reactivity risk and reduce IMC formation. Additionally, welding in a vacuum or inert gas environment can help in reducing the formation of IMCs and improve the weld quality.

7.4.4 FSW of Aluminum-to-Steel

A paradigm shift in joining technology has been brought about by the use of lightweight, dissimilar metals for industrial applications. Every year the necessity for enhanced fuel efficiency and

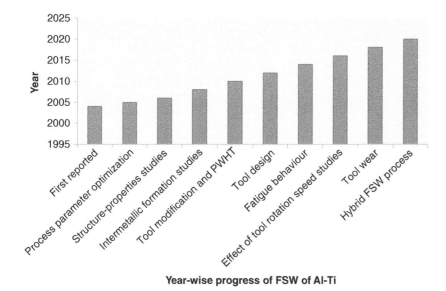

Year-wise progress of FSW of Al-Ti

Figure 7.9 Year-wise progress of FSW of Al–Ti.

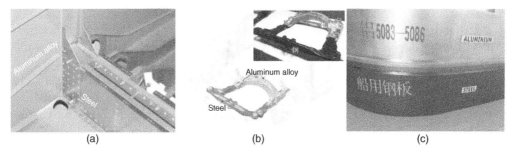

(a)   (b)   (c)

Figure 7.10 Commonly utilized Al–steel joint hybrid structures in various sectors. (a) Aerospace, (b) automobile, and (c) shipbuilding industry. Source: Wan and Huang [17]/Springer Nature.

reduced emissions are highlighted. Research-minded automotive industries are now aiming for lightweight materials to enhance fuel efficiencies and vehicle emission reduction [16]. Figure 7.10 shows the use of Al–steel joins in the aerospace, automobile, and shipbuilding sectors. Consequently, these material combinations are difficult-to-weld due to their incompatibility using traditional fusion welding methods. Different fusion and solid-state welding techniques are reported in the literature. Unfortunately, the fusion welding processes are non-recommended due to thermal and solidification issues. To overcome these issues, friction-based joining methods are proven to join Al–steel material due to low heat input and solid-state nature. The following friction-based joining methods were reported in the literature: friction welding, FSW, FSSW, FS spot fusion joining, hybrid FSW, FS scribe joining, FS brazing, FS dovetailing, friction bit joining, friction melt bonding, FS extrusion, and FS-assisted diffusion welding. The year-wise progression of FSW of Al–steel material joining is shown in Fig.7.11.

The following parameters are crucial for obtaining sound joints for FSW of Al–steel alloy joints: tool design, TPO, process parameters, material position, reducing welding defects, and IMC formation. It is suggested that for butt joint positioning, steel and Al are kept at AS and RS, respectively, while for lap joint positioning, Al is placed on top of steel material. Similarly, it is suggested that

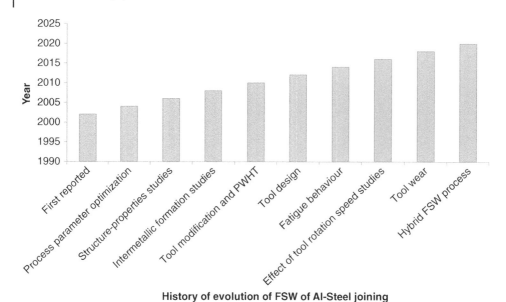

History of evolution of FSW of Al-Steel joining

Figure 7.11 Year-wise progression of FSW of aluminum-to-steel material joining.

TPO is kept at the Al side due to tool wear, partial deformation of steel, and large plastic deformation of Al, causing dispersion and mixing of steel particles in the Al matrix [18]. The other parameters related to the tool also contribute to the successful Al–steel joints, such as the diameter of the shoulder, the ratio of shoulder-to-pin diameter, and shoulder and pin geometries. Figure 7.12 shows the microstructure of FSW of Al–steel joints.

The following IMCs, $FeAl_2$, $FeAl_3$, $Al_{13}Fe_4$, and Al_5Fe_2, are stated in the literature for FSW of Al–steel joining. The interface's IMC layer thickness is reported to vary from <0.1 to58.1 μm. Better joint strength was achieved at the IMC layer thickness <5 μm. However, an increase in the thickness of IMC deteriorates the properties of the joints. The thickness of IMCs is controlled by TRS, WS, shoulder diameter, and shoulder-to-pin diameter ratio, respectively. The different hardness values ~250–739 HV is reported in the literature. The variations in the hardness values are due to variations in the volume fraction of IMCs [18].

Optimizing the welding parameters, including tool design, TRS, WS, weld sequence, and the use of interlayers or transition materials, is essential to address these issues. Interlayers can reduce IMC formation and improve bonding between dissimilar metals. It is also recommended to

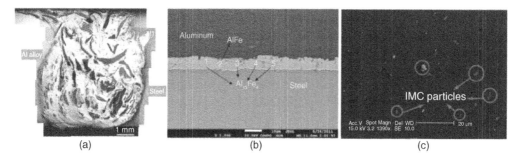

Figure 7.12 Microstructure of FSW of Al–steel joints, (a) complex WZ, (b) Al–steel interface with IMC layer, and (c) presence of IMCs in the WZ. Source: Mehta 2019 [18]/Springer Nature.

perform pre-weld surface preparation and post-weld treatments to improve the joint's performance and reduce the risk of corrosion. Additionally, using shielding gases or a vacuum during the welding operation can minimize the risk of contamination and improve the quality of the joint.

7.5 Recent Developments in Tool Design and Tool Materials

Several developments in FSW of dissimilar metal joining have occurred in recent years, particularly in tool design and materials. Some of the recent developments are:

1) *Tool materials:* The new tool material development has shown enhanced wear resistance and longer tool life. Advanced tools like tungsten-based tools (tungsten rhenium (W-Re), tungsten carbide (WC)), and polycrystalline cubic boron nitride (PCBN) [3] are particularly useful for welding dissimilar materials with high melting points.
2) *Hybrid tool designs:* Hybrid tools which combine different materials have been developed to improve tool life and reduce tool wear. These tools can withstand higher temperatures and stresses, making them suitable for welding high-strength dissimilar materials, such as Al–Cu, Al–Ti, Al–Mg, and Al–Steels.
3) *Multi-component tool designs:* To improve welding process control and weld quality, multicomponent tools like dual-shoulder or multi-shoulder tools have been created. By enhancing material flow and mixing during welding, these tools can produce welds that are more uniform and have superior mechanical qualities.
4) *Coatings and surface treatments:* It was created to improve tool wear resistance and reduce IMC formation. These coatings can also improve heat transfer and reduce the risk of sticking or galling.
5) *Smart tool technology:* It incorporates sensors and monitoring systems built into the tool to enhance process control and give welding operators real-time feedback. These devices can monitor temperature, force, and torque variations and change the welding parameters as necessary to increase the uniformity and quality of the welds.

The dependability, quality, and effectiveness of FSW's dissimilar metal joining have been increased by recent advancements in tool design and materials, making it more practical for a wider range of industrial applications. For the FSW of Al–Ti joining, other issues including the machining of the tool materials, availability, and cost must also be taken into account.

7.6 Parameter Optimization

To produce high-quality, defect-free welds, the FSW of dissimilar metal joining is a challenging process that needs careful parameter optimization. Some of the crucial variables that must be optimized for FSW of dissimilar metal joining include the following:

1) *Tool geometry:* It may have a substantial impact on the weld quality and joint strength. The tool geometry should be optimized for sufficient material flow and mixing between dissimilar metals while minimizing defects, such as voids or excessive flash.
2) *Tool rotational speed:* Throughout the welding process, it affects the heat input and material flow. It should be tuned to produce proper plastic deformation and metal mixing while reducing the risk of excessive heat input, which might result in IMC formation or material degradation.

3) *Weld speed:* It impacts the welding process's heat dissipation time and material flow rate. It should be optimized to balance proper material mixing and heat input while reducing the risk of excessive heat input, which can cause material degradation.

4) *Axial force:* It is applied to the tool that affects the material flow and deformation during welding. It should be optimized to achieve adequate material mixing and deformation while minimizing the risk of excessive force, leading to material damage or excessive tool wear.

5) *Interlayer material:* Using interlayers or transition materials can significantly improve the weld quality and joint strength in dissimilar metal joining. The interlayer material should be optimized to achieve adequate bonding and prevent IMC formation while minimizing the risk of thermal mismatch or material degradation.

Using a combination of modeling and experimental methodologies, these parameters should be optimized systematically and experimentally. This could lead to better weld quality and joint performance by identifying the best parameter values for FSW of dissimilar metals.

7.7 Common Defects that Occur in FSW of Dissimilar Metal Joining

The complex process of FSW of dissimilar metal joining might result in a number of welding joint defects. The following are some of the most frequent defects that might appear during FSW of dissimilar metal joining:

1) *Lack of fusion:* It occurs when there is insufficient bonding between dissimilar metals due to insufficient heat input or inadequate mixing. This can result in reduced joint strength.

2) *Voids:* These are pockets of gas that are trapped during the welding process due to insufficient pressure or inadequate material flow. Voids can lead to reduced joint strength and increased susceptibility to cracking.

3) *Incomplete mixing:* It occurs when there is inadequate mixing between dissimilar metals, resulting in an uneven distribution of material and the formation of unmixed regions. This can lead to reduced joint strength and increased susceptibility to cracking.

4) *IMC formation:* The formation of IMCs can occur due to the reaction between dissimilar metals during the welding process. These compounds can be brittle, reducing joint strength, and increasing susceptibility to cracking.

5) *Flash:* The excess material is extruded from the welding zone during welding. Excessive flash can lead to a weakened joint and cause problems during subsequent machining or finishing operations.

6) *Cracking:* It can occur due to a range of factors, including inadequate heat input, material flow, and bonding, as well as the presence of defects, such as voids and IMCs. It can decrease joint strength substantially and may result in premature failure of the welded joint.

For FSW of dissimilar metal joining, the welding parameters and tool design should be optimized to reduce these defects. Interlayers or transition materials may be used to enhance the bonding of dissimilar metals, as well as modifying the TRS, WS, axial force, and tool shape. Furthermore, cutting-edge monitoring and control systems can find and eliminate defects throughout the welding process.

7.8 Future Recommendations for Dissimilar Metal Joining

The research on FSW of joining dissimilar metal joining has been ongoing for several years. Some of the key research findings and developments in this area are:

1) *Tool design and material optimization:* Researchers have explored various tool designs, such as tapered, stepped, and threaded, to improve material flow and reduce defects in the welded joint in case of dissimilar Al alloys, Al–Cu system, Al–Ti system and in case of Al–steel system various tool designs, such as threaded and tri-flute has been used. Additionally, advanced tool materials, such as WC, diamond, and PCBN, have been investigated to improve tool life and reduce wear during welding.
2) *Weld parameter optimization:* The effect of various welding parameters, such as TRS, WS, and axial force, on the quality of the welded joint has been extensively studied. Researchers have identified optimal parameter ranges for FSW of dissimilar metal joining that can result in improved joint strength and reduced defects.
3) *Interlayer development:* Using an interlayer material between dissimilar metal joining can help in facilitating bonding and reduce the formation of IMCs. Researchers have explored the use of various interlayer materials, such as Cu and Mg (for dissimilar Al alloys), Zn and Ni (for Al–Cu system), Cu and Ni (for Al–Ti and Al–steel system), to enhance the weld joint quality. The selection of interlayer in dissimilar metal joining is crucial as it is directly correlated to the structure-properties of the joint. While selecting the interlayer, the following points need to be considered: determination of formation enthalpy (ΔH) and Gibbs energy of compound (ΔG), estimation of mechanical property through compound's ductility, another processing aspect, such as accessibility and cost, need to be considered, and finally, a feasibility analysis using test experiments to optimize weld parameters [19, 20].
4) *Structure-properties characterization:* FSW's structure properties of dissimilar metal joints have been extensively studied. Researchers have used techniques, such as electron microscopy, X-ray diffraction, and mechanical testing, to understand the influence of various variables on the structure properties of the joint.
5) *Application-specific testing:* The performance of FSW of dissimilar metal joints in real-world applications has also been studied. Researchers have evaluated the performance of FSW joints in various industries, such as aerospace, automotive, biomedical, and shipbuilding. The application-specific testing procedure to estimate the performance of the joint under different conditions is also an ongoing area.

Overall, the research on FSW of dissimilar metal joining has led to significant improvements in the quality and reliability of the welded joint, enabling the use of this process in a broader range of applications.

Acknowledgments

The authors like to sincerely thank the All India Council for Technical Education (AICTE), New Delhi, for their kind financial support under the research promotion scheme project no. 8-117/FDC/RPS (Policy-1)/2019-20.

References

1 Gadakh, V.S. and Badheka, V.J. (2022). Sustainability of fusion and solid-state welding process in the era of industry 4.0. In: *Handbook of Smart Materials, Technologies, and Devices*, 1637–1654. https://doi.org/10.1007/978-3-030-84205-5_113.

2 Mishra, R.S. and Ma, Z.Y. (2005). Friction stir welding and processing. *Mater. Sci. Eng.: R: Rep. 50* (1–2): 1–78. https://doi.org/10.1016/j.mser.2005.07.001.

3 Kumar, N., Yuan, W., and Mishra, R.S. (2015). Introduction. In: *Friction Stir Welding of Dissimilar Alloys and Materials*, 1–13. Butterworth-Heinemann https://doi.org/10.1016/b978-0-12-802418-8.00001-1.

4 Bang, H.S., Bang, H.S., Jeon, G.H. et al. (2012). Gas tungsten arc welding assisted hybrid friction stir welding of dissimilar materials Al6061-T6 aluminum alloy and STS304 stainless steel. *Mater. Des. 37*: 48–55. https://doi.org/10.1016/j.matdes.2011.12.018.

5 Gadakh, V.S., Badheka, V.J., and Mulay, A.S. (2021). Solid-state joining of aluminum to titanium: A review. *Proc. Inst. Mech. Eng., Part L: J. Mater.: Des. Appl. 235* (8): 1757–1799. https://doi.org/10.1177/14644207211010839.

6 Amancio-Filho, S.T., Sheikhi, S., dos Santos, J.F., and Bolfarini, C. (2008). Preliminary study on the microstructure and mechanical properties of dissimilar friction stir welds in aircraft aluminium alloys 2024-T351 and 6056-T4. *J. Mater. Process. Technol. 206* (1–3): 132–142. https://doi.org/10.1016/j.jmatprotec.2007.12.008.

7 Ghosh, M., Husain, M.M., Kumar, K., and Kailas, S.V. (2013). Friction stir-welded dissimilar aluminum alloys: microstructure, mechanical properties, physical state. *J. Mater. Eng. Perform. 22* (12): 3890–3901. https://doi.org/10.1007/s11665-013-0663-3.

8 Murr, L.E. (2010). A review of FSW research on dissimilar metal and alloy systems. *J. Mater. Eng. Perform. 19* (8): 1071–1089. https://doi.org/10.1007/s11665-010-9598-0.

9 Al-Badour, F., Merah, N., Shuaib, A., and Bazoune, A. (2014). Thermo-mechanical finite element model of friction stir welding of dissimilar alloys. *Int. J. Adv. Manuf. Technol. 72* (5–8): 607–617. https://doi.org/10.1007/s00170-014-5680-3.

10 İpekoğlu, G. and Çam, G. (2014). Effects of initial temper condition and postweld heat treatment on the properties of dissimilar friction-stir-welded joints between AA7075 and AA6061 aluminum alloys. *Metall. Mater. Trans. A 45* (7): 3074–3087. https://doi.org/10.1007/s11661-014-2248-7.

11 Mehta, K.P. and Badheka, V.J. (2016). A review on dissimilar friction stir welding of copper to aluminum: process, properties, and variants. *Mater. Manuf. Processes 31* (3): 233–254. https://doi.org/10.1080/10426914.2015.1025971.

12 Al-Roubaiy, A.O., Nabat, S.M., and Batako, A.D.L. (2014). Experimental and theoretical analysis of friction stir welding of Al-Cu joints. *Int. J. Adv. Manuf. Technol. 71* (9–12): 1631–1642. https://doi.org/10.1007/s00170-013-5563-z.

13 Saeid, T., Abdollah-zadeh, A., and Sazgari, B. (2010). Weldability and mechanical properties of dissimilar aluminum-copper lap joints made by friction stir welding. *J. Alloys Compd. 490* (1–2): 652–655. https://doi.org/10.1016/j.jallcom.2009.10.127.

14 Xue, P., Xiao, B.L., Wang, D., and Ma, Z.Y. (2011). Achieving high property friction stir welded aluminium/copper lap joint at low heat input. *Sci. Technol. Weld. Joining 16* (8): 657–661. https://doi.org/10.1179/1362171811Y.0000000018.

15 Akbari, M., Abdi Behnagh, R., and Dadvand, A. (2012). Effect of materials position on friction stir lap welding of Al to Cu. *Sci. Technol. Weld. Joining 17* (7): 581–588. https://doi.org/10.1179/1362171812Y.0000000049.

16 Haghshenas, M. and Gerlich, A.P. (2018). Joining of automotive sheet materials by friction-based welding methods: a review. *Eng. Sci. Technol. Int. J. 21* (1): 130–148. https://doi.org/10.1016/j.jestch.2018.02.008.

17 Wan, L. and Huang, Y. (2018). Friction stir welding of dissimilar aluminum alloys and steels: a review. *Int. J. Adv. Manuf. Technol. 99* (5–8): 1781–1811. https://doi.org/10.1007/s00170-018-2601-x.

18 Mehta, K.P. (2019). A review on friction-based joining of dissimilar aluminum-steel joints. *J. Mater. Res. 34* (1): 78–96. https://doi.org/10.1557/jmr.2018.332.

19 Fang, Y., Jiang, X., Mo, D. et al. (2019). A review on dissimilar metals' welding methods and mechanisms with interlayer. *Int. J. Adv. Manuf. Technol. 102* (9–12): 2845–2863. https://doi.org/10.1007/s00170-019-03353-6.

20 Shah, L.H., Gerlich, A., and Zhou, Y. (2018). Design guideline for intermetallic compound mitigation in Al-Mg dissimilar welding through addition of interlayer. *Int. J. Adv. Manuf. Technol. 94* (5–8): 2667–2678. https://doi.org/10.1007/s00170-017-1038-y.

8

Friction Stir Welding of Aluminum and Its Alloy

Palani Sivaprakasam[1], Kolar Deepak[2], Durairaj Raja Joseph[3], Melaku Desta[1], Putti Venkata Siva Teja[4], and Murugan Srinivasan[5]

[1] Department of Mechanical Engineering, CEME, Addis Ababa Science and Technology University, Addis Ababa, Ethiopia
[2] Department of Mechanical Engineering, Vardhaman College of Engineering, Jawaharlal Nehru Technological University, Hyderabad, Telangana, India
[3] Department of Aerospace Engineering, School of Aeronautical Sciences, HITS, Hindustan Institute of Technology & science, Chennai, India
[4] Department of Mechanical Engineering, Dhanekula Institute of Engineering & Technology, Jawaharlal Nehru Technological University, Kakinada, Ganguru, Andhra Pradesh, India
[5] Department of Mechanical Engineering, Mahendra Engineering College, Anna university, Namakkal, India

8.1 Introduction

FSW is a sophisticated technique designed for continuously joining materials. In FSW, a tool is cylindrical, shouldered, and has a profiled probe that is rotated while being smoothly put into the joint contour that exists connecting two plates or sheet materials. During the welding process, the components will be fastening to prevent sliding away from one another. Heat is generated through the process of frictional surface between the hard welding tool and the materials being joined. This heat softens or melts the latter and permits the tool to be moved along the length of the weld line. It is generally agreed upon that this method represents one of the metal joining's great developments [1]. FSW of materials which have melting point at a relatively high temperature like nickel base alloys, steel, stainless steel, and so forth, according to papers from the recent decade, actually demand a higher level of sophistication in tool materials and process control of FSW [2].

FSW is employed to join a wide range of material types as well as various material assemblages. The technique for connecting aluminum and its alloys has been the focus of the majority of FSW research and development efforts. Aluminum alloys from the 2xxx, 5xxx, 6xxx, 7xxx, and 8xxx series were effectively FSWed to achieve high-quality welds while remaining within the necessary parametric limits [2]. Aluminum and steel alloys must be joined when using different alloys and the Honda Company has completed this welding effectively in a vehicle suspension system developed for mass manufacturing [3]. Ti alloys, Al alloys, polymers, Mg alloys, and other dissimilar materials can be joined very effectively by FSW. FSW has recently attracted significant scientific and technological interest in a number of industries, including aerospace, railroad, clean energy, and automotive [4].

Developing unwanted transition phases, porosity, and voids in the weld zone are difficulties in joining aluminum alloys. Thermal and residual stresses are also produced as a result of variations

Friction Stir Welding and Processing: Fundamentals to Advancements, First Edition.
Edited by Sandeep Rathee, Manu Srivastava, and J. Paulo Davim.

in thermal coefficients of additives and the metal matrix. Brazing has not been utilized as a way for welding Al alloy with MMCs, even though several researchers have concentrated on welding of Al-MMCs. This is due to the expensive cost of filler as well as a lack of accuracy in the welding process [5]. FSW is recognized as an improved alternative approach for the production of aluminum joints in a timely and dependable fashion [6]. Over the past decade, scholars have studied FSW and its effects on material strength [7–10]. Furthermore, FSW is an effective method for producing aluminum joints that offer accurate welding with low filler costs and high material strength.

The FSW approach was utilized in order to investigate the corrosion patterns exhibited by two different alloys (AA6061 and AA8011) under a range of corrosion circumstances meant to mimic marine applications. Corrosion study reveals that the weld area specimens are more corrosion resistant than base Al alloy specimens. The salt medium caused significantly less weight loss than other solutions. Analysis using a SEM reveals extensive surface corrosion on the base metal AA8011 [11]. Recent research has demonstrated that the inclusion of nanoparticles can enhance the joint's weld characteristics in FSW. FSW of Al alloys (5083 and 6082) with nanoparticle (carbide and oxide) addition is evaluated different from one another critically.

In addition, the impacts of nanomaterial fortification on the weld properties and microstructure are analyzed, with effects of welding processes on the particles. The dispersion of the reinforcing nanoparticles influences how the joint properties are determined to be significantly influenced by friction stir welding conditions. In addition, the strewed nanoparticles improve joint properties and facilitate in grain refinement. The microstructures and weld properties of Al alloys and dissimilar Al alloys welds are greatly influenced by the variety, quantity, nanoparticles (size), as well as the welding variables [12].

In addition to these benefits, FSW has significant drawbacks like lower speed welding of 6000-series aluminum alloys that are 5 mm thick on commercially available machines [13]. Moreover, a backing bar is needed because this welding procedure necessitates a firm clamping of the workpieces. The other issue is dealing with the keyhole at each weld's ends. Nonetheless, worldwide research and development efforts are reducing these restrictions [13]. The current state of the art in FSW of aluminum and its alloys, as well as the fundamentals of the FSW process and its variables, are examined. Mechanical properties of FSW, additive mixed FSW of Al alloy, testing and characterization, such as tensile, hardness, and microstructure analyses, as well as industrial applications, are also explored.

8.2 Fundamentals of FSW

FSW is a formation that happens as a result of thermomechanical deformation of the workpieces as the resultant heat rises over the solidus temperature of the workpieces. Since the creation of this joining technology, significant weight savings have been achieved in applications where it is important [14, 15]. Figure 8.1 shows the FSW process.

8.3 FSW of Aluminum and Its Alloy

There are several uses of aluminum alloys, such as transportation, maritime, and aerospace industries, as structural components. The various series of aluminum alloy are shown in Figure 8.2 [17]. Both the aluminum alloys (AA8011 and AA6082-T6) that were taken into account for this research have potential uses in aerospace. The aerospace sector has a significant opportunity to connect

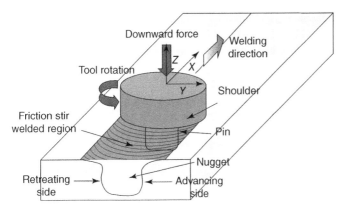

Figure 8.1 Image of the FSW in schematic form. Source: Mishra et al. [16]/with permission of Elsevier.

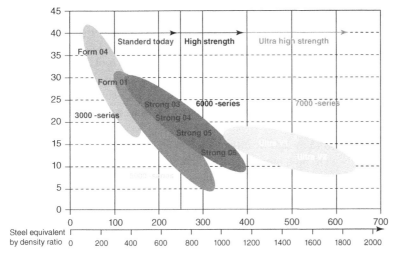

Figure 8.2 High-strength aluminum plates [Automotives]. Source: Ref. [17]/with permission of Elsevier.

different aluminum alloys to increase performance. Because to their low cost, low weight, and exceptional strength, aluminum alloys continue to be the material of choice [18].

The joints made using the SFSP method have better tensile characteristics than those made using the NFSP method. It was also found UTS and % elongation at the joint can be improved with the application of SFSP. The possibility of applying tool torque as a process factor in the thermo-mechanical circumstances created during FSW was examined in several papers. Torque, as opposed to force, should be used as a process control measure, according to Longhurst et al. [19]. The authors claim that torque control enables the production of flawless welds and allows the factors to be adjusted to modify in the surface of materials. A temperature control system was created by Bachmann et al. [20] using an analytical torque model. This monitoring model generated weld joints that have superior quality and greater uniformity over the joint's whole length.

Leito et al. [21] found that torque measured while welding might be attributed to the control factors, subsequently a clearly established empirical connection, when welds without flaws were produced. When applying process parameters that resulted in the manufacture of welds with faults, it was unable to establish a clear correlation among torque advancement and those factors while the torque results were practically aleatory. Similar to this, Galvo et al. [22] revealed that by examining the torque advancement in welding, it is feasible to identify the production of flaws caused by the realization of sizable amounts of intermetallics from beneath the tool. A connection between the torque variations that results from welding and the appearance of surface flaws was also found [23].

From all of these studies, it is feasible to draw the conclusion that modeling and understanding how process parameters affect torque can be useful tools for managing processes and spotting welding errors as well as choosing process parameters for various applications.

8.4 Influences of Process Parameters

8.4.1 Tool Profile

For FSW, a tool is used that is necessary to the perform [24] and includes of a shoulder and a pin, both of which are vital to the welding process. The shoulder provides much of the required heat, while the pin is in control of moving the plasticized substance [25]. Due to the tool's effect on the grain size, microstructure homogeneity, and joint flow, the joint is more likely to be structurally sound. The FSW technique primarily depends on the shoulder and pin of the tool to create strong welds. In Figure 8.3 [11], one can see the many different pin shapes employed in FSW tools to successfully produce weld joints. Consistent force is provided by the shoulder contact surface, the diving pin combines the borders of the materials and homogenous flow, and the tool stirring actions gives strength.

Groove and plunge depth were found to significantly affect tensile characteristics, as reported by Guo et al. [26]. The joint's TS were 86% more than base metal under conditions. The taper threaded FSW tool had stronger tensile, percentage elongation, and flexure strengths than comparable profile tools [27]. The effect of tool–pin shape on the AA1100 aluminum alloy joint was examined by Tamadon et al. [28]. The investigation revealed that the cylindrical threaded tool produced the best tensile strength because of a significant level of flexibility generated by greater contact of the tool with the welded substance. A further comparison was investigated for the application of pins with different configurations: square cylindrical/cam, tapered/cam. Using a cylindrical cam and square pins, faultless Al–Cu couplings were manufactured. With the square pin, however, greater mechanical qualities were achieved [29].

Comparing the highest result for the tapered-threaded junction to the use of the cylindrical tool, greater values of around 37% were obtained. It was determined based on the specifications that the majority of the samples were linked correctly with a rotating speed of 475 rpm. The material was improperly mixed in joints created at a welding speed of 300 mm/min. The welding speed levels were 150–475 mm/min used, with a rotational speed of 475 rpm producing the best joint quality [30].

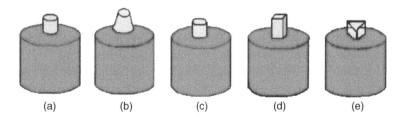

Figure 8.3 FSW tools pin profiles. (a) Straight, (b) Tapered, (c) Threaded, (d) Square, (e) Triangular. Source: Alfattani et al. [11]/ with permission of Elsevier.

The FSW of Al alloys (5083 and 7068) were planned to evaluate TS and Vickers hardness.

The three different kinds of tools are triangular, straight cylindrical, and tapered. The investigation's process factors include axial load of 3–6 kN, rotating speed of 800–1400 rpm, welding speed of 30–60 mm/min, and plate thickness of 5–8 mm. The welding zone's improved hardness value and UTS are evidence that tool profile effects are effectively put to use [31].

Fuji et al. [32] employed a triangular prism and probes (columnar) threaded and without threads for FSW butt welding of Al alloys. It has been concluded that the most effective configuration approach was a columnar tool without threads; if metal deformation resistance was low, whereas more resistant to alloys, it was proved to be an appropriate answer only by reducing the rotation speed. Wei et al. [33] performed test on plates made of Al–Li alloy, by varying the transverse speed of the welding pins and the force applied to them. Since four improper speed tunnel type faults were discovered, they reported that the optimal rotational and transverse speed to produce plates with no grooves was welded. When pin axial pressures were excessively high, NZ-wide material depressions and wavy burrs were observed. A groove type defect was seen when pin axial pressures were too low.

In addition, Patil et al. [34] studied how changing welding transverse speeds affected the quality of welds in AA6082-O plates, assuming a constant pin rotating speed and pin profiles. Results showed that the tensile strength of joints made using a taper screw thread pin was higher than that of joints made with a tri-flute pin at both slow and fast welding speeds. Elangovan et al. [35] examined AA6061 aluminum alloy plates to determine how the axial pressure of the pins and the profiles of the pins affected the location of the functionally significant weld (FSW) zones. Specifically, they showed that there is a correlation between pin shape and the pressure that should be applied to the pins. The ideal pin pressure for a straight and tapered cylinder was around 6 kN. In contrast, it was found that, regardless of the pin axial pressure, welds produced using a threaded cylindrical pin profile, as well as welds produced using a square pin profile and a triangular pin profile, were all sound and defect-free. When compared to joints made with a straight cylindrical pin profile, those made with a square or triangular pin profile exhibited greater tensile properties.

Zhang et al. [36] examined how different heat treatments affected the quality of FSW joints and how they could be improved in aluminum alloys. Because the HAZ is the weakest section of a typical FWSed precipitation hardened Al alloy series, special attention was paid to it. It has been claimed that the FSWed Al alloy welded plates can be optimized by a combination of an appropriate condition at low temperature and cooling.

According to a number of reports, there are optimal FSW settings for welding plates of incompatible alloys. Both the initial plate circumstances and the post-FSW dissimilar alloy plate conditions were unique. The FSW of Al alloy sheets (age-hardened) is likely to include certain metallurgical difficulties. Al-copper/Al-magnesium-plate scheme identified a few unusual metallurgical issues that are worth mentioning here [37]. It is well known that low melting eutectic formed by the constituent element during fusion welding of dissimilar Al alloys is very difficult to work with and often leads to cracking. As a result, a significant loss in hardness is anticipated to occur on the heat-treatment alloy commencing the BM across the NZ, but this is not the case with the non-age-hardening alloy plate. On average, the stated hardness reduction is greater on the AS than on the RS (RS). Several methods exist for dealing with this metallurgical asymmetry and inhomogeneity, including plate homogenization performed after welding, double-sided friction stir welding (DS-FSW), pins that deviate only slightly from the welding line during their translation, and adjustments to shoulder diameter and pin configuration/direction relative to the joining plates.

The importance of the pin depth into the sheet thickness as a crucial welding parameter is also highlighted by Tutar et al. [38]. They saw an appreciable increase in tensile strength as the pins were driven deeper into the plate's thickness and significant to FSW zones. These mathematical models demonstrated a rise in tensile strength with an increase in axial force and higher pin speed, after which the material strength began to deteriorate. As a result, material flaws and local locations of incomplete welding occur in the AS because it does not go far enough to the RS at the fastest welding speeds. As the pin's rotational speed and axial force increase, the ductility does as well, but the opposite is true for the pin's transverse speed.

8.4.2 Rotational Speed

Yin et al. [39] studied the influence of tool offset, material positions, and rotational and traverse speed on the weld joints, microstructure, and mechanical characteristics of different alloys. With tool RS of 1000 rpm and traverse speeds of 50–150 mm/min, they claimed that sound joints could be produced without tool offset. When compared to the foundation material, the joint's TS were 75% higher. The welds' quality was not significantly controlled by tool rotation speeds, according to Kalemba and Dymek [40]. Due to their proneness to softening, porosity, hot cracking, and due to the heat cycle's involvement, massive structures made of high-strength aluminum–lithium alloys are linked with significant challenges when joined by arc welding. For connecting these alloys, the FSW technique is incredibly effective [41, 42]. Rotational speed is among the criteria with the greatest influence on the outcomes of TS tests [43, 44].

8.4.3 Tool Tilt Angle

The tilt angles that are typically employed in practice range from 0 degrees to 3°, where a tilt angle of 0° indicates that the tool is at a 90° angle to the work piece. TTA regulates the flow of materials all over the joining process, which in turn has an effect on the amount of heat produced. Friction and the decomposition of plastic both contribute to the production of heat in FSW. A tilted tool creates a greater forging effect more intense material flow than 0° TTA [45]. The TTA of 1° to 3° was sufficient to produce a weld junction without any defects. But sound joints were welded with a 2° TTA [46]. It has been proven that as TTA rises from 1 to 2, the strength of the joint increases before decreasing. With the exception of the TTA of 3°, which displays brittle failure, the fractographic examination reveals a blend of failure for all the specimens. Overall, the study recommends using 1–2° to obtain a strong weld of this specific type of AMC [47].

8.4.4 Axial Force

The amount of energy input to FSW process, the formation of heat, the progression of the microstructure, and the joint qualities are all under the control of the process factors, such as the tool rotations, the welding speed, the axial load, and the tool geometry [48]. The tool forces involved in a linear weld, particularly the bear force, have received the most attention since they directly cause the workpiece penetration depth in these welds. Five joints were created using the cast aluminum alloy A319, while welding speeds and tool rotation were kept constant. A research was conducted on A319 FSW aluminum alloys to find out how axial force affected microstructure, tensile strength, and hardness. More tensile strength in A319 alloy was found in the joint created with a 6 kN axial stress [49]. Force signals are commonly measured to be the most reliable source of data for locating flaws in the FSW process [50, 51]. This is so because the welding force, which varies according to the material type, welding operations, stirring pin form, and workpiece thickness, is produced by the physical connection with the workpiece and the tool. Shrivastava et al. [52] modeled the FSW force and investigated the relationship among the force and the flow of material near the stirring device. Additionally, it has been suggested that faults could be found using the frequency spectrum of forces. According to the study by Sahu et al. [53], the axial force more accurately envisages the tensile strength than the temperature signals obtained from thermocouples (K-type).

Hepeng Jia et al. [54] investigated high-speed FSW of aluminum alloy (6061-T6). Lower spindle axial force is directly proportional to improved weld material fluidity. Similar to this, as welding speed is increased, welding heat is reduced and the welded material's fluidity is reduced, resulting in an increase in spindle axial force. Also, as the plunge depth is increased, the area of the tool shoulder that comes into contact with the material will do so more frequently, increasing the frictional force that will raise the spindle axial force.

8.4.5 Welding Speed

Liu et al. investigated FSW procedure that involved changing the welding speed from 50–200 mm/min while maintaining a constant rotational speed of the tool [55]. While welding speed increased, heat also did, which led to an increase in UTS. The nugget zone's grain size increased with faster welding, according to the results. Mishra and Ma [16] investigated a number of FS welding process components. The mechanical and fatigue properties of FSW joints were strongly impacted by the shape of the tool pin and welding speed. The feed rate and tool pin form, which determine the weld joint's soundlessness, were found to have a significant impact on the microstructural and fatigue properties of the material. The various process parameters considered for FSW of Al alloy are presented in Table 8.1.

8.5 Testing and Characterization of FSW of Al and Its Alloy

8.5.1 Tensile Tests

The tensile test performed for FSWed AMCs which had the least B_4C level, possessed the maximum tensile strength (172 MPa) [65]. This is explained by Al's high plasticity and density [66]. The tensile strength diminishes in FSWed Al-MMCs when B_4C content rises and SiC content falls. This is explained by the uniform distribution of small Al–SiC particles. Al–SiC particle agglomeration during solidification, which was exacerbated by higher SiC percentages, was the cause of the decline in composites' strength. Al-MMCs have high yield strength because their matrix had a lot of dislocations. With a maximum elongation of 7%, YS of 106.61 MPa and UTS of 188.77 MPa were found in DFSW [67]. A lower elongation but a higher ultimate tensile strength when subjected to heavy loads were found in Al 6082 alloy. Under tensile testing circumstances, it can withstand a heavier load than the AA2014 [68]. It was discovered that the heat-affected zone (HAZ) on the advancing side, which had the least qualities related to the other locations, was where the specimens' defects were located [69].

8.5.2 Hardness Test

The maximum hardness value was produced by an increase in axial force, welding, and rotation speed. The smallest tool rotation dictated the minimal hardness values in dissimilar-FSW [31]. A microhardness of about 90 HV was gained with AA6082 advancing side NFSP, which was higher than the microhardness of FSWed [18]. The hardness increase was caused by the entire solution and fresh precipitation in the stir zone [70]. The application of current resulted in an increase in

Table 8.1 FSW process parameters.

S. no	Materials used	Rotational speed (rpm)	Welding speed mm/min	Tilt angle (°)	Axial force (kN)	Tool profile	Plunge depth (mm)	Key findings	References
1.	AA6061-T6	850–1450	30–110	1-5	—	SC,CT,ST,TT, SS	—	SC tool, 3° TA, RS 1150 rpm, and WS 30 mm/min can produce an excellent joint	[56]
2	AA5083 and AA2024	800–1600	40–80	—	15-35	TH, TS, SC, CG, PS	—	TH tool, moderate level of RS, WS & AF for optimum conditions	[57]
3	AA2014-T6	900	90	0-4	—	TT	—	AF 14.42 kN, 2° TA provide better joint efficiency	[46]
4	AA6082-T6	800–1250	14.5–31	—	—	ST	—	Higher RS and low WS have significant increase in temperature	[58]
5	AA5058-Polycarbonate	960–1940	45 and 90	2	—	ST	0.3	RS 1600 rpm, WS 45 mm/min strongest joint	[59]
6	AA5005-St-52	920–1200	54 and 90	2	—	ST	0.1, 0.3, and 0.6	RS 1200 rpm, WS 90 mm/min, with PD 0.3 mm for sound weld	[60]
7	AA6061 and AA6082	800–1400	40–60	0-2	—	SC		RS 1178.2 rpm, WS 45.92 mm/min, and TA 1.89° for optimum conditions	[61]
8	AA5083 and AA7068	800–1400	30–60	—	3-6	SC, ST, STT	—	RS 1200 rpm, WS 30 mm/min, and AF 3 kN for higher tensile strength	[31]
9	AA2024-T3 and AA7075-T6	900–1200	20–60	—	—	—	—	RS 1050 rpm, WS 40 mm/min produce quality weld	[62]
10	AA 6061-T6	800–1200	30–60	—	—	SC	—	RS 1200 rpm, WS 30 mm/min for optimum parameters	[63]
11	Al-TiC	800–1600	100–100	—	—	SC	—	RS 1200 rpm, WS 30 mm/min for best wear resistance	[64]

RS: Rotational speed, AF: Axial force, WS: Welding speed, TA: Tilt angle, TP: Tool profile, SC: Simple cylindrical, CG: Tapered cylinder with grooves, ST: Simple tapered, TT: Tapered with threads, SS: Simple square. STT: Simple triangular tool, TH: Tapered hexagon, TS: Tapered square, PS: Paddle shape, CT: Cylindrical with threads.

the total hardness of the weld zones, particularly the weld zones at the AS, with respect to the conventional FSWed joint. There was a 2.74–7.38% point increase in the joint's tensile strength between 100 and 400 A [71]. The microhardness values of nano alumina-added FSW composites were higher than titanium oxide. This is due to uniform and grain refinement of nano alumina [72]. The study's findings show that as transverse speeds increase, microhardness and strength increase as well. The microstructure revealed that with increasing transverse speeds, the grain size became smaller in FSW of AA6061-T651 plates [73].

8.5.3 Microstructure Evolution

Dynamic recrystallization is typically anticipated to be inhibited by Al and its alloys propensity to display extremely high dynamic recovery rates. Even so, it has been repeatedly noted in the literature [74–76] that new equiaxed grains might occur in FSW welds composed of Al alloys. Figure 8.4 shows different weld zones of FSW of aluminum alloy. Grain that is large and lengthy, which are classic of the morphology observed in flat rolled plate, is visible in the base metal's microstructure (Figure 8.4a).

Although the grain formation in the HAZ (Figure 8.4b) has the identical rolling morphology as the base metal, there is evidence of a little coarsening of the grains due to the influence of heat transfer from the WNZ. Comparatively to the WNZ, the material in the TMAZ is subjected to lower temperatures, strain rates, strains, and as a result, recrystallization cannot take place under these circumstances (Figure 8.4c,d). Heavy plastic shearing and intensive heat creation coupled to produce WNZ, which is described by a dispersion of equiaxed fine grains. This implies that the plasticized substance in the weld nugget underwent dynamic recrystallization (Figure 8.4e) [77].

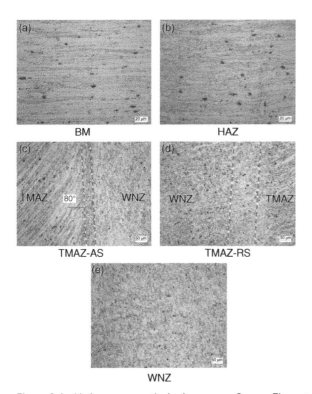

Figure 8.4 Various zone-optical microscopes. Source: Zhao et al. [77]/with permission of Elsevier.

8.6 Additive Mixed Friction Stir Process of Al and Its Alloy

The manufacturing of composite using FSP is a growing solid-state method. Successful surface characteristics adaptation occurs as solid-state plastic deformation, enhancing and modifying the surface. During the early stages of friction stir processing, aluminum and other lightweight metal matrices were the first applications. However, the encouraging outcomes of these lightweight metals have drawn researchers' attention to additional nonferrous, ferrous metals, and polymers.

According to studies, reinforcing particles have improved joint and surface qualities in terms of mechanical properties [14, 15]. The AA6082 joints were given a toughening and hardening enhancement by the addition of Si and Ti Carbide nanoparticles. The qualities are influenced by the kind of carbide nanoparticle. For instance, the toughness and elongation of the joint with Si carbide were superior to those of the joints strengthened with Ti carbide.

In addition, when aluminum alloys come in contact with particles that are stiffer, this leads to significant wear [78]. When stiff particles such as oxides (Al_2O_3, TiO_2,) and Si, Ti-carbides or their mixture were introduced to the aluminum alloy base matrix [76, 79], this issue was resolved. The mechanical and tribological properties could be enhanced by nanoparticles' addition in Al alloy.

Saeidi et al. [80] discovered that upon inclusion of nano Al_2O_3, the length of the joint line of FSW (dissimilar) joints significantly improved corrosion resistance and fracture behavior. Fernandez et al. [81] performed various particle incorporation techniques on the wear resistance behavior of FSP-produced $AA6061$-T6/SiC/Al_2O_3 composites. In addition, researchers attempted to construct bulk surface composites using FSP utilizing a choice of particle pre-deposition techniques [16, 82, 83]. The initial investigation was performed on the FSP production of an Al/SiCp surface composite [16]. Regarding mechanical and metallurgical performance, they discovered a significant improvement in surface qualities. Dixit et al. [82] have employed small perforations to introduce nitinol below the surface of the plate (Al alloy 1100) in order to create FSP-based surface composites. They found that the greater spreading of particles in the aluminum matrix enhances mechanical characteristics. Functionally graded MMCs is made by particles' inclusion with varying layouts to the consumable tool's tip via FSP [83].

8.7 Applications

This welding technique is ideal for prefabricating flat sections made of aluminum alloys, such as shell of railroad cars, decks, shell plating, and bulkheads, in the shipbuilding and railway industries. It enables one to cut production costs and project completion times [84]. FSW has been adopted for lightweight structures by a wide variety of automotive manufacturers, including "Audi [85], Ford [86], Mazda [87], as well as Tesla." The FSW technique was utilized in the construction of aircrafts by "Airbus (or EADS) (A340, A350, A380, and A400M) [88], Boeing, Bombardier, Lockheed Martin, BAE Systems, Embraer (Legacy 450 and 500), as well as Eclipse Aerospace (in Eclipse 500)" in order to join various components without the use of rivets for the purpose of lighter construction. Apple Inc. utilized FSW technology in order to achieve their goal of making their "2012 Apple iMac" 40% more thin. Significant advancements in the automobile industry are predicted by advanced welding of complex geometries. It has found use in many sectors, including aerospace, automotive, maritime, and railway. Because temperature during FSW plays a key role, it is possible to thoroughly understand the process through thermal analysis, which also enables energy and money savings. However, because the measuring tools are inaccurate, doing an experimental study of thermal behavior is difficult. Moreover, determining the tool's perpendicular movement over a curved surface was equally another challenge.

8.8 Conclusions

- This paper discusses the current state of FSW, FSW of aluminum and its alloy, including the fundamentals.
- The different factors associated with FSW, such as tool profile, rotational speed, axial force, tool tilt angle and welding speed, are discussed.
- The mechanical properties of associated with FSW using Al its alloy dependent on the combined effect of various process parameters.
- In addition, additive mixed friction stir process of Al and its alloy is also discussed.
- Finally, the major application of FSW process in industry is also discussed.

References

1 Troughton, M. (2009). *Handbook of Plastics Joining A Practical Guide 2009*, 15–35. Norwich, NY, USA: William Andrew Inc. https://doi.org/10.1016/B978-0-8155-1581-4.50016-0.
2 Packer, S., 2013, January. Tool geometries and tool materials for friction stir welding high melting temperature materials. In Proceedings of the 1st International Joint Symposium on Joining and Welding (pp. 473-476). Woodhead Publishing.
3 Worldwide, H. (2012). *Honda Develops New Technology to Weld Together Steel and Aluminum and Achieves World's First Application to the Frame of a Mass production Vehicle.*
4 Meng, X., Huang, Y., Cao, J. et al. (2021). Recent progress on control strategies for inherent issues in friction stir welding. *Prog. Mater Sci.* 115: 100706.
5 Lisheng, Z.U.O., Xianrui, Z.H.A.O., Zeyang, L.I. et al. (2020). A review of friction stir joining of SiCp/Al composites. *Chin. J. Aeronaut. 33* (3): 792–804.
6 Palanivel, R., Mathews, P.K., Dinaharan, I., and Murugan, N. (2014). Mechanical and metallurgical properties of dissimilar friction stir welded AA5083-H111 and AA6351-T6 aluminum alloys. *Trans. Nonferrous Met. Soc. China 24* (1): 58–65.
7 Salih, O.S., Ou, H., Sun, W., and McCartney, D.G. (2015). A review of friction stir welding of aluminium matrix composites. *Mater. Des. 86*: 61–71.
8 Karthikeyan, P. and Mahadevan, K. (2015). Investigation on the effects of SiC particle addition in the weld zone during friction stir welding of Al 6351 alloy. *Int. J. Adv. Manuf. Technol. 80*: 1919–1926.
9 Bozkurt, Y. and Boumerzoug, Z. (2018). Tool material effect on the friction stir butt welding of AA2124-T4 Alloy Matrix MMC. *J. Mater. Res. Technol. 7* (1): 29–38.
10 Parikh, V.K., Badgujar, A.D., and Ghetiya, N.D. (2019). Joining of metal matrix composites using friction stir welding: a review. *Mater. Manuf. Processes 34* (2): 123–146.
11 Alfattani, R., Yunus, M., Mohamed, A.F. et al. (2021). Assessment of the corrosion behavior of friction-stir-welded dissimilar aluminum alloys. *Materials 15* (1): 260.
12 Vimalraj, C. and Kah, P. (2021). Experimental review on friction stir welding of aluminium alloys with nanoparticles. *Metals 11* (3): 390.
13 Amini, A., Asadi, P., and Zolghadr, P. (2014). Friction stir welding applications in industry. *Adv. Frict. Stir Weldin. Process.* 671–722.
14 Bahrami, M., Dehghani, K., and Givi, M.K.B. (2014). A novel approach to develop aluminum matrix nano-composite employing friction stir welding technique. *Mater. Des. 53*: 217–225.

15 Scudino, S., Liu, G., Prashanth, K.G. et al. (2009). Mechanical properties of Al-based metal matrix composites reinforced with Zr-based glassy particles produced by powder metallurgy. *Acta Mater. 57* (6): 2029–2039.

16 Mishra, R.S. and Ma, Z.Y. (2005). Friction stir welding and processing. *Mater. Sci. Eng.: R: Rep. 50* (1-2): 1–78.

17 Aluminium Alloys in the Automotive Industry: A Handy Guide. Available online

18 Mabuwa, S. and Msomi, V. (2020). Comparative analysis between normal and submerged friction stir processed friction stir welded dissimilar aluminium alloy joints. *J. Mater. Res. Technol. 9* (5): 9632–9644.

19 Prado, R.A., Murr, L.E., Shindo, D.J., and Soto, K.F. (2001). Tool wear in the friction-stir welding of aluminum alloy 6061+ 20% Al_2O_3: a preliminary study. *Scr. Mater. 45* (1): 75–80.

20 Dialami, N., Cervera, M., Chiumenti, M., and de Saracibar, C.A. (2017). A fast and accurate two-stage strategy to evaluate the effect of the pin tool profile on metal flow, torque and forces in friction stir welding. *Int. J. Mech. Sci. 122*: 215–227.

21 Dialami, N., Cervera, M., Chiumenti, M., and Segatori, A. (2019). Prediction of joint line remnant defect in friction stir welding. *Int. J. Mech. Sci. 151*: 61–69.

22 Dialami, N., Chiumenti, M., Cervera, M. et al. (2017). Enhanced friction model for Friction Stir Welding (FSW) analysis: Simulation and experimental validation. *Int. J. Mech. Sci. 133*: 555–567.

23 Dialami, N., Chiumenti, M., Cervera, M., and de Saracibar, C.A. (2013). An apropos kinematic framework for the numerical modeling of friction stir welding. *Comput. Struct. 117*: 48–57.

24 Agapiou, J.S. and Carlson, B.E. (2020). Friction stir welding for assembly of copper squirrel cage rotors for electric motors. *Procedia Manuf. 48*: 1143–1154.

25 Zhao, Y.H., Lin, S.B., Qu, F.X., and Wu, L. (2006). Influence of pin geometry on material flow in friction stir welding process. *Mater. Sci. Technol. 22* (1): 45–50.

26 Guo, C., Shen, Y., Hou, W. et al. (2018). Effect of groove depth and plunge depth on microstructure and mechanical properties of friction stir butt welded AA6061-T6. *J. Adhes. Sci. Technol. 32* (24): 2709–2726.

27 Gupta, M.K. (2020). Effects of tool profile on mechanical properties of aluminium alloy Al 1120 friction stir welds. *J. Adhes. Sci. Technol. 34* (18): 2000–2010.

28 Tamadon, A., Baghestani, A., and Bajgholi, M.E. (2020). Influence of WC-based pin tool profile on microstructure and mechanical properties of AA1100 FSW welds. *Technologies 8* (2): 34.

29 Sharma, N., Siddiquee, A.N., Khan, Z.A., and Mohammed, M.T. (2018). Material stirring during FSW of Al–Cu: effect of pin profile. *Mater. Manuf. Processes 33* (7): 786–794.

30 Janeczek, A., Tomków, J., and Fydrych, D. (2021). The influence of tool shape and process parameters on the mechanical properties of AW-3004 aluminium alloy friction stir welded joints. *Materials 14* (12): 3244.

31 Jayaprakash, S., Siva Chandran, S., Sathish, T. et al. (2021). Effect of tool profile influence in dissimilar friction stir welding of aluminium alloys (AA5083 and AA7068). *Adv. Mater. Sci. Eng. 2021*: 1–7.

32 Fujii, H., Cui, L., Maeda, M., and Nogi, K. (2006). Effect of tool shape on mechanical properties and microstructure of friction stir welded aluminum alloys. *Mater. Sci. Eng., A 419* (1-2): 25–31.

33 Wei, S., Hao, C., and Chen, J. (2007). Study of friction stir welding of 01420 aluminum–lithium alloy. *Mater. Sci. Eng., A 452*: 170–177.

34 Patil, H.S. and Soman, S.N. (2010). Experimental study on the effect of welding speed and tool pin profiles on AA6082-O aluminium friction stir welded butt joints. *Int. J. Eng. Sci. Technol. 2* (5): 268–275.

35 Elangovan, K., Balasubramanian, V., and Valliappan, M. (2008). Influences of tool pin profile and axial force on the formation of friction stir processing zone in AA6061 aluminium alloy. *Int. J. Adv. Manuf. Technol.* *38*: 285–295.

36 Liu, H.J., Zhang, H.J., Huang, Y.X., and Lei, Y.U. (2010). Mechanical properties of underwater friction stir welded 2219 aluminum alloy. *Trans. Nonferrous Met. Soc. China 20* (8): 1387–1391.

37 Koilraj, M., Sundareswaran, V., Vijayan, S., and Rao, S.K. (2012). Friction stir welding of dissimilar aluminum alloys AA2219 to AA5083–optimization of process parameters using Taguchi technique. *Mater. Des. 42*: 1–7.

38 Tutar, M., Aydin, H., Yuce, C. et al. (2014). The optimisation of process parameters for friction stir spot-welded AA3003-H12 aluminium alloy using a Taguchi orthogonal array. *Mater. Des. 63*: 789–797.

39 Yin, K., Cao, L., and Wang, N. (2019). Mechanical properties and residual stresses of 5083 to AM60B dissimilar friction stir welding with different process parameters. *J. Adhes. Sci. Technol. 33* (23): 2615–2629.

40 Kalemba, I. and Dymek, S. (2016). Microstructure and properties of friction stir welded aluminium alloys. *Weld. Int. 30* (1): 38–42.

41 Velichko, O.V., Ivanov, S.Y., Karkhin, V.A. et al. (2016). Methods of overlapping closed joints in friction stir welding of thin wall structures made of aluminium alloys. *Weld. Int. 30* (10): 797–801.

42 Lukin, V.I., Ioda, E.N., Skupov, A.A., and Ovchinnikov, V.V. (2016). Friction stir welding of V-1461 and V-1469 high strength aluminium–lithium alloys. *Weld. Int. 30* (7): 552–556.

43 Labus Zlatanovic, D., Balos, S., Bergmann, J.P. et al. (2021). Influence of tool geometry and process parameters on the properties of friction stir spot welded multiple (AA 5754 H111) aluminium sheets. *Materials 14* (5): 1157.

44 MohammadiSefat, M., Ghazanfari, H., and Blais, C. (2021). Friction stir welding of 5052-H18 aluminum alloy: modeling and process parameter optimization. *J. Mater. Eng. Perform. 30*: 1838–1850.

45 Zhai, M., Wu, C., and Su, H. (2020). Influence of tool tilt angle on heat transfer and material flow in friction stir welding. *J. Manuf. Processes 59*: 98–112.

46 Rajendran, C., Srinivasan, K., Balasubramanian, V. et al. (2019). Effect of tool tilt angle on strength and microstructural characteristics of friction stir welded lap joints of AA2014-T6 aluminum alloy. *Trans. Nonferrous Met. Soc. China 29* (9): 1824–1835.

47 Acharya, U., Roy, B.S., and Saha, S.C. (2021). On the role of tool tilt angle on friction stir welding of aluminum matrix composites. *Silicon 13*: 79–89.

48 Sivabalan, S., Sridhar, R., Parthiban, A., and Sathiskumar, G. (2021). Experimental investigations of mechanical behavior of friction stir welding on aluminium alloy 6063. *Mater. Today Proc. 37*: 1678–1684.

49 Maurya, S.K., Kumar, R., Mishra, S.K. et al. (2022). Friction stir welding of cast aluminum alloy (A319): effect of process parameters. *Mater. Today Proc. 56*: 1024–1033.

50 Mishra, D., Gupta, A., Raj, P. et al. (2020). Real time monitoring and control of friction stir welding process using multiple sensors. *CIRP J. Manuf. Sci. Technol. 30*: 1–11.

51 Boldsaikhan, E., Corwin, E.M., Logar, A.M., and Arbegast, W.J. (2011). The use of neural network and discrete Fourier transform for real-time evaluation of friction stir welding. *Appl. Soft Comput. 11* (8): 4839–4846.

52 Shrivastava, A., Pfefferkorn, F.E., Duffie, N.A. et al. (2015). Physics-based process model approach for detecting discontinuity during friction stir welding. *Int. J. Adv. Manuf. Technol. 79*: 605–614.

53 Sahu, S.K., Mishra, D., Pal, K., and Pal, S.K. (2021). Multi sensor based strategies for accurate prediction of friction stir welding of polycarbonate sheets. *Proc. Inst. Mech. Eng., Part C: J. Mech. Eng. Sci. 235* (17): 3252–3272.

54 Jia, H., Wu, K., Sun, Y., and Hu, F. (2023). Experimental research on process optimization for high-speed friction stir welding of aluminum alloy. *JOM* 1–13.

55 Liu, H.J., Zhang, H.J., and Yu, L. (2011). Effect of welding speed on microstructures and mechanical properties of underwater friction stir welded 2219 aluminum alloy. *Mater. Des. 32* (3): 1548–1553.

56 Shehabeldeen, T.A., Abd Elaziz, M., Elsheikh, A.H. et al. (2020). A novel method for predicting tensile strength of friction stir welded AA6061 aluminium alloy joints based on hybrid random vector functional link and henry gas solubility optimization. *IEEE Access 8*: 79896–79907.

57 Abd Elaziz, M., Shehabeldeen, T.A., Elsheikh, A.H. et al. (2020). Utilization of random vector functional link integrated with marine predators algorithm for tensile behavior prediction of dissimilar friction stir welded aluminum alloy joints. *J. Mater. Res. Technol. 9* (5): 11370–11381.

58 Lambiase, F., Paoletti, A., and Di Ilio, A. (2018). Forces and temperature variation during friction stir welding of aluminum alloy AA6082-T6. *Int. J. Adv. Manuf. Technol. 99*.

59 Derazkola, H.A. and Elyasi, M. (2018). The influence of process parameters in friction stir welding of Al-Mg alloy and polycarbonate. *J. Manuf. Processes 35*: 88–98.

60 Aghajani Derazkola, H. and Khodabakhshi, F. (2019). Intermetallic compounds (IMCs) formation during dissimilar friction-stir welding of AA5005 aluminum alloy to St-52 steel: Numerical modeling and experimental study. *Int. J. Adv. Manuf. Technol. 100*: 2401–2422.

61 Kumar, R., Dhami, S.S., and Mishra, R.S. (2021). Optimization of friction stir welding process parameters during joining of aluminum alloys of AA6061 and AA6082. *Mater. Today Proc. 45*: 5368–5376.

62 Padmanaban, R., Balusamy, V., and Vaira Vignesh, R. (2020). Effect of friction stir welding process parameters on the tensile strength of dissimilar aluminum alloy AA2024-T3 and AA7075-T6 joints. *Materialwiss. Werkstofftech. 51* (1): 17–27.

63 Rathinasuriyan, C. and Kumar, V.S. (2021). Optimisation of submerged friction stir welding parameters of aluminium alloy using RSM and GRA. *Adv. Mater. Process. Technol. 7* (4): 696–709.

64 Akinlabi, E.T., Mahamood, R.M., Akinlabi, S.A., and Ogunmuyiwa, E. (2014). Processing parameters influence on wear resistance behaviour of friction stir processed Al-TiC composites. *Adv. Mater. Sci. Eng.*, *2014*.

65 Ali, K.S.A., Mohanavel, V., Vendan, S.A. et al. (2021). Mechanical and microstructural characterization of friction stir welded SiC and B4C reinforced aluminium alloy AA6061 metal matrix composites. *Materials 14* (11): 3110.

66 Laska, A. and Szkodo, M. (2020). Manufacturing parameters, materials, and welds properties of butt friction stir welded joints–overview. *Materials 13* (21): 4940.

67 Sidhu, R.S., Kumar, R., Kumar, R. et al. (2022). Joining of dissimilar Al and Mg metal alloys by friction stir welding. *Materials 15* (17): 5901.

68 Praneetha, K., Apoorva, M., Laxmi, T.P. et al. (2022). Experimental investigation on aluminium alloy AA6082 and AA2014 using the friction stir welding. *Mater. Today Proc. 62*: 3397–3404.

69 Nakowong, K. and Sillapasa, K. (2021). Optimized parameter for butt joint in friction stir welding of semi-solid aluminum alloy 5083 using Taguchi technique. *J. Manuf. Mater. Process. 5* (3): 88.

70 Cavaliere, P., De Santis, A., Panella, F., and Squillace, A. (2009). Effect of welding parameters on mechanical and microstructural properties of dissimilar AA6082–AA2024 joints produced by friction stir welding. *Mater. Des. 30* (3): 609–616.

71 Chen, S., Zhang, H., Jiang, X. et al. (2019). Mechanical properties of electric assisted friction stir welded 2219 aluminum alloy. *J. Manuf. Processes 44*: 197–206.

72 Singh, T., Tiwari, S.K., and Shukla, D.K. (2022). Novel method of nanoparticle addition for friction stir welding of aluminium alloy. *Adv. Mater. Process. Technol. 8* (1): 1160–1172.

73 Napitupulu, R.A., Simanjuntak, S.L., Manurung, C., et al. (2019, April). Friction stir welding of aluminium alloy 6061-t651. *IOP Conference Series: Materials Science and Engineering* (Vol. 508, No. 1, p. 012064). IOP Publishing.

74 Cho, J.H., Kim, W.J., and Lee, C.G. (2014). Texture and microstructure evolution and mechanical properties during friction stir welding of extruded aluminum billets. *Mater. Sci. Eng., A 597*: 314–323.

75 Bayazid, S.M., Farhangi, H., Asgharzadeh, H. et al. (2016). Effect of cyclic solution treatment on microstructure and mechanical properties of friction stir welded 7075 Al alloy. *Mater. Sci. Eng., A 649*: 293–300.

76 Singh, T., Tiwari, S.K., and Shukla, D.K. (2020). Effect of nano-sized particles on grain structure and mechanical behavior of friction stir welded Al-nanocomposites. *Proc. Inst. Mech. Eng., Part L: J. Mater.: Des. Appl. 234* (2): 274–290.

77 Zhao, Y., Yang, Z., Domblesky, J.P. et al. (2019). Investigation of through thickness microstructure and mechanical properties in friction stir welded 7N01 aluminum alloy plate. *Mater. Sci. Eng., A 760*: 316–327.

78 Fujii, H., Cui, L., Nakata, K., and Nogi, K. (2008). Mechanical properties of friction stir welded carbon steel joints—Friction stir welding with and without transformation. *Weld. World 52*: 75–81.

79 Singh, T., Tiwari, S.K., and Shukla, D.K. (2020). Effects of Al_2O_3 nanoparticles volume fractions on microstructural and mechanical characteristics of friction stir welded nanocomposites. *Nanocomposites 6* (2): 76–84.

80 Saeidi, M., Barmouz, M., and Givi, M.K.B. (2015). Investigation on AA5083/AA7075+ Al_2O_3 joint fabricated by friction stir welding: characterizing microstructure, corrosion and toughness behavior. *Mater. Res. 18*: 1156–1162.

81 María, Abreu Fernández, C., Rey, R.A., Julia Cristóbal Ortega, M. et al. (2018). Friction stir processing strategies to develop a surface composite layer on AA6061-T6. *Mater. Manuf. Processes 33* (10): 1133–1140.

82 Dixit, M., Newkirk, J.W., and Mishra, R.S. (2007). Properties of friction stir-processed Al 1100–NiTi composite. *Scr. Mater. 56* (6): 541–544.

83 Gandra, J., Miranda, R., Vilaca, P. et al. (2011). Functionally graded materials produced by friction stir processing. *J. Mater. Process. Technol. 211* (11): 1659–1668.

84 Gesella, G. and Czechowski, M. (2017). The application of friction stir welding (FSW) of aluminium alloys in shipbuilding and railway industry. *J. KONES 24* (2): 85–90.

85 Audi of America News Channel (2011). High-tech production of aluminium bodies, February.

86 The Aluminum Association (2008). Airbus to use friction stir welding.

87 Freeman, J., Moore, G., Thomas, B., and Kok, L. (2006). Advances in FSW for commercial aircraft applications. *Proceedings of the 6th International Symposium on Friction Stir Welding, Toronto, ON, Canada*, October (pp. 10–13).

88 Femandez, F. (2010). Friction stir welding applied on mid size aircraft. *8th FSW Symposium*, May.

9

Mechanical Characterization of FSWed Joints of Dissimilar Aluminum Alloys of AA7050 and AA6082

Mohd Sajid[1], Gaurav Kumar[1], Husain Mehdi[2], and Mukesh Kumar[1]

[1] Department of Mechanical Engineering, VCE Dr. A.P.J. Abdul Kalam Technical University, Lucknow, Uttar Pradesh, India
[2] Department of Mechanical Engineering, Meerut Institute of Engineering and Technology, Dr. A.P.J. Abdul Kalam Technical University, Lucknow, Uttar Pradesh, India

9.1 Introduction

FSW is a solid-state welding method created by The Welding Institute (TWI) in the UK [1] and is a noteworthy new development in welding. FSW can join plates and sheet materials, including Mg, Al, Cu, and steel, using a nonconsumable tool. The most potential FSW use is in Al-alloys. There is no need for filler metal or shielding gas when joining different Al-alloys, particularly those that traditional fusion welding processes often cannot join. Solid-state dynamic recrystallization is used in the welds to enable superplastic deformation. Because they are lightweight, Al-alloys are used more frequently in the industry. Al-alloys are challenging to fuse using traditional fusion welding procedures, and the joint quality may be enhanced because of welding flaws such as distortion, porosity, and fractures [1]. FSW has lately been chosen as a dependable technique for maintaining the alloy's characteristics when welding occurs in the solid state. This method connects several material combinations, including brass, aluminum, copper, magnesium, and other materials. In the procedures of connecting materials, the FSW tool is crucial [2]. Due to their exceptional corrosion resistance, castability, and demand for improved strength-to-weight ratios in lightweight constructions, Al-alloys are being used more and more often. Previous research examined how FSW process factors affected the joining of different Al-alloys, such as those in the 2xxx to 8xxx range. A little research has been done on welding parameters' impact on FSWed joints of 7xxx and 6xxx series [3–6]. The consequence of the FSP on the TIG weldments of AA7075 and AA6061 with different filler materials was analyzed, and it revealed that the metallurgical characteristics of TIG weldments were enhanced after applying the FSP [7–12]. The UTS and average weld temperature of various joints were found to have risen due to welding tool deviations on the RS of the AA7075 in the FSWed AA7075 and AA6061 joints. The influence of TS and materials' position on the metallurgical characteristics and material flow was analyzed. They demonstrated that the AA6061 situated at the AS and numerous vortex centers forming precipitously in the stir zone (SZ) significantly increased the effectiveness of the material combination [13]. The AA6061 plate was positioned on the AS, and they revealed the maximum joint efficiency. The susceptibility and microstructural properties of FSWed joints of AA6056 and AA7075 were analyzed [14] and revealed that the positioning of base plates in the FSW of different alloys might result in more pronounced temperature asymmetry. The differing joint qualities, which predominantly depend on the base material's properties and processing parameters (RTS, TS, and material locations), might be significantly impacted by this material flow between AS and RS, stress, and asymmetry in temperature. According to reports, a small material amount from the leading edge experiences chaotic flow. Good joints can be produced without flaws only when the weld contact is on AS [15]. High rotational speeds can be used with better materials while compromising lower surface integrity [16, 17]. More effective material mixing was achieved when the AA6xxx was positioned on RS and the softer AA5xxx was on the AS [18, 19]. Maximum joint efficiency was perceived when AA5xxx was positioned on the AS. On the AS, joints made with the soft AA5xxx Al alloy showed marginally lower longitudinal ultimate tensile strength (UTS) than contrariwise ones [20]. Another research looked at the FSW of the alloys A356 and AA6061. Because the dominating material in the welded region generally originated from the harder RS, greater longitudinal UTS of the FSWed joints was observed when a somewhat soft material was positioned at the AS [21]. A specific forecast can only be made after research due to the strong relationship between the influence of the relative positions of the base plates. The impact of several process variables, such as RTS, TS, and pin profile, on the mechanical properties of the FSWed joints of AA5083 and AA2024 was analyzed.

In contrast, the tool pin design with a straight cylinder has the lowest UTS and % strain. However, after achieving a maximum value, the UTS began to decline because of the increased RTS and TS. In the current study, an effort has been made to comprehend how TS and RTS affect the mechanical characterization of FSWed joints of AA7050 and AA6082. Different FSW tests are carried out under various welding circumstances to achieve this.

9.2 Materials and Methods

The FSW was accomplished on 6 mm thick AA7050 and AA6082 plates, respectively, whose mechanical characteristics and chemical composition are described in Tables 9.1 and 9.2. Before welding, all the base plates were sliced into 160×80 mm. Utilizing an FSW machine, single-pass FSW was completed. The dimension of the FSW tool is demonstrated in Figure 9.1. For each

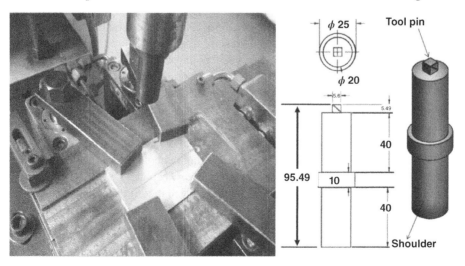

Figure 9.1 Experimental setup and FSP tool specification.

Table 9.1 Mechanical characteristics of AA7050 and AA6082.

Material	UTS (N/mm²)	YS (N/mm²)	Strain (%)	Hardness (HV)
AA7050	474	312.52	12.6	112
AA6082	226	142.65	24.8	72.5

Table 9.2 Chemical composition of Al-alloys.

Base metal	Mg	Zn	Fe	Mn	Cu	Zr	Cr	Si	Al
AA7050	2.1	5.9	0.14	0.08	1.8	0.09	0.05	0.09	Balance
AA6082	0.92	0.12	0.35	0.75	0.09	0.01	0.18	1.2	Balance

Table 9.3 Processing parameter for FSW.

Sample no	RTS (rev/m)	TS (mm/min)	Tilt angle (°)
1	800	90	0
2	900		
3	1000		
4	1100		

experiment, the axial load and tool tilt angle were adjusted to 7.2 kN and 0°, respectively. Several experiments were conducted to identify the welding parameters that produced reliable welds. Table 9.3 is a list of the welding parameters that were used. While AA7050 was in RS, AA6082 was in the AS. An optical microscope was employed for microstructure analysis. AA7050 and AA6082 were etched using a reagent of 150 ml H_2O, 6 ml HCl, 6 ml HF, and 3 ml HNO_3. Vickers hardness was also assessed at the load of 100 gm for 15 seconds on the FSWed region of all the specimens. After that, the tensile samples were prepared along the transverse direction of the weld as per ASTM E8 standard.

9.3 Results and Discussion

9.3.1 Tensile Strength

The processing parameters of FSW affect the tensile strength and microhardness of AA6082 and AA7050 weldments. Based on the analysis, the FSW parameter range was determined. The UTS of welded joints was perceived under the uniaxial direction. Three tensile samples were made for each parameter, and the mean UTS of these three specimens was taken. Table 9.4 displays the mechanical characteristics and effectiveness of the welded joints between AA7050 and AA6082. When the RTS increases at constant TS, the UTS increases, and UTS rises with an increase in RTS. The highest strength (187.75 N/mm²) was perceived at RTS of 1100 rev/min with a TS of 90 mm/min, while the lowest UTS (127.68 N/mm²) was perceived at RTS of 800 rev/min with a TS of 90 mm/min.

Table 9.4 Tensile properties of the welded joint.

Specimen no	RTS (rev/m)	TS (mm/min)	UTS (N/mm²)	Strain (%)	Microhardness at SZ (HV)
1	800	90	127.68	13.94	78.23
2	900		164.21	18.31	73.69
3	1000		173.25	18.93	76.25
4	1100		187.75	20.67	86.71

As the RTS rose, the equiaxed and fine grain structure that resulted from the increasing heat input into the welded joints enhanced the tensile characteristics. The top surfaces of the base metal may have too much stirred welded material when the RTS is higher than 1100 rev/m, which might lead to tiny voids in the SZ. A temperature rise, a coarsening of the grain size, and a slower cooling rate at a temperature higher than desirable may all work together to lower the UTS of the FSWed joints at high RTS. The revolving tool created frictional heat, and at low RTS, the base metal would not have produced enough plasticized flow. Certain defects were also found during the material flow around the welded joints in AS [23]. The FSWed joint has a higher % strain than the other different processing parameters at 1100 rev/m and a TS of 90 mm/min. The true stress–strain diagram of weldments along the different welding parameters is revealed in Figure 9.2.

The highest joint efficiency (83.08%) was perceived at RTS of 1100 rev/m, TS of 90 mm/min, and the lowest joint efficiency (56.50%) was perceived at RTS of 800 rev/m, TS of 90 mm/min. This is due to a lower RTS having a narrower temperature distribution, a subpar square pin stirring action, and a subpar tool shoulder consolidation of the FSWed materials [24].

9.3.2 Microhardness

The microhardness profile with different process parameters is shown in Figure 9.3. Due to aging precipitations on AA7050, the microhardness value diminished from the parent material (AA7050) to HAZ. The microhardness was carried out from the AA6082 to the AA7050, comprising the welded zone. Three hardness values were measured at each location, and the mean value was taken. Because of the equiaxed and very fine grain structure seen at the SZ, the hardness values were greater than the TMAZ and HAZ. The reduced hardness value was detected due to

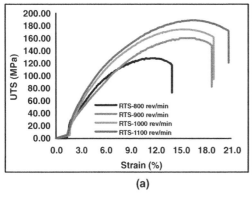
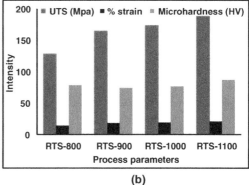

(a)   (b)

Figure 9.2 (a) Stress–strain curve of the FSWed joints of AA6082 and AA7050, (b) comparison of mechanical properties with process parameters.

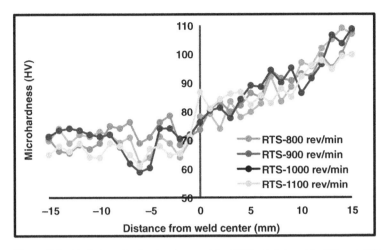

Figure 9.3 Variation of microhardness of the FSWed joints with different RTS.

precipitates and coarse grains in the HAZ, whereas at TMAZ, the falling trend of hardness was observed due to the dissolution of the precipitate [25], as revealed in Figure 9.3. Because of the refined grains and high heat input, the microhardness value at the SZ enhanced as the RTS increased [26]. The hardness value at TMAZ increases because of plastic deformation imposed by the shoulder in this zone which initiates the refinement of grains. In SZ, the hardness value is increased as compared to the base metal AA6082 and less than the second base metal (AA7050) due to recrystallization and dissolution of precipitates. It was perceived that the microhardness at SZ was increased at high RTS due to the refinement of grain structure and found small and equi-axed grain structures in the SZ. Hardness is influenced by dislocation density and phase dispersion. Less substantial is the influence of hardness levels on the tensile strength of the weldments [27]. The solidification and cooling rate of the weldments dramatically inflated the hardness value in the center and bottom of the FSWed joints. An important consideration while examining the metallurgical phase is the hardness value. The minimal hardness value (73.69 HV), as shown in Figure 9.3, was perceived at RTS of 900 rev/m and TS of 90 mm/min; however, the more significant microhardness value (86.71 HV) was perceived at RTS of 1100 rev/min and TS of 90 mm/min.

9.3.3 Microstructural Evaluation

Figure 9.4a demonstrates the optical microscopy of the welded joint, while Figure 9.4b–d reveals the microstructure at RTS of 900, 1000, and 1100 rev/min at SZ observing the expected grain structure of the weldments of AA7050 and AA6082. The TS has less influence on the welded joints. It was expected that the higher RTS, the lower the resultant efforts were. The amount of time required at the mixture temperature for the materials to soften must be guaranteed by the TS. Figure 9.4 depicts the influence of RTS (800–1100) rev/m with a constant TS of 90 mm/min on the microstructure of AA7050 and AA6082 FSWed joints. The thermocouple detected temperature ranges of 395–432 °C in the SZ for all the specimens. The temperature in the SZ is likely to have been greater than TMAZ and HAZ regions [28]. The fine grain size was found at an RTS of 1100 rev/m as compared to 800 rev/m. The grain size was measured by the Image J software and perceived the average grain size in the SZ was 9.2 µm at 1100 rev/m, whereas 21.52 µm and 14.65 µm grain sizes were observed at RTS of 800 rev/m and 900 rev/m, respectively. The microstructure of the weldment was made up of base metal, TMAZ, and HAZ regions. The TMAZ and HAZ region comprised coarse

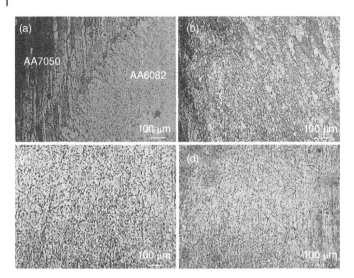

Figure 9.4 Optical microstructure of the welded joint at SZ: (a) welded region, (b) Specimen 1, (c) Specimen 3, (d) Specimen 4.

grain structure compared to SZ. This microstructure varied depending on the FSWed joint's process parameters.

When the microstructure was investigated at RTS of 800 rev/m with constant TS, a large gap between the grain structure was observed. When the microstructure at RTS of 1100 rev/m with TS of 90 mm/min was examined, less porosity ensued at the welded regions. Whereas, when the microstructure at RTS of 1100 rev/m was examined, fine grain structure was observed at the SZ as shown in Figure 9.4. This condition demonstrated that increasing the RTS, reduced grain size. Low plasticity temperature and a drop in temperature were predicted to induce an increase in welding defects [29]. The welding quality of samples at RTS of 1100 rev/m with TS of 90 mm/min was very well observed, with very few welding defects.

9.3.4 Fractured Surface Analysis

Figure 9.5 reveals the formation of tiny and equiaxed dimples due to intense plastic deformation during the welding, with subsequent dissolution of the precipitates, together with the DRX of the base metals. The primary metal became more ductile due to fractures spreading throughout it. As a result, the fragmented region shown in Figure 9.5 will develop dimples on its surface. The main cause of the portion was an interface that had ruptured, showing cleavage and dimples. In the high RTS, a ductile fracture feature, and fine and uniform dimples were evident. During the tension test, a cup-cone failure was discovered around the samples' perimeter at 45° to the tensile axis [30], indicating ductile fracture. Micro-void coalescence, nucleation, and dimple coalescence were the primary driving forces behind fractography. The dimples get narrower as the FSP passes rise [31] in the high RTSs. The more homogeneous and uniform dimples shown in Figure 9.5 contributed to the improvement in ductility and strength. The microstructure of the FSWed joints was modified and redistributed at a high tool rotating speed. Since the dendrite structures were removed during the redistribution, the interdendritic fracture was avoided, and microvoids were distributed throughout the base metal.

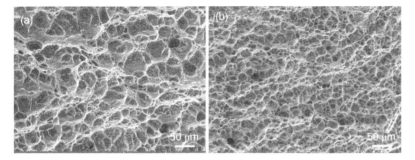

Figure 9.5 SEM images of tensile fractured specimen, (a) Specimen 1, (b) Specimen 4.

9.3.5 Conclusions

Based on the positively fabricated AA7050 and AA6082 joints using FSW, the influence of FSW parameters on their metallurgical characterization was investigated. The following conclusions were drawn:

- The FSW parameters on the AA7050 and AA6082 joints converted the elongated dendritic structures to equiaxed and refined ones.
- The joint's efficiency of the FSWed joints of AA7050 and AA6082 was found higher at RTS of 1100 rev/m with a TS of 90 mm/min.
- The highest UTS of 187.75 N/mm^2 was perceived at RTS 1100 rev/m, TS 90 mm/min, while minimum tensile stress of 127.68 N/mm^2 was perceived at RTS of 800 rev/m with a TS of 90 mm/min.
- When the microstructure at RTS of 1100 rev/m with different TS was examined, refined and equiaxed grain structures were observed in the SZ. This condition demonstrated that increasing the RTS and TS reduced grain size. Insufficient plasticity temperature and a drop in temperature were predicted to induce an increase in welding defects.
- The reduced hardness value was detected due to precipitation and coarsening of grains in the HAZ region. Whereas at TMAZ, the declining trend of hardness was observed due to the dissolution of the precipitate.

References

1 Khan, N.Z., Siddiquee, A.N., Khan, Z.A., and Shihab, S.K. (2015). Investigations on tunneling and kissing bond defects in FSW joints for dissimilar aluminum alloys. *J. Alloys Compd.* 648: 360–367.
2 Srinivasulu, P., Rao, G.K.M., and Gupta, M.S. (2015). Evaluation of bending strength of friction stir welded AA 6082 aluminum alloy butt joints. *Int. J. Adv. Res. Sci. Eng.* 4 (1): 1262–1270.
3 Li, B. and Shen, Y. (2012). A feasibility research on friction stir welding of a new-typed lap–butt joint of dissimilar Al alloys. *Mater. Des.* 34: 725–731.
4 Palanivel, R., Koshy Mathews, P., Murugan, N., and Dinaharan, I. (2012). Effect of tool rotational speed and pin profile on microstructure and tensile strength of dissimilar friction stir welded AA5083-H111 and AA6351-T6 aluminum alloys. *Mater. Des.* 40: 7–16.

5 da Silva, A.A.M., Arruti, E., Janeiro, G. et al. (2011). Material flow and mechanical behaviour of dissimilar AA2024-T3 and AA7075-T6 aluminum alloys friction stir welds. *Mater. Des.* 32: 2021–2027.

6 Hatamleh, O. and DeWald, A. (2009). An investigation of the peening effects on the residual stresses in friction stir welded 2195 and 7075 aluminum alloy joints. *J. Mater. Process. Technol.* 209: 4822–4829.

7 Mehdi, H., Mabuwa, S., Msomi, V., and Yadav, K.K. (2022). Influence of friction stir processing on the mechanical and microstructure characterization of single and double V-groove tungsten inert gas welded dissimilar aluminum joints. *J. Mater. Eng. Perform.* https://doi.org/10.1007/s11665-022-07659-7.

8 Mabuwa, S., Msomi, V., Mehdi, H., and Saxena, K.K. (2022). Effect of material positioning on Si-rich TIG welded joints of AA6082 and AA8011 by friction stir processing. *J. Adhes. Sci. Technol.* https://doi.org/10.1080/01694243.2022.2142366.

9 Hashmi, A.W., Mehdi, H., Mabuwa, S. et al. (2022). Influence of FSP parameters on wear and microstructural characterization of dissimilar TIG welded joints with Si-rich filler metal. *Silicon* 14: 11131–11145. https://doi.org/10.1007/s12633-022-01848-8.

10 Hashmi, A.W., Husain Mehdi, R.S., Mishra, P.M. et al. (2022). Mechanical properties and microstructure evolution of AA6082/Sic nanocomposite processed by multi-pass FSP. *Trans. Indian Inst. Met.* https://doi.org/10.1007/s12666-022-02582-w.

11 Salah, A.N., Mabuwa, S., Mehdi, H. et al. (2022). Effect of multipass FSP on Si-rich TIG welded joint of dissimilar aluminum alloys AA8011-H14 and AA5083-H321: EBSD and microstructural evolutions. *Silicon* 14: 9925–9941. https://doi.org/10.1007/s12633-022-01717-4.

12 Mehdi, H. and Mishra, R.S. (2021). Effect of friction stir processing on mechanical properties and heat transfer of TIG welded joint of AA6061 and AA7075. *Def. Technol.* 17 (3): 715–727. https://doi.org/10.1016/j.dt.2020.04.014.

13 Cole, E.G., Fehrenbacher, A., Duffie, N.A. et al. (2014). Weld temperature effects during friction stir welding of dissimilar aluminum alloys 6061-T6 and 7075-T6. *Int. J. Adv. Manuf. Technol.* 71: 643–652.

14 Guo, J.F., Chen, H.C., Sun, C.N. et al. (2014). Friction stir welding of dissimilar materials between AA6061 and AA7075 Al alloys effects of process parameters. *Mater. Des.* 56: 185–192.

15 Bala Srinivasan, P., Dietzel, W., Zettler, R. et al. (2005). Stress corrosion cracking susceptibility of friction stir welded AA7075–AA6056 dissimilar joint. *Mater. Sci. Eng., A* 392: 292–300.

16 Kumar, K. and Kailas, S.V. (2010). Positional dependence of material flow in friction stir welding: analysis of joint line remnant and its relevance to dissimilar metal welding. *Sci. Technol. Weld. Joining* 15: 305–311.

17 Steuwer, A., Peel, M.J., and Withers, P.J. (2006). Dissimilar friction stir welds in AA5083– AA6082: the effect of process parameters on residual stress. *Mater. Sci. Eng., A* 441: 187–196.

18 Silva, A.A.M., Arruti, E., Janeiro, G. et al. (2011). Material flow and mechanical behaviour of dissimilar AA2024-T3 and AA7075-T6 aluminium alloys friction stir welds. *Mater. Des.* 32: 2021–2027.

19 Park, S.-K., Hong, S.-T., Park, J.-H. et al. (2010). Effect of material locations on properties of friction stir welding joints of dissimilar aluminium alloys. *Sci. Technol. Weld. Joining* 15: 331–336.

20 Aval, H.J., Serajzadeh, S., and Kokabi, A.H. (2011). Thermo-mechanical and microstructural issues in dissimilar friction stir welding of AA5086–AA6061. *J. Mater. Sci.* 46: 3258–3268.

21 Nait Salah, A., Mehdi, H., Mehmood, A. et al. (2021). Optimization of process parameters of friction stir welded joints of dissimilar aluminum alloys AA3003 and AA6061 by RSM. *Mater. Today Proc.* 56 (4): 1675–1684. https://doi.org/10.1016/j.matpr.2021.10.288.

22 Sundaram, N.S. and Murugan, N. (2010). Tensile behavior of dissimilar friction stir welded joints of aluminium alloys. *Mater. Des.* 31: 4184–4193.

23 Azimzadegan, T. and Serajzadeh, S. (2010). An Investigation into microstructures and mechanical properties of AA7075-T6 during friction stir welding at relatively high rotational speeds [J]. *J. Mater. Eng. Perform.* 19 (9): 1256–1263.

24 Mehdi, H. and Mishra, R.S. (2022). Modification of microstructure and mechanical properties of AA6082/ZrB2 processed by multipass friction stir processing. *J. Mater. Eng. Perform.* https://doi.org/10.1007/s11665-022-07080-0.

25 Several, P. and Jaiganesh, V. (2014). A detailed investigation on the role of different Tool Geometry in Friction Stir Welding of various Metals & their Alloys. *Proceedings of the International Colloquium on Materials, Manufacturing & Metrology – ICMMM*, August 8–9, pp 103–107.

26 Ahmadnia, M., Seidanloo, A., Teimouri, R. et al. (2015). Deter-mining Influence of ultrasonic-assisted friction stir welding parameters on mechanical and tribological properties of AA6061 joints. *Int. J. Adv. Manuf. Technol.* 78 (9–12): 2009–2024.

27 Mehdi, H. and Mishra, R.S. (2020). Influence of friction stir processing on weld temperature distribution and mechanical properties of TIG-welded joint of AA6061 and AA7075. *Trans. Indian Inst. Met.* https://doi.org/10.1007/s12666-020-01994-w.

28 Cavaliere, P. and Squillace, A. (2005). High temperature deformation of friction stir processed AA7075 aluminum alloy. *Mater. Charact.* 55: 136–142.

29 Mahoney, M.W., Rhodes, C.G., Flintoff, J.G. et al. (1998). Properties of friction-stir-welded 7075 T651 aluminum [J]. *Metall. Mater. Trans. A* 29: 1955–1964.

30 Mehdi, H., Gaurav, S., Kumar, T., and Sharma, P. (2017). Mechanical characterization of SA-508Gr3 and SS-304L steel weldments. *Int. J. Adv. Prod. Ind. Eng.* 2 (1): 41–46.

31 Strombeck, A.V., Santos, J.F.D., Torster, F., et al. (1999). Fracture toughness behaviour of FSW joints on aluminum alloys. *Proceedings of the First International Symposium on Friction Stir Welding, Paper No. S9-P1*, California, USA, June 1999, TWI Ltd.

10

Sample Preparation and Microstructural Characterization of Friction Stir Processed Surface Composites

Manu Srivastava[1], Sandeep Rathee[2], Shazman Nabi[2], and Atul Kumar[3]

[1] Department of Mechanical Engineering, PDPM Indian Institute of Information Technology, Design and Manufacturing Jabalpur, India
[2] Department of Mechanical Engineering, National Institute of Technology Srinagar, Jammu & Kashmir, India
[3] School of Mechanical Engineering, Vellore Institute of Technology, Vellore, India

10.1 Introduction

Composite fabrication via FSP is gaining tremendous popularity in metals (nonferrous and ferrous alloys) and even in polymers. During the composite fabrication process by FSP, drastic microstructural changes take place resulting in associated improvement in mechanical as well as other surface properties [1–3]. This is owing to severe plastic deformation (SPD) caused by FSP and addition of reinforcement particles. Evolution of mechanical and other properties as well as several functional characteristics developed in friction stir processed (FSPed) composites can be understood and related to FSP parameters only through an effective regime of microstructural characterization. Changes in material properties of material bear direct correlation with microstructural modifications. This also paves the way for process optimization, identification and analysis of imperfections, defects, and all undesirable phases. Thus, a detailed understanding of effective microstructural characterization practice is key in analyzing microstructural evolution during surface composites (SCs) fabrication through FSP.

Many texts on materials science and engineering focus on material processing routes like casting, hot and cold working, but very few texts deal with microstructure study procedures required for microstructural evolution of SCs fabricated via FSP. Each of these processes typically affects the microstructure of the product differently; consequently, causing varied mechanical properties and an associated peculiar response by products to every service condition. This makes microstructural analysis and its correlation with mechanical properties an important characteristic. In order to draw meaningful conclusions on evolved mechanical properties of fabricated SCs, correct microstructural examination practice is required. This chapter addresses this aspect in detail. In this chapter, standard microstructural characterization techniques are presented. Aspects such as the microstructure sample preparation procedure with brief introduction of equipment used in microstructural characterization are covered. A few illustrative examples are added toward the end of each section of this chapter with an aim to facilitate understanding of concepts.

10.2 Sample Preparation for Microscopic Analysis of Metals, Alloys, and Composites

Microscopy involves the study of surface and in-depth features like morphology, grain size and structure, presence of various phases, and dispersion of reinforcement particles (in case of MMCs). Generally, microcopy is done with the interaction of some means of the probe with the prepared sample. Commonly used probes are visible light, beams of high-energy electrons, and X-rays which form the basis of optical microscopy (OM), electron microscopy, and X-ray diffraction, respectively. These techniques provide two-dimensional imaging. The selection of the type of method (microscopy) is out of the scope of the present text and mainly depends on the end requirements of the user. Following is the list of four main classes of microscopy methods employed for obtaining the microstructural features.

1) Stereo-zoom microscope
2) Optical microscopy (OM)
3) Scanning electron microscopy (SEM)
4) Transmission electron microscopy (TEM)

OM is one of the simplest and oldest microscopy technologies as compared to SEM and TEM. OM is mainly utilized to study various microstructural features above 1 µm size. However, below this size, SEM and TEM techniques are utilized. Optical micrography technique is most commonly used in research and in a teaching university laboratory or in an industrial research and development lab. This section introduces detailed discussion on sample preparation and various aspects of OM during MMCs' microstructural examination.

Identification of microstructural features present in a sample being viewed under microscope requires a thorough knowledge of metallurgy, structure, and morphology of the alloys being analyzed. But, the correctly prepared sample is a critical prerequisite for microscopy. Correct specimen preparation for OM is a challenge for the identification and interpretation of various microconstituents and phases. Issues in sample preparation vary with each material type and the conditions of its processing. This section aims to provide a comprehensive methodology for sample preparation required for microstructural study. Since the results of microstructural examination depend on quality of the samples prepared, the utmost care and attention are required during their preparation. For metals and their composites, sample preparation mainly consists of grinding, polishing, and etching. These steps are explained in the following section.

10.2.1 Method of Sampling

To examine the microstructure of FSPed plate, the samples are machined from it. Depending on the objective of what need to be studied, the samples may be cut out from transverse direction or they may be sliced along a plane parallel to the plate surface. There may be several other directions from which the samples may be taken depending on the objective of the study, but most commonly transverse section samples are frequently studied for analysis of various FSP zones. Sliced samples, however, may be useful for such purposes as to study the material flow. We discuss here the commonly chosen direction, i.e. transverse direction. Chosen specimens are cut from the object perpendicular to direction of FSP route either by sawing, wire EDM, or by using an abrasive cutting wheel aligned with water cooling or water-soluble cutting oil. For cutting of hard material, abrasive or diamond wheels are used. One such transverse sample is shown in Figure 10.1.

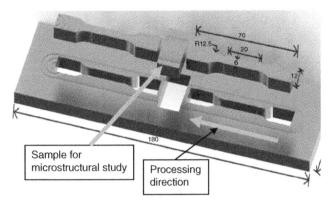

Figure 10.1 Direction of sample cutting for microstructural study [23] Rathee et al., 2018 / Springer Nature.

The samples so machined from the FSPed plate are then processed as described in the following section.

10.2.2 Initial Preparation

Care must be taken during machining that both the machined surfaces are flat, even, and mutually parallel. Uneven surfaces of samples, if any, need to be finished to make them flat, even, mutually parallel, and smooth. This is accomplished by:

- Grinding the samples on a belt grinding having coarse grit, e.g. 100 mesh, a typical belt grinder is shown in Figure 10.2.
- This can also be done manually by filing.

10.2.3 Mounting of Samples

The samples for micrography need to be polished using a standard metallurgical procedure. These samples are usually thin and have irregular shape which makes them difficult to evenly handle during polishing. For the sake of ease of holding and handling during grinding or polishing, the machined samples are mounted in a polymer mold of suitable size. The samples can be mounted either by hot mounting or by cold mounting process, the details are given below:

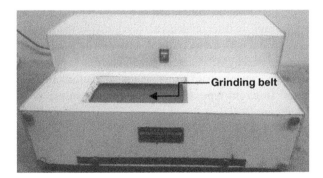

Figure 10.2 Typical belt grinder.

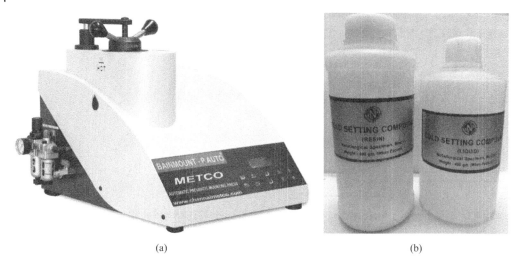

(a)   (b)

Figure 10.3   (a) Typical arrangement for hot mounting, (b) items for cold mounting.

- Hot mounting can be done using hot mounting press at a temperature and pressure of about 160° C and 10–30 N/mm^2, respectively, in a thermosetting plastic like acrylic resin, phenolic resin, or bakelite.
- Certain materials are sensitive to temperatures which prevail during hot mounting temperature; in such cases, hot mounting procedure is undesirable. Cold mounting is done generally for such specimens which get affected using hot mounting process. It can be done using cold setting compounds (refer Figure 10.3) easily available in the market like epoxy or polyester resins.

10.2.4   Polishing of Samples

The samples mounted in molds are first ground on belt grinder once again to remove any leftover layer of mounting material that could have remained on the surface being polished. Subsequent to belt grinding, the samples are polished which involves polishing of specimen on waterproof abrasive papers of varying grit sizes. The grit sequence of abrasive papers ranges from the coarser grade say of 100 grit to the fine grit size of 600 and above depending upon the type of material and micro-constituent feature being studied. For example, polishing up to 1000 grit size may be sufficient for some types of steels while aluminum needs polishing on finer grades. On each girt size, the polishing is performed under ample supply of cooling water with even polishing force and for a specified time. Between successive stages, the sample is rotated by 90°. Each stage of polishing is designed to remove the marks created in the previous stage. Polishing can either be done on automatic polishing machine or on manual polishing discs. Different types of polishing systems in use (viz. automatic, or manual type) are shown in the Figure 10.4.

10.2.5   Method of Polishing

There are a variety of methods available for polishing. These are mechanical, chemical, and electrochemical. Mechanical polishing is the most commonly used method of polishing for

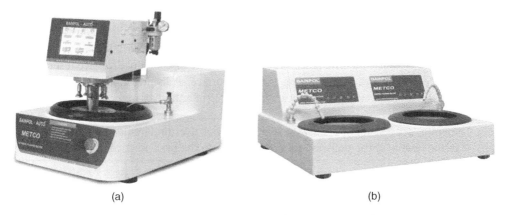

Figure 10.4 (a) Automatic polishing machine; (b) manual disc polishing machine.

metals and alloys. The steps adopted in mechanical polishing procedure of metals and alloys are as given below:

1) The specimen is polished generally on rotating disc with any chosen orientation on a coarser beginning grit of 100, 150, or 200. However, polishing can be manually done by hands without the use of rotating discs. There is a prescribed down hold polishing force which has a specified value for every material. Normally, for soft materials less force is used and vice versa. The speed of rotation and polishing time are also specified and these parameters of down hold force, disc rotation speed, and polishing time need to be kept constant during all successive polishing stages.

2) In the next step, successively fine grits with equal increment are used and the sample is polished by changing orientation generally by 90° in every stage. Before starting with next finer grade, the specimen is usually turned to about 90° (as shown in Figure 10.5). Angle or orientation of specimen is changed so that scratch marks caused during the previous stage are eliminated and get replaced with new ones. The down hold force, speed of disc, and time of polishing are kept fixed for every stage. The change in angle is generally kept as 90°; however, it can be varied based on the prior knowledge and working experience of the personnel involved.

3) Process mentioned at step 2 is repeated till the final finish using finest grit is achieved.

> Note: For polishing of specimen on machine, care should be taken about its speed. It should be kept in mind that higher speeds remove more material but result in inferior surface finish. An important point to be kept in mind is proper cleaning of hands and specimen after completion of each polishing stage with water to ensure removal of grit particles.

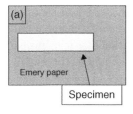 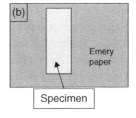

Figure 10.5 (a) Polishing of specimen at first paper; (b) polishing of specimen at subsequent grade.

10.2.6 Fine Polishing

After polishing of samples on finest grit abrasive papers, the samples are polished on discs covered with velvet cloth carrying paste/slurry of submicron size polishing medium like alumina, magnesia, or diamond with lubricant. The polishing cloth should be of good quality since it serves dual function of holding abrasive and applying it to the surface. Polishing cloth should be tightly stretched over the disc and clamped or stuck with the help of adhesives to the wheel.

> Note: The entire process mentioned varies with material, user requirement, and the person's skills. Also, the quality of the polished surface depends on the user practices.

10.3 Etching

Etching is the process of revealing the structure of material using suitable etchants. An etchant in its most common actions selectively dissolves specific phase(s) from a surface, thereby exposing the structure by creating contrast that can be viewed through optics. It is a matter of common knowledge that a lot of microstructural observations cannot be directly obtained from the polished surface. Then, the surfaces must be treated with etchant to reveal structural characteristic like grain boundaries, grains, slip lines, and phase boundaries. Various available etching techniques include chemical etching, electrochemical etching, vacuum cathodic etching, mechanical treatments, and thermal treatments. Among these available techniques, chemical and electrochemical etchings are most popular and commonly used. The etchant serves either one or all of the following requirements:

- It may reveal the structure by attacking grain boundaries.
- It may reveal the structure by discoloring of grains of different orientations.
- It may reveal the structure by staining different phases in case of alloys having more number of phases.

The mechanism of actions of etchant simply reveals that a unique chemical reagent (or etchant) may be needed for surfaces containing different phases. Even for the same material, a different etchant may be selected depending on what actually need to be viewed. Special texts are available which present comprehensive information on various combinations of reagents which can be used for various metals and alloys [4, 5].

10.3.1 Methods of Etching

The reagents, agents, and etchants use a complex combination of various chemicals including concentrated acid and thus they may be extremely toxic. It is therefore, necessary that the etching must be carried out in a well-ventilated space and the chemicals must be stored securely. The etching area must have ample supply of running water readily available. Protections, such as masks and gloves, and sample handling instruments, such as forceps, should always be used. Among several techniques, three methods are most commonly used to accomplish etching of samples. These are: Swab, Immersion, and Electrolytic etching methods. Depending on the material and etchant combination, the etchant is required to be applied for specific duration that may involve several seconds. As soon as etching application is over, the etched surface should be thoroughly washed and dried. Etching reveals the following type of information:

- – Grain boundaries and grain size
- – Different phases
- – Different zones
- – Presence of various constituents

– Information regarding coatings
– Homogeneity, morphology, and distribution of constituents.

Note: For better results, etching should be done just after the fine polishing. And soon after etching, it should be optically examined.

10.4 Microstructural Evolution

The term microstructural refers to the analysis of microscale structural constituents of macroscale landscape of a surface. Once the samples are prepared, polished, and etched, the next step is to obtain the relevant results in the form of macrographs, microstructural images, and correlation of these with the measured properties of the fabricated part/composite, etc. The capturing and analysis of microstructural images are mainly done using optical microscope. A typical OM is as shown in Figure 10.6.

Analysis is discussed in detail in the following section as a case of various processing zones of SCs fabricated by FSP.

After the etching of specimens, almost all of the materials possess structural features that are the characteristics of the material also bear the hallmark of processing the material has been subjected

Figure 10.6 Optical microscope and related parts. Source: Image courtesy: NIT Srinagar, India.

Figure 10.7 Micrograph showing cross-sectional view of AA6063/SiC surface composite [6] Rathee et al., 2018 / Taylor & Francis.

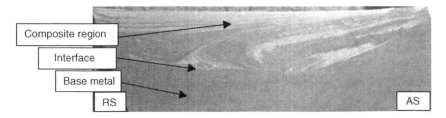

Figure 10.8 Micrographic image of AA6063/SiC surface composite captured via SEM [7] Rathee et al., 2018 / with permission from ELSEVIER / CC BY-NC-ND 4.0.

to. These features may be studied with unaided eyes or with low magnification microscopes that are generally called as stereo-zoom microscopes. This study generally refers to macrostructural study. In macrostructural analysis of SCs, the macrographs are taken and studied to observe the reinforcement particle distribution, shape and size of SZ, and soundness of the processed region. A typical macrograph is as shown in Figure 10.7. The bright region shows the SiC powder dispersion in metal matrix, while light/dark region shows the metal matrix. This was achieved with the help of stereo-zoom microscope. Macrostructural study can also be performed by SEM at low magnification. Figure 10.8 shows a macrographic image of aluminum matrix SC taken using SEM.

10.4.1 Microstructural Zones

Characteristically, transverse sections region of a typical FSPed comprises of four distinct microstructural zones that evolve as a result of various degrees of plastic deformation and thermal effects experienced in each zone. This state of plastic deformation and heat directly represent the effects of FSP process parameters. These distinguishing zones are demonstrated in Figure 10.9. These four zones are named as:

a) Stir zone (SZ);
b) Thermomechanically affected zone (TMAZ);
c) Heat-affected zone (HAZ); and
d) Base material/unaffected parent material.

As shown in Figure 10.9, these zones are clearly distinguished through grain size, grain orientation, and morphology, among several other features. The zones are also separated by a clearly

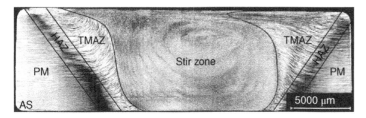

Figure 10.9 Various FSPed zones obtained after FSW [8].

defined zone boundary. All these have been exposed with clarity by application of appropriate microstructural procedure. Peculiar characteristics also result in different mechanical properties of each characteristic zone. The characteristics of each zone are discussed in the subsequent sections.

a) **Stir zone (SZ):** This corresponds to a fully recrystallized area previously occupied by rotating tool pin. Immense grain refinement is evident from the micrograph (refer Figure 10.9). This refinement resulted from the SPD affected by stirring action of the tool. This zone is also referred to as dynamically recrystallized zone. The grain refinement and dynamic recrystallization (DRX) as evident from the microstructure is because of simultaneous application of SPD and frictional heating during FSP. SZ is a complex region to analyze as the grains in this region are morphologically similar yet structurally different containing sub-domains. The top layer particles are shoulder driven, while bulk particles are of pin driven. These minute structural details can only be analyzed through very sophisticated techniques such as TEM.

It is evident that the knowledge of action of mechanical processing (like FSP) on the material being processed bear close relation to resulting microstructure. This enables one to conclude on evolution of various zones citing evidence from micrographs. The ultrafine grain refinement in microstructure during composite fabrication using FSP is attributed to vigorous stirring causing intense plastic deformation succeeded by DRX because of available heat [9–14]. Since material is intensely plastically deformed, its grains get fragmented resulting in numerous low-angle misoriented grain boundaries further leading to increased number of nucleating sites. Consequently, during DRX, transformation of low-angle grain boundaries into higher angle and fresh grain nucleation at favorable sites initiates resulting in creation of fine equiaxed granular structure (this, however, can be confirmed through TEM). Additionally, degree of recrystallization is enhanced owing to the presence of reinforcing material as its surface acts as multiple heterogeneous nucleation sites. Higher counts of initial recrystallized grains is directly proportional to fineness in ultimately achieved microstructure [15].

b) **Thermomechanically affected zone (TMAZ):** As the name implies, it is the area which is affected by heat and mechanical deformation. The process mechanics illustrate that material in this zone undergoes plastic deformation to lesser degree as compared to SZ and the temperature is also less. In some materials like aluminum where considerable plastic strain occurs without recrystallization, there is generally a visual boundary between SZ and TMAZ. A typical macrograph showing the transition between SZ and TMAZ is shown in Figure 10.9. Although TMAZ experiences plastic deformation, recrystallization in this zone does not occur much. This can be seen in micrograph shown in Figure 10.10b.

c) **Heat-affected zone (HAZ):** This region is closest to the processed region in which the material experiences effect of heat but no plastic deformation occurs in this region. However, effects of heat are large enough to cause grain growth; some metallurgical changes (e.g. sensitization in case of some stainless steels) and changes in some microstructural and/or mechanical properties are observed which can be accounted to thermal affects only. In HAZ, the grain size increases due to the thermal effects.

d) **Unaffected material or parent metal:** This is the region remote from the processed region with no deformation, although it may have experienced a thermal cycle from the processed region, but is not affected by the heat in terms of microstructure or mechanical properties. The microstructural characteristics of different zones are summarized in Table 10.1.

The phenomena of DRX and its effect on the grain structure in case of composites fabricated via FSP can be understood with the help of an example. Figure 10.10 shows the micrographs of

Table 10.1 Microstructural characteristics in different zones after FSP.

Different zones	Static/dynamic recrystallization	Static/dynamic recovery	Grain refinement/ growth	Precipitate nucleation, growth, or dissolution
SZ	√	√	Grain refinement	√
TMAZ	√	√	Grain growth	√
HAZ	. . . .	Static recovery	Grain growth	√
BM	No microstructural changes take place			

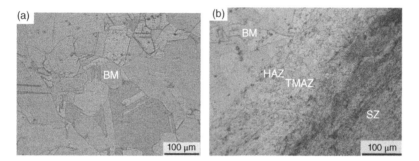

Figure 10.10 Microstructural images of Cu/SiC surface composite: (a) BM; (b) grain structure difference between SZ, TMAZ, and BM [16].

Cu/SiC SCs fabricated via FSP. Figure 10.10a shows the grain structure of unaffected base metal while Figure 10.10b shows the microstructural difference between BM, TMAZ, and SZ. It can be clearly seen that grains in TMAZ are broken due to plastic deformation, but recrystallization phenomena do not takes place while in case of SZ, recrystallization takes place due to sufficient heat and plastic deformation.

Another aspect related to grain size enhancement or reduction during FSP is the heat input. Another factor, i.e. absence or presence of reinforcement particle, has a very dominant effect on the evolved grain dimensions during FSP. In the case of FSP without reinforcement particles, generation of heat facilitates grain growth [17]. There are two contributing effects in the latter case, both of which have contradictory impact on grain size. First, coarsen granular structure is obtained owing to grain growth at elevated heat owing to increased speed of rotation and reduction in traverse speed [18, 19]. Second, reinforcement particles may obstruct grain boundary expansion, thereby restricting granular growth (due to pinning effect) [11]. A broadly accepted fact is the dominance of pinning effect over heat effect in MMCs, thus resulting in microstructural refinement owing to grain nucleation and reinforcement pinning effect which restricts grain growth [9, 20–22].

With massive technical and technological advancement, numbers of equipment that can be utilized for microstructural study have considerably improved in terms of resolution and details. SEM and TEM are used to study the microstructural aspects at higher magnifications. SEM is mainly utilized to study the surface features like dispersion of reinforcement particles, bonding and interface details of BM, and reinforcement particles, while TEM is utilized to enhance the depth of investigation. TEM is generally used to look through the sample where some features such as dislocations, ultrafine grains, sub grain boundaries, twins, precipitates, and crystal lattice that cannot be recorded using SEM.

> Note: Before starting the preparation of samples for SEM and TEM analyses, the users should exactly know what type of information they intend to study. For example, if they want to take microimages using SEM with energy dispersive spectroscopy (EDS) on a nonconductive sample, the conductive technique should be very important.

10.4.2 Sample Preparation for SEM Analysis

The common steps in sample preparation of MMCs/SCs for plate/solid samples are as follows:

- The sample preparation procedure for SEM analysis is quiet similar to sample preparation for OM analysis (refer Section 10.2).
- SEM analysis can be done with or without etching depending upon the type of investigation. For example, if dispersion of reinforcement particles is studied in MMCs, then the analysis can be done without etching (refer Figure 10.11).
- The size of sample for SEM analysis should be kept small depending upon the size of sample holder of SEM. The size of sample holders generally varies in circular shape of diameter ranging from 0.4″, 0.5″, 1″, 1.25″, etc. The size of samples to be prepared should be in line with size of sample holders.
- For irregular-shaped samples, separate jaw holders can also be used (Figure 10.12).

This chapter illustrates that micrography is an important means to expose various macroscale and microscale features present in processed regions. Several details are made visible through these techniques which can be analyzed and interpreted by associated knowledge of materials science and process mechanics. While knowledge of science and mechanics is purely theoretical, the macrography/micrography bears the evidence of what has happened in the processed regions. These macrostructural/microstructural analyses form an important basis for structure property correlation.

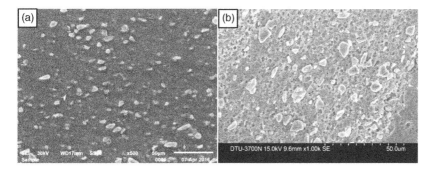

Figure 10.11 Microstructural images captured using SEM: (a) distribution of SiC particles in AA 6063/SiC SCs (without etching condition) [23] Rathee et al., 2018 / Taylor & Francis; (b) distribution of SiC particles in AA 6063/SiC SCs (etching condition) [7] Rathee et al., 2018 / with permission from ELSEVIER / CC BY-NC-ND 4.0.

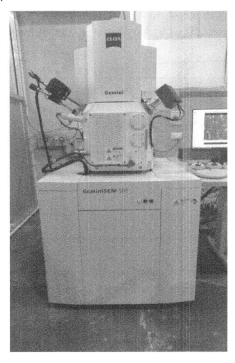 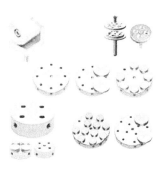

Figure 10.12 Typical SEM and sample holders. Source: Image Courtesy: NIT Srinagar, India.

Acknowledgment

The authors, Dr Manu Srivastava and Dr Sandeep Rathee thank the Science & Engineering Research Board (SERB) for its financial assistance under the project (vide sanction order no. SPG/2021/003383) to perform this work.

References

1 Rathee, S., Maheshwari, S., Noor Siddiquee, A., and Srivastava, M. (2018). A review of recent progress in solid state fabrication of composites and functionally graded systems via friction stir processing. *Critical Reviews in Solid State and Materials Sciences* **43** (4): 334–366.
2 Rathee, S., Maheshwari, S., and Noor Siddiquee, A. (2018). Issues and strategies in composite fabrication via friction stir processing: a review. *Materials and Manufacturing Processes* **33** (3): 239–261.
3 Rathee, S., Maheshwari, S., Siddiquee, A.N., and Srivastava, M. (2017). Effect of tool plunge depth on reinforcement particles distribution in surface composite fabrication via friction stir processing. *Defence Technology* **13** (2): 86–91.
4 Walker, P. and Tarn, W.H. (1991). *Handbook of Metal Etchants*. CRC Press.
5 Petzow, G. (1999). *Metallographic Etching*. ASM International.
6 Rathee, S., Maheshwari, S., Siddiquee, A.N., and Srivastava, M. (2017). Analysis of microstructural changes in enhancement of surface properties in sheet forming of al alloys via friction stir processing. *Materials Today: Proceedings* **4** (2, Part A): 452–458.

7 Rathee, S., Maheshwari, S., Noor Siddiquee, A., and Srivastava, M. (2018). Distribution of reinforcement particles in surface composite fabrication via friction stir processing: Suitable strategy. *Materials and Manufacturing Processes* **33** (3): 262–269.

8 Doude, H.R., Schneider, J.A., and Nunes, A.C. (2014). Influence of the tool shoulder contact conditions on the material flow during friction stir welding. *Metallurgical and Materials Transactions A* **45** (10): 4411–4422.

9 Morisada, Y., Fujii, H., Nagaoka, T. et al. (2006). Effect of friction stir processing with SiC particles on microstructure and hardness of AZ31. *Materials Science and Engineering: A* **433** (1–2): 50–54.

10 Morisada, Y., Fujii, H., Nagaoka, T. et al. (2006). MWCNTs/AZ31 surface composites fabricated by friction stir processing. *Materials Science and Engineering: A* **419** (1–2): 344–348.

11 El-Rayes, M.M. and El-Danaf, E.A. (2012). The influence of multi-pass friction stir processing on the microstructural and mechanical properties of Aluminum Alloy 6082. *Journal of Materials Processing Technology* **212** (5): 1157–1168.

12 Karthikeyan, L., Senthilkumar, V.S., and Padmanabhan, K.A. (2010). On the role of process variables in the friction stir processing of cast aluminum A319 alloy. *Materials & Design* **31** (2): 761–771.

13 Srivastava, M., Rathee, S., Maheshwari, S., and Siddiquee, A.N. (2018). Influence of multiple-passes on microstructure and mechanical properties of Al-Mg/SiC surface composites fabricated via underwater friction stir processing. *Materials Research Express* **5** (6): 066511.

14 Srivastava, M., Rathee, S., Maheshwari, S., and Siddiquee, A.N. (2019). Optimisation of friction stir processing parameters to fabricate AA6063/SiC surface composites using Taguchi technique. *International Journal of Materials and Product Technology* **58** (1): 16–31.

15 Barmouz, M., Asadi, P., Givi, M.K.B. et al. (2011). Investigation of mechanical properties of Cu/SiC composite fabricated by FSP: Effect of SiC particles' size and volume fraction. *Materials Science and Engineering: A* **528** (3): 1740–1749.

16 Akramifard, H.R., Shamanian, M., Sabbaghian, M. et al. (2014). Microstructure and mechanical properties of Cu/SiC metal matrix composite fabricated via friction stir processing. *Materials & Design* **54**: 838–844.

17 Faraji, G. and Asadi, P. (2011). Characterization of AZ91/alumina nanocomposite produced by FSP. *Materials Science and Engineering: A* **528** (6): 2431–2440.

18 Azizieh, M., Kokabi, A.H., and Abachi, P. (2011). Effect of rotational speed and probe profile on microstructure and hardness of AZ31/Al2O3 nanocomposites fabricated by friction stir processing. *Materials & Design* **32** (4): 2034–2041.

19 Abbasi Gharacheh, M., Kokabi, A.H., Daneshi, G.H. et al. (2006). The influence of the ratio of "rotational speed/traverse speed" (ω/v) on mechanical properties of AZ31 friction stir welds. *International Journal of Machine Tools and Manufacture* **46** (15): 1983–1987.

20 Asadi, P., Faraji, G., and Besharati, M. (2010). Producing of AZ91/SiC composite by friction stir processing (FSP). *The International Journal of Advanced Manufacturing Technology* **51** (1–4): 247–260.

21 Morisada, Y., Fujii, H., Nagaoka, T. et al. (2007). Fullerene/A5083 composites fabricated by material flow during friction stir processing. *Composites Part A: Applied Science and Manufacturing* **38** (10): 2097–2101.

22 Izadi, H. (2012). *Proceedings of the 9th International Conference on Trends in Welding Research*. ASM.

23 Rathee, S., Maheshwari, S., Siddiquee, A.N. et al. (2017). *Investigating effects of groove dimensions on microstructure and mechanical properties of AA6063/SiC surface composites produced by friction stir processing. Transactions of the Indian Institute of Metals* 1–8.

11

Microstructural Characterization and Mechanical Testing of FSWed/FSPed Samples

Prem Sagar[1], Sushma Sangwan[1], and Mohankumar Ashok Kumar[2]

[1] Department of Mechanical Engineering, The Technological Institute of Textile Sciences, Maharshi Dayanand University, Rohtak, Haryana, India
[2] Department of Mechanical Engineering, Government College of Engineering, Anna University, Coimbatore, Tamil Nadu, India

11.1 Introduction

Greenhouse gases (GHGs) are essential for maintaining the planet's climate at a level that is suitable for human survival, as they not only cause infectious diseases but also increase and aggravate cardiovascular and respiratory problems in humankind. In the first quarter of 2022, Europe's (EU) economy GHG emissions totaled 1029 million tons of carbon dioxide (CO_2) equivalents (a 7% and 6% increase compared with the same quarter of 2020 and 2021) [1], of which 27% of total EU-27 GHG emissions came from the transport sector. Research shows that one of the major causes of GHG emissions is the transport industry [2]. Transportation-related emissions should be lowered by 94 percent from 2005 levels by 2050 to achieve the National climate objectives of the EU-2030 and Paris climate goals [2, 3]. Study proves that in addition to improvement of aerodynamics and efficiency, reduction of vehicle weight is indeed a pivotal approach to reducing carbon dioxide discharge [4]. According to a linear regression analysis of curb weight vs. CO_2 emissions, CO_2 emissions can be reduced by 8% when a vehicle's weight is reduced by 10% [5]. Research activities related to lightweight materials development have surged rapidly in recent years to manufacture lightweight vehicles with reduced emissions of CO_2.

Materials like magnesium (Mg) and aluminum (Al) are the kinds of materials that are most frequently employed to meet the goal of using lightweight materials. Magnesium and aluminum both have low densities compared to steel; as a result, steel is being largely replaced by these materials. Generally, these Mg and Al metal matrix alloys are hard-to-weld materials due to the presence of flaws such as void and micro-macro cracks [6]. Thus, expanding the use of these metals calls for reliable joining and welding techniques [7]. Friction stir welding (FSW) is a solid-state joining methodology designed and established by The Welding Institute of the United Kingdom [8, 9]. The illustration of basic FSW process is presented in Figure 11.1.

FSW is a modified kind of welding that works well for joining difficult-to-join materials since it allows for the joining of these materials without melting them, yet forming an effective joint. Particularly in the case of metals that were previously difficult to weld, such as magnesium AZ

Friction Stir Welding and Processing: Fundamentals to Advancements, First Edition.
Edited by Sandeep Rathee, Manu Srivastava, and J. Paulo Davim.

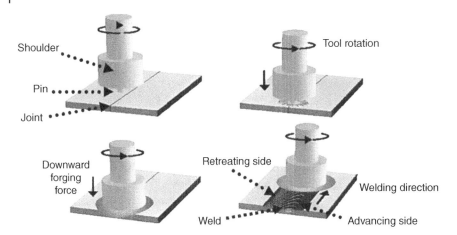

Figure 11.1 Illustration of basic FSW process. Source: Reproduced with permission Aissani et al. [10] Taylor & Francis.

series alloys like AZ31B and AZ61, and aluminum (2xxx and 7xxx series), FSW sparked a revolution [11]. Moreover, FSW made it possible to successfully join metals that are different from one another. Attributed to its enormous potential, the fabricating industry quickly adopted FSW technology, which is today utilized on a global scale.

Many problems in other conventional welding techniques, such as solidification and liquefaction, void, and porosity, can be circumvented with FSW since the mass of the material is not melting. FSW is a solid-state joining method that may produce junctions with high quality and strength with little distortion. This method may work with an extensive array of material lengths and thicknesses and produce the often-used butt and lap joints. FSW seems to be the best approach to welding with efficient joints and lesser utilization of energy [11]. In FSW, two components are joined together using a revolving, nonconsumable tool. The tool is made up of three parts: the pin, the shoulder, and the shank.

While welding, the tool shoulder rubs against the base metals and causes the required frictional heating. The shank is made up of the tool and the tool's shoulder. The tool pin enters the joint where the workpieces need to be joined during welding. The rotation of the pin paired with a downward force causes this plunge to happen. One of the crucial steps in the FSW process is the plunge, which starts the thermomechanical conditions needed to begin the weld. The tool shoulder provides the necessary friction heating once the pin has fully inserted itself into the workpiece, softening the material (below its melting point temperature). The workpiece deforms plastically because of tool penetration, and friction causes further heat to be produced and dynamic recrystallization. FSW can significantly refine the microstructure of alloys with coarse grains since it is a powerful technique for severe plastic deformation [12]. According to Xiao et al. [13], considerable microstructural alteration, such as the fundamental disintegration of eutectic networks and noteworthy grain refinement, can cause FSW to dramatically improve the mechanical characteristics of base alloys.

In connection with the above-stated benefits of FSW, FSP which works on the same principle of FSW, is an impressive technique to develop lightweight metal matrix composite (MMC) materials [14, 15]. These materials bear superior properties as compared to the base matrix. To a larger extent, the challenge associated with developing lightweight materials is the associated manufacturing process. Any adopted process should have the potential of uniformly distributing the reinforcement particulates and avoid forming clusters in the base matrix. Among all, state-of-the-art FSP technology is considered to be far ahead in achieving the homogeneous and uniform dispersal of the secondary phase particulates [16]. FSP is a green fabrication method since it uses less heat and produces no harmful fumes, gases, or radiation during the processing process. In order to achieve the specifically increased characteristics in MMCs, the dispersion of hard nano particulates in the base matrix is essential. Various manufacturing techniques [17–21] have been adopted by researchers to homogenously disperse secondary phase particulates in the base matrix and to fabricate bulk metal composites. Each one of these composite fabrication techniques, change the material's phase from solid to liquid. FSP is solid-state processing (which does not have a phase change process) and is considered best among all for the homogeneous and uniform distribution of the particulates. The stirring action of FSP (as shown in Figure 11.2) has been efficiently adopted to distribute secondary phase particulates in the base matrix, obtaining impressive grain refinement and producing MMC materials with enhanced properties.

Every material produced via the FSP technique is claimed to have enhanced microstructural and mechanical characteristics. Henceforth, it became essential to probe the gain in microstructural and mechanical properties of the material developed via the FSP technique.

So as per the above discussion, material welded via FSW and the materials developed via FSP need to be examined for the development of microstructural and mechanical properties. Microstructural characterization and mechanical testing of FSW samples involve several steps to evaluate the quality and performance of the weld joint. Here is a general outline of the process:

Sample Preparation: Obtain representative samples from the friction stir weld joint. The samples can be extracted perpendicular or parallel to the weld direction. The samples should be prepared carefully to minimize any damage or alteration to the microstructure.

Metallographic Preparation: Prepare the samples for metallographic analysis. This involves cutting, mounting, grinding, and polishing the samples to a mirror-like finish. The goal is to obtain a flat, damage-free surface that can be analyzed under a microscope.

Microstructural Analysis: Examine the microstructure of the friction stir weld joint using optical microscopy (OM) or scanning electron microscopy (SEM). OM provides a general overview of the weld region, while SEM offers higher magnification and detailed information about the microstructural features.

a) *Grain Structure*: Assess the grain structure in the weld zone, including grain size, grain morphology, and grain boundaries. Compare the weld zone with the base material to identify any changes in the microstructure caused by the FSW process.

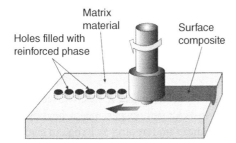

Figure 11.2 Surface composite fabrication via FSP utilizing the zig-zag hole method. Source: Iwaszko and Sajed [22]/MDPI/ CC BY 4.0.

b) *Defects and Inclusions*: Look for any defects, such as voids, cracks, or porosity, that may have formed during the welding process. Additionally, check for the presence of any inclusions or impurities that could affect the mechanical properties.

c) *Phase Identification*: Identify the different phases present in the weld region, particularly if the base material contains multiple phases or if any phase transformations occurred during welding.

d) *Hardness Testing*: Measure the hardness of the weld joint using microhardness testing techniques, such as Vickers or Knoop hardness. This provides an indication of the material's strength and can help identify any variations in hardness across the weld.

e) *Tensile Testing*: Perform tensile testing on the FSW samples to evaluate their mechanical properties, such as ultimate tensile strength, yield strength, and elongation. This testing can be conducted using standard tensile testing machines.

f) *Impact Testing (optional)*: If impact resistance is a critical factor for the application, perform impact testing, such as Charpy or Izod testing, to assess the weld joint's ability to withstand sudden loading or dynamic events.

g) *Other Mechanical Testing (optional)*: Depending on the specific requirements, additional mechanical tests, such as bending, fatigue, or creep testing, can be conducted to further evaluate the weld joint's performance under specific loading conditions.

h) *Data Analysis*: Analyze the obtained microstructural and mechanical testing results to determine the weld quality, mechanical properties, and any correlations between the microstructure and mechanical performance. Compare the results with relevant standards or specifications to assess the weld's suitability for the intended application.

It is important to note that the specific techniques, equipment, and parameters used for microstructural characterization and mechanical testing may vary depending on the materials being welded, the equipment available, and the specific requirements of the study or application [23].

The following are the detailed various methodologies used to check the basic metallurgical and mechanical characteristics.

11.2 Microstructural Characterization

The FSW/FSP technique produces three distinct microstructural zones: the nugget zone (NZ), the thermomechanical affected zone (TMAZ), and the heat-affected zone (HAZ) [24–27]. Plastic deformation is absent in HAZ, this zone is only affected by thermal energy but shows no significant change in microstructural properties. On the other hand, TMAZ, which is next to NZ, is plastically deformed by the developed thermal energy. The NZ is the highly affected zone; it is largely influenced by the developed heat energy. Attributed to this high-temperature NZ goes under the highest plastic deformation, which consequently gives development generally consisting of refined grains owing to the full dynamic recrystallization (refer Figure 11.3).

Coarse and fine grains in the microstructure of base alloys characterize base alloys. The weld NZ after FSW has fine grains that are equiaxed. Since grains in HAZ only experience a heat cycle rather than plastic deformation, grain growth takes place without undergoing a change in shape. The grains in TMAZ exhibit a considerable change in morphology over this period, suggesting that the tool may have caused plastic deformation [28]. In addition, the disintegration of precipitates in

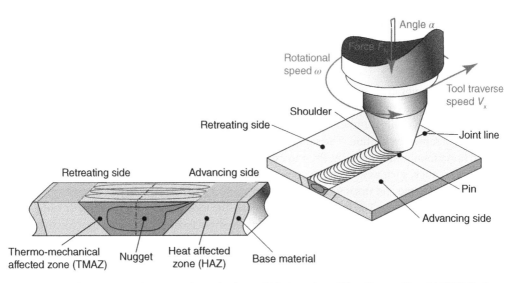

Figure 11.3 Microstructural zones and terminology of friction stir welding. Source: Hossfeld [28]/Springer Nature/CC BY 4.0.

the TMAZ has been similar but to a lesser extent, whereas it has been noted that precipitation in the HAZ has significantly coarsened [29, 30].

Throughout the welding process, grains have dissolved roughly half of the alloy's precipitates into the base metal matrix. The continual dynamic recrystallization process is the pivotal mechanism for the creation of tiny equiaxed grains and continuous grain boundary misorientation along the weld line in the stirred zone [31]. As a result, studying grain evolution is a crucial phenomenon. The majority of property evolution is influenced by grain size and shape.

The microstructure characterization of FSWed and FSPed samples is usually examined by standard metallurgical methodologies. The basic five methods of examining the microstructural features of the FSWed and FSPed samples are widely adopted. These techniques are:

a) Optical microscopy (OM) examination
b) X-ray diffraction (XRD) analysis
c) Scanning electron microscopy (SEM)
d) Transmission electron microscopy (TEM)
e) Electron backscatter diffraction (EBSD) examination

11.2.1 Optical Microscopy Examination

OM is a widely used technique in various fields, including biology, materials science, forensics, and many others. It involves using visible light to observe and analyze samples at the microscopic level. Here is an overview of OM examination:

- Light Source: Optical microscopes typically use a light source, such as a halogen lamp or LED, to illuminate the sample. The light passes through a condenser lens, which focuses it onto the sample.

- Objective Lens: The light transmitted through the sample enters the objective lens, which is the primary lens responsible for magnifying the image. Objective lenses have different magnification powers and numerical apertures, allowing the user to select the desired level of magnification and resolution.
- Eyepiece: The magnified image formed by the objective lens is further magnified by the eyepiece. The eyepiece contains a lens that brings the image to the viewer's eye, allowing them to observe and analyze the sample.
- Stage and Sample Preparation: The sample is placed on a stage, which is a platform that can be moved horizontally and vertically. Prior to examination, the sample may undergo various preparation techniques, such as staining, sectioning, or mounting on a glass slide, depending on the nature of the sample and the intended analysis.
- Focus and Magnification: The microscope user adjusts the focus by moving the stage or using fine focus knobs to bring the desired area of the sample into sharp focus. By changing the objective lens, the user can achieve different levels of magnification to observe different features of the sample.
- Brightfield and Phase Contrast: Brightfield microscopy is the most common technique, where the sample appears dark against a bright background. In phase contrast microscopy, a special condenser and objective lens are used to enhance the contrast between different parts of the sample, particularly transparent or translucent specimens.
- Polarized Light: Polarized light microscopy utilizes polarizers and specialized filters to examine samples that interact with polarized light, such as minerals, crystals, or certain biological tissues. It provides information about the sample's optical properties and structural characteristics.
- Fluorescence Microscopy: Fluorescence microscopy uses specific fluorescent dyes or markers that emit light of different wavelengths when excited by a particular wavelength of light. This technique enables the visualization of specific structures or molecules within the sample, such as fluorescently labeled proteins or DNA.
- Image Capture and Analysis: In modern OM, digital cameras are often used to capture images of the sample. These images can be further processed, analyzed, and enhanced using image processing software. Measurements, annotations, and quantitative analysis can be performed on the acquired images.
- Limitations: OM has certain limitations, such as limited resolution due to the diffraction of light, which restricts the level of detail that can be observed. Additionally, the depth of field can be shallow, limiting the ability to focus on multiple planes simultaneously. These limitations can be overcome by using advanced techniques such as confocal microscopy or electron microscopy.
- Om provides valuable insight into the structure, composition, and behavior of microscopic samples. It continues to be a fundamental tool in scientific research, medical diagnostics, quality control, and numerous other applications. The FSWed and FSPed samples contain fine equiaxed grains, especially at the stir zone (SZ). The investigation of the SZ is done under an optical microscope. Before examining the samples, all the samples went through standard metallographic procedures as described by ASTM standards. Basically, this procedure follows the polishing and grinding of the samples followed by the cleaning of samples with a suitable etchant.

The specimen prepared via FSW and FSP processes can be examined under the magnification range of the optical microscope from 10× to 100×. The main feature of FSWed and FSPed processes is the evolution of fine grains in the SZ. With the use of the OM technique, one can examine the size and morphology of grains along with the grain boundary and orientation of these grains. Lakshminarayanan, A. K et al. [31] impressively present the change in the microstructure of the base metal when processed via FSW. The typical OM images of the FSWed in SZ and other zones are shown in Figure 11.4.

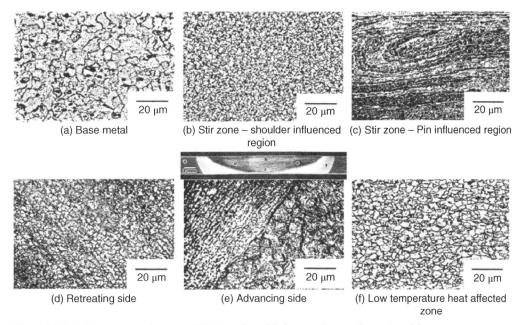

Figure 11.4 Microstructure images of friction stir welded zones. Source: Reproduced from Lakshminarayanan et.al. [31]/© with permission from Elsevier.

11.2.2 X-ray Diffraction (XRD) Analysis

XRD analysis is a powerful technique used to determine the atomic and molecular structure of crystalline materials. It is based on the principle that when X-rays interact with a crystalline sample, they undergo constructive and destructive interference, resulting in a characteristic diffraction pattern. Here is an overview of XRD analysis:

- Instrumentation: XRD analysis requires a specialized instrument called an X-ray diffractometer. The diffractometer consists of an X-ray tube that generates X-rays and a detector that measures the intensity of the diffracted X-rays. The sample is mounted on a goniometer, which allows it to be rotated to different angles.
- Sample Preparation: Crystalline samples are typically prepared as small, powdered particles. The sample should be finely ground and homogenized to ensure a representative analysis. For some applications, thin film samples or single crystals may be used.
- Bragg's Law: XRD analysis is based on Bragg's Law, which relates the angle of incidence, the wavelength of X-rays, and the spacing between crystal lattice planes. According to Bragg's Law, constructive interference occurs when $2d \sin \theta = n\lambda$, where d is the spacing between lattice planes, θ is the angle of incidence, n is an integer, and λ is the wavelength of X-rays.
- Diffraction Pattern: The X-ray beam is directed onto the sample, and the diffracted X-rays are detected. As the sample is rotated, different crystallographic planes diffract X-rays at different angles, resulting in a diffraction pattern. The pattern consists of a series of peaks corresponding to the angles at which constructive interference occurs.
- Data Analysis: The diffraction pattern is recorded and analyzed to extract information about the crystal structure. The position, intensity, and shape of the diffraction peaks provide valuable data. The positions of the peaks correspond to the lattice spacing, and their intensities reflect the

arrangement of atoms in the crystal. The shape and width of the peaks can reveal information about the crystal size, defects, or strain.

- Indexing and Structure Determination: To determine the crystal structure, the diffraction pattern is compared to known crystal structures in a database. This process, called indexing, involves matching the observed diffraction angles with the theoretical angles for different crystal systems. Once indexing is achieved, the crystal structure can be determined by solving mathematical equations and refining the parameters using specialized software.
- Applications: XRD analysis is widely used in materials science, geology, chemistry, and various other fields. It is used to study the crystalline structure of materials, identify unknown compounds, determine phase composition, analyze crystal defects, measure crystal size and strain, and investigate phase transitions.
- XRD analysis is a nondestructive technique that provides detailed information about the atomic and molecular arrangement within crystalline materials. It has played a crucial role in advancing our understanding of materials and their properties, enabling the development of new materials and technologies.

In order to examine material's crystallographic structure, chemical composition, and physical characteristics, a nondestructive method known as XRD analysis (XRD) can give precise details. It is based on the constructive interference of crystalline samples and monochromatic X-rays. For the specimens developed via FSWed and FSPed samples, various crystallographic structures, chemical compositions, and physical properties of a material are essential to be known. XRD analysis supports determining the basic chemical composition of the developed material. In order to examine the chemical composition of the material after FSP, Hernández-García et al. [32] performed an XRD analysis and present it in Figure 11.5.

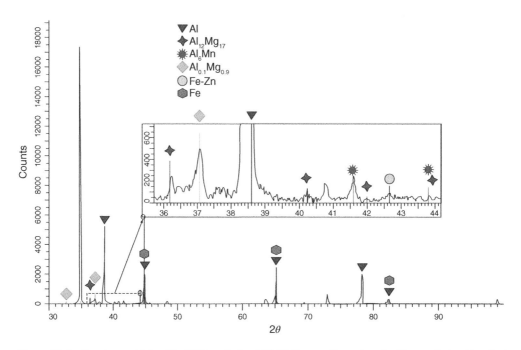

Figure 11.5 Spectrum obtained by XRD analysis of SZ 32. Source: Reproduced with permission Hernández-García et al. [32]/Springer Nature.

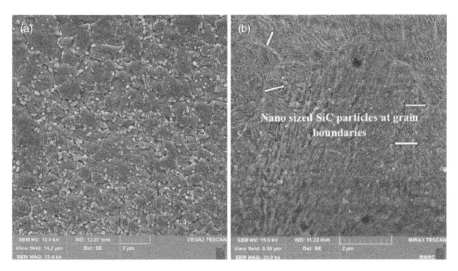

Figure 11.6 FESEM examination of steel/SiC base composite fabricated via FSP. Source: Reproduced with permission from Fotoohi et al. [33]/© Springer Nature.

11.2.3 Scanning Electron Microscopy (SEM)

SEM or Field emission scanning electron microscopy (FESEM) analysis is a powerful analytical technique to perform analysis on a wide range of materials, at high magnifications, and to produce high-resolution images. Especially in the case of composites processed via the FSP technique, it is essential to find the reinforcement distribution. SEM analysis helps in investigating the homogenous distribution of the reinforcement particles throughout the base matrix. In addition, energy-dispersive X-ray spectroscopy analysis usually taken in conjunction with SEM helps in determining various peaks of the presentation materials at a particular location. FESEM analysis results to examine the uniform distribution of silicon carbide particles in base metal steel are presented in Figure 11.6. Fotoohi Nezhad Khales et al. [33] conclude that silicon carbide particles dispersed uniformly in the base matrix of the steel metal. This uniform distribution of secondary phase particulates directly contributes to the enhancement of mechanical and tribological characteristics of the base steel metal alloy. Further to investigate the formation of clusters or any agglomeration of the secondary phase particulates, FESEM analysis plays a key role. Usually, FESEM analysis may be conducted 10–300,000X to determine the various cracks and porosity, if available.

11.2.4 Transmission Electron Microscopy (TEM)

Once after the preparation of the specimen via FSW or FSP, the basic texture of the material undergoes various changes. All these changes need to be analyzed via suitable methods to examine the real evolution. TEMs provide topographical, morphological, compositional, and crystalline information. Transmission electron microscopes are capable of imaging at a significantly higher resolution than light microscopes, owing to the smaller de Broglie wavelength of electrons. This enables the instrument to capture fine detail – even as small as a single column of atoms, which is thousands times smaller than a resolvable object seen in a light microscope. The images allow viewing samples on a molecular level, making it possible to analyze structure and texture. TEM instruments have multiple operating modes including conventional imaging, scanning TEM imaging (STEM), diffraction, spectroscopy, and combinations of these. Even within conventional imaging, there are many fundamentally different ways that contrast is produced, called "image contrast mechanisms." Contrast can arise from position-to-position differences in the thickness or density ("mass-thickness contrast"), atomic number ("Z contrast," referring to the common abbreviation Z for atomic number), crystal structure or orientation ("crystallographic contrast" or "diffraction contrast"), the slight quantum-mechanical phase shifts that individual atoms produce in electrons that pass through them ("phase contrast"), the energy lost by electrons on passing through the sample ("spectrum imaging"), and more. Each mechanism tells the user a different kind of information, depending not only on the contrast mechanism but on how the microscope is used – the settings of lenses, apertures, and detectors. What this means is that a TEM is capable of returning an extraordinary variety of nanometer- and atomic-resolution information including size, morphology, and atom bonding. For this motivation, TEM is considered a key tool for materials fields.

11.2.5 Electron Backscatter Diffraction (EBSD) Examination

With its versatility in mapping orientation, crystal type, and perfection over a wide range of step sizes, EBSD, a powerful microstructural analyzer tool, allows crystallographic information to be obtained from small volumes of material in a scanning electron microscope (SEM). These kinds of investigations have been conducted utilizing spatially resolved acoustic spectroscopy (SRAS), which examines elastic waves rather than a diffraction event, as well as X-ray, neutron, and/or electron diffraction in a transmission electron microscope. The technique that is chosen depends on a number of variables, including the spatial resolution, the area or volume that is being investigated, and whether the data are static or dynamic. EBSD is a technique used to analyze the crystallographic properties of materials at a microscopic scale. It is commonly employed in the field of materials science and is particularly useful for studying crystalline materials such as metals, ceramics, and semiconductors.

In an EBSD examination, a focused beam of high-energy electrons is directed onto the surface of a sample. The electrons interact with the atoms in the material, causing them to scatter. Some of the scattered electrons will backscatter, meaning they are reflected back toward the detector. The pattern of backscattered electrons contains information about the crystal structure and orientation of the material.

The backscattered electrons are detected using a specialized detector, typically a phosphor screen or a CCD camera. The detector records the intensity and position of the electrons, which is then processed to extract the crystallographic information. The electron patterns obtained from different regions of the sample are analyzed to determine the crystal orientation, grain boundaries, and other microstructural features.

a) The EBSD technique can provide valuable information about the grain size and shape, crystallographic texture, and phase identification of the material. It is often used to study the microstructure of metals and alloys, characterizes the deformation behavior of materials, investigate phase transformations, and analyze the quality of materials in manufacturing processes.

b) Overall, the EBSD examination offers a powerful tool for understanding the crystallographic properties of materials, enabling researchers to gain insights into the structure–property relationships and optimize material performance in various applications.

An SEM with an EBSD detector that has at least a phosphor screen, a small lens, and a low-light charge-coupled device (CCD) camera is used to conduct EBSD experiments. For quick measurements, the CCD chip has a native resolution of 640×480 pixels; for slower, more sensitive measurements, the CCD chip resolution can go up to 1600×1200 pixels. Commercially available EBSD systems normally come with one of two distinct CCD cameras. The main benefit of high-resolution detectors is that they are more sensitive, allowing for a more thorough analysis of the data contained in each diffraction pattern. The diffraction patterns are binned for texture and

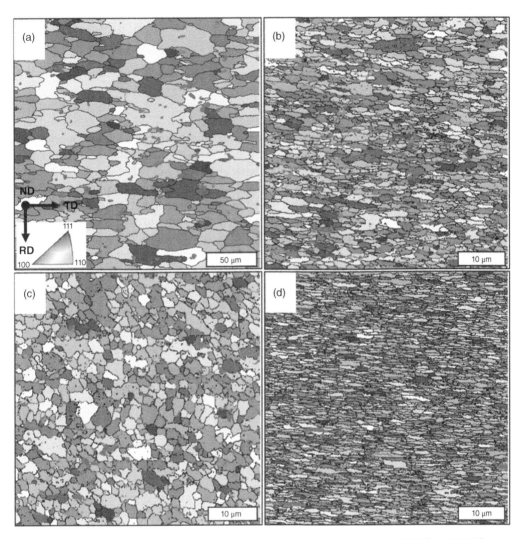

Figure 11.7 EBSD maps for grain boundaries and interfaces for (a) BM, (b) HAZ, (c) TMAZ, and (d) NZ. Source: Reproduced from Acuña et al. [34]/© Springer Nature.

orientation measurements to minimize their size and speed up processing. Patterns can be indexed at up to 1800 patterns/second by contemporary CCD-based EBSD systems. This enables very rapid and rich microstructural maps to be generated. Acuña et al. [34] investigate the microstructure of the aluminum-based composite; the obtained results are shown in Figure 11.7, which suggests that significant changes occurred in the microstructure of the base metal after FSP.

11.3 Mechanical Testing

FSW and FSP directly contribute to the evolution of mechanical properties. There are a number of these properties where the effect of FSP and FSW can be visualized [35]. The basic mechanical properties that can be examined to investigate the real gain in the basic mechanical characteristic are:

a) Microhardness
b) Tensile strength

c) Compressive strength
d) Nanoindentation
e) Fatigue testing

11.3.1 Microhardness

Hardness test methods use an indenter probe that is displaced into FSWed and FSPed surfaces under a specific load. The indentation typically has a defined dwell time. In traditional mechanical testing, the size or depth of indentation is measured to determine hardness. Microhardness testing, with applied loads under 10 N, is typically used for smaller FEWed and FSPed samples and plated surfaces. In FSWed and FSPed samples, usually examining microhardness is considered along a horizontal line underneath at some distance from the top surface and at some equal intervals between adjacent points. As many as values can be taken for a distance to either side of the central line passing through the SZ. At every indentation point, more values may be taken and the mean value was considered. It is also evident from many studies that examining microhardness values is pivotal for any friction stir welded or friction stir processed samples [36, 37]. It can be seen from Figure 11.8 that after employing FSP, the microhardness of the developed specimen increased significantly when compared to base monolithic alloy.

11.3.2 Tensile Strength

A destructive engineering and materials science test known as tensile testing involves applying controlled tension to a sample until it completely fails. One of the most used methods for

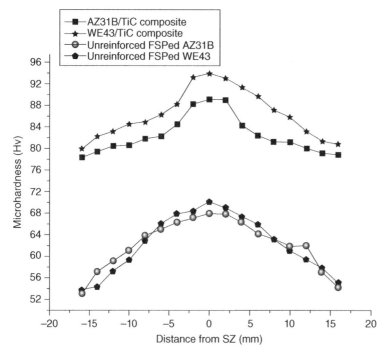

Figure 11.8 Microhardness values enhancement after FSP. Source: Reproduced with permission Singla et al. [37]/Taylor & Francis.

mechanical testing is the tensile testing methods. It is used to determine a material's strength as well as how far it can be stretched before breaking. The yield strength, ultimate tensile strength, ductility, strain hardening properties, Young's modulus, and Poisson's ratio are all assessed using this test method, many research work reported the same [38–40]. In addition, in order to determine ductility and fracture mechanism, the tensile testing methods are very much important for materials prepared via FSW or FSP.

11.3.3 Compressive Strength

Composite compression testing techniques offer a way to apply a compressive load to the material without risking buckling. For composite materials that are in the form of a relatively thin, flat rectangular test specimen, such as laminate panels, compression tests are conducted. For specimens fabricated via FSP or joined via FSW, it became very much essential to investigate their compressive strength. As most of the specimens prepared via FSP or FSW have their potential application in the field where these materials are subjected to repeated loading. Hence, it is advisable for the researcher to work in the field of joining or developing materials via FSP or FSW.

11.3.4 Nano-indention

The nano-hardness of FSWed and FSPed samples can be measured using a Hysitron triboindenter. The triboindenter is equipped with a Berkovich diamond three-sided pyramid tip in a quasi-static indentation mode, having a radius of curvature of nearly 150 nm. At a suitable peak load, an indentation may be created on the polished surface of specimens with a fixed loading time and unloading time and up to desired depth.

11.3.5 Fatigue Testing

The design of FSWed and FSPed composite materials may include many instances of fatigue testing. Nonetheless, a focus on methodologies for small-specimen-level fatigue testing points to the necessity of ongoing test method standardization. A test specimen or a structure is subjected to cyclic loading during a fatigue test. The applied load is cycled between the defined maximum and minimum levels until a fatigue failure occurs, unlike monotonic tests where loading increases till failure. The output of the fatigue test is considered a run-out if the specified number of loading cycles exceeds and the failure does not occur. An S-N diagram, which plots the amplitude of the alternating stress against the number of cycles to failure, N, often represents the findings of several fatigue tests that were carried out using a variety of cyclic stress levels. The same minimum stress to maximum stress ratio – also known as the cyclic stress ratio, or R – is used for all of the fatigue tests represented in the plot. Furthermore, stiffness and strength decrease in composite materials and structures brought on by earlier cyclic loading are investigated using fatigue testing. A composite structure may undergo fatigue testing at various stages of its design, using test pieces of various sizes and complexity. Small-specimen and simple element-level tests are used to examine the fatigue behavior of composite materials. In the case of materials developed via FSW and FSP, the examination of fatigue is very much important. Especially, the materials developed via FSP and FSW find their application in the field automotive and aerospace industry where materials undergo periodic fatigue, and the investigation of fatigue properties became more vital.

11.4 Conclusions

FSP/FSW both have promising aspect whether in joining similar or different materials and fabricating new composites. Being a solid-state process, it offers various advantages, especially in obtaining defect-free joints or materials. In addition, both of these processes significantly contribute to enhancing microstructural and mechanical characteristics. As FSW and FSP have applications in aerospace and automobile engineering, where passenger safety is the utmost priority of the manufacturing industry. In order to achieve supreme quality materials, examining microstructural and mechanical properties with due care became more important. The major challenge in investigating these properties is the selection of testing methodology and the zone from where the specimen is extracted. However, more microstructural and mechanical testing techniques may be investigated in order to build a group of advanced materials via FSW and FSP, and the results should be compared with the current approaches.

References

1 Andrew, R.M. (2021). Towards near real-time, monthly fossil CO_2 emissions estimates for the European Union with current-year projections. *Atmos. Pollut. Res.* 12 (12): 101229. https://doi.org/10.1016/j.apr.2021.101229.

2 Fontaras, G., Zacharof, N.-G., and Ciuffo, B. (2017). Fuel consumption and CO_2 emissions from passenger cars in Europe – Laboratory versus real-world emissions. *Prog. Energy Combust. Sci.* 60: 97–131. https://doi.org/10.1016/j.pecs.2016.12.004.

3 Meireles, M., Robaina, M., and Magueta, D. (2021). The effectiveness of environmental taxes in reducing CO_2 emissions in passenger vehicles: the case of mediterranean countries. *Int. J. Environ. Res. Public Health* https://doi.org/10.3390/ijerph18105442.

4 Kawajiri, K., Kobayashi, M., and Sakamoto, K. (2020). Lightweight materials equal lightweight greenhouse gas emissions?: a historical analysis of greenhouse gases of vehicle material substitution. *J. Cleaner Prod.* 253: 119805. https://doi.org/10.1016/j.jclepro.2019.119805.

5 Lutsey, N. (2010). *Review of Technical Literature and Trends Related to Automobile Mass-reduction Technology*. Institute of Transportation Studies, UC Davis, Institute of Transportation Studies, Working Paper Series.

6 Rose, A.R., Manisekar, K., and Balasubramania, V. (2012). Influences of tool rotational speed on tensile properties of friction stir welded AZ61A magnesium alloy. *Proc. Inst. Mech. Eng., Part B: J. Eng. Manuf.* 226 (4): 649–663. https://doi.org/10.1177/0954405411399352.

7 Cao, X., Jahazi, M., Immarigeon, J.P. et al. (2006). A review of laser welding techniques for magnesium alloys. *J. Mater. Process. Technol.* 171: 188–204. https://doi.org/10.1016/j.jmatprotec.2005.06.068.

8 Mishra, R., De, P. and Kumar, N. (2014) Friction Stir Welding and Processing: Science and Engineering, Friction Stir Welding and Processing: Science and Engineering. https://doi.org/10.1007/978-3-319-07043-8.

9 Staron, P., Koçak, M., and Williams, S. (2002). Residual stresses in friction stir welded Al sheets. *Appl. Phys. A* 74: s1161–s1162. https://doi.org/10.1007/s003390201830.

10 Aissani, M., Gachi, S., Boubenider, F. et al. (2010). Design and optimization of friction stir welding tool. *Mater. Manuf. Processes* 25 (11): 1199–1205. https://doi.org/10.1080/10426910903536733.

11 Davim, J. P. (Ed.). (2021). Welding Technology. *Mater. Form. Mach. Tribol.* https://doi.org/10.1007/978-3-030-63986-0

12 Ma, Z.Y. (2008). Friction stir processing technology: a review. *Metall. Mater. Trans. A* 39 A (3): 642–658. https://doi.org/10.1007/s11661-007-9459-0.

13 Xiao, B.L. et al. (2011). Enhanced mechanical properties of Mg–Gd–Y–Zr casting via friction stir processing. *J. Alloys Compd.* 509 (6): 2879–2884. https://doi.org/10.1016/j.jallcom.2010.11.147.

14 Sharma, S., Handa, A., Singh, S.S. et al. (2019). Synthesis of a novel hybrid nanocomposite of AZ31Mg-Graphene-MWCNT by multi-pass friction stir processing and evaluation of mechanical properties. *Mater. Res. Express* 6 (12): https://doi.org/10.1088/2053-1591/ab54da.

15 Esmaily, M., Svensson, J.E., Fajardo, S. et al. (2017). Fundamentals and advances in magnesium alloy corrosion. *Prog. Mater Sci.* 89: 92–193. https://doi.org/10.1016/j.pmatsci.2017.04.011.

16 Peron, M., Torgersen, J., and Berto, F. (2017). Mg and its alloys for biomedical applications: exploring corrosion and its interplay with mechanical failure. *Metals* https://doi.org/10.3390/met7070252.

17 Guan, H. et al. (2022). A review of the design, processes, and properties of Mg-based composites. *Nanotechnol. Rev.* 11 (1): 712–730. https://doi.org/10.1515/ntrev-2022-0043.

18 Sunil, B.R., Reddy, G.P.K., Patle, H. et al. (2016). Magnesium based surface metal matrix composites by friction stir processing. *J. Magnesium Alloys* 4 (1): 52–61. https://doi.org/10.1016/j.jma.2016.02.001.

19 Azizieh, M., Larki, A.N., Tahmasebi, M. et al. (2018). Wear behavior of AZ31/Al$_2$O$_3$ magnesium matrix surface nanocomposite fabricated via friction stir processing. *J. Mater. Eng. Perform.* 27 (4): 2010–2017. https://doi.org/10.1007/s11665-018-3277-y.

20 Moustafa, E.B. (2021). Hybridization effect of BN and Al$_2$O$_3$ nanoparticles on the physical, wear, and electrical properties of aluminum AA1060 nanocomposites. *Appl. Phys. A* 127 (9): 724. https://doi.org/10.1007/s00339-021-04871-5.

21 Moustafa, E.B. et al. (2020). Microstructural, mechanical and thermal properties evaluation of AA6061/Al$_2$O$_3$-BN hybrid and mono nanocomposite surface. *J. Mater. Res. Technol.* 9 (6): 15486–15495. https://doi.org/10.1016/j.jmrt.2020.11.010.

22 Iwaszko, J. and Sajed, M. (2021). Technological aspects of producing surface composites by friction stir processing – a review. *J. Compos. Sci.* https://doi.org/10.3390/jcs5120323.

23 Prado, R., Murr, L.E., Shindo, D.J. et al. (2001). Tool wear in the friction-stir welding of aluminum alloy 6061+20% Al$_2$O$_3$: a preliminary study. *Scr. Mater.* 45: 75–80. https://doi.org/10.1016/S1359-6462(01)00994-0.

24 Benavides, S., Li, Y., Murr, L. et al. (1999). Low-temperature friction-stir welding of 2024 aluminum. *Scr. Mater.* 41 (8): 809–815. https://doi.org/10.1016/S1359-6462(99)00226-2.

25 Su, J.-Q., Nelson, T.W., Mishra, R., and Mahoney, M. (2003). Microstructural investigation of friction stir welded 7050-T651 aluminium. *Acta Mater.* 51: 713–729. https://doi.org/10.1016/S1359-6454(02)00449-4.

26 Yang, B., Yan, J., Sutton, M.A., and Reynolds, A.P. (2004). Banded microstructure in AA2024-T351 and AA2524-T351 aluminum friction stir welds. *Mater. Sci. Eng., A* 364: 55–65. https://doi.org/10.1016/S0921-5093(03)00532-X.

27 Sutton, M., Reynolds, A.P., Yang, B. et al. (2003). Mode I fracture and microstructure for 2024-T3 friction stir welds. *Mater. Sci. Eng., A* 354: 6–16. https://doi.org/10.1016/S0921-5093(02)00078-3.

28 Hossfeld, M. (2022). Shoulderless Friction Stir Welding: a low-force solid state keyhole joining technique for deep welding of labile structures. *Prod. Eng.* 16 (2): 389–399. https://doi.org/10.1007/s11740-021-01083-x.

29 Arora, K., Pandey, S., Schaper, M. et al. (2010). Microstructure evolution during friction stir welding of aluminum alloy AA2219. *J. Mater. Sci. Technol.* 26: 747–753. https://doi.org/10.1016/S1005-0302(10)60118-1.

30 Buffa, G., Fratini, L., and Shivpuri, R. (2007). CDRX modelling in friction stir welding of AA7075-T6 aluminum alloy: Analytical approaches. *J. Mater. Process. Technol.* 191: 356–359. https://doi.org/10.1016/j.jmatprotec.2007.03.033.

31 Lakshminarayanan, A.K. and Balasubramanian, V. (2010). An assessment of microstructure, hardness, tensile and impact strength of friction stir welded ferritic stainless steel joints. *Mater. Des.* 31 (10): 4592–4600. https://doi.org/10.1016/j.matdes.2010.05.049.

32 Hernández-García, D. et al. (2017). Friction stir welding of dissimilar AA7075-T6 to AZ31B-H24 alloys. *MRS Adv.* 2: 1–9. https://doi.org/10.1557/adv.2017.609.

33 Fotoohi, M., Sajjadi, S., and Kamyabi-Gol, A. (2020). Multipass friction stir processing of steel/SiC nanocomposite: assessment of microstructure and tribological properties. *J. Mater. Eng. Perform.* 29: https://doi.org/10.1007/s11665-020-04947-y.

34 Acuña, R., Cristóbal, M.J., Abreu, C.M. et al. (2019). Microstructure and wear properties of surface composite layer produced by friction stir processing (FSP) in AA2024-T351 aluminum alloy. *Metall. Mater. Trans. A* 50: https://doi.org/10.1007/s11661-019-05172-6.

35 Sagar, P., Handa, A., and Kumar, G. (2022). Metallurgical, mechanical and tribological behavior of reinforced magnesium-based composite developed Via Friction stir processing. *Proc. Inst. Mech. Eng., Part E: J. Process Mech. Eng.* 236 (4): 1440–1451. https://doi.org/10.1177/09544089211063099.

36 Sagar, P. and Handa, A. (2020). Role of tool rotational speed on the tribological characteristics of magnesium based az61a/tic composite developed via friction stir processing route. *J. Achiev. Mater. Manuf. Eng.* 101 (2): 60–75. https://doi.org/10.5604/01.3001.0014.4921.

37 Singla, S., Sagar, P., and Handa, A. (2023). Magnesium-based nanocomposites synthesized using friction stir processing: an experimental study. *Mater. Manuf. Processes* https://doi.org/10.1080/10426914.2023.2195909.

38 Sagar, P., Singla, S., and Handa, A. (2023). Fabrication and characterization of magnesium-based WE43/TiC nanocomposite material developed via friction stir processing and study of significant parameters. *J. Eng. Mater. Technol.* 145 (3): 1–11. https://doi.org/10.1115/1.4062321.

39 Sagar, P. and Handa, A. (2021). Prediction of wear resistance model for magnesium metal composite by response surface methodology using central composite design. *World J. Eng.* https://doi.org/10.1108/WJE-08-2020-0379.

40 Sagar, P., Kumar, G., and Handa, A. (2023). Progressive use of nanocomposite hydrogels materials for regeneration of damaged cartilage and their tribological mechanical properties. *Proc. Inst. Mech. Eng., Part N: J. Nanomater. Nanoeng. Nanosyst.* 239779142311514. https://doi.org/10.1177/23977914231151487.

12

Comparative Analysis of Microstructural and Mechanical Characteristics of Reinforced FSW Welds

Tanvir Singh

Department of Mechanical Engineering, St. Soldier Institute of Engineering & Technology, Punjab Technical University, Jalandhar, Punjab, India

12.1 Introduction

With remarkable thermal mechanical and physiological characteristics, metal matrix-reinforced composites based on aluminum are closely studied by the aerospace, transportation, and marine industries [1]. Mouritz [2] emphasized the importance of materials for the best design cum their operation with proper maintenance over the entire life span right from the starting phase of design to the product's end life. However, Chawla and Chawla [3] suggested making the best next-generation engineering materials. However, Kaczmar [4] claimed that there were several challenges in constructing the best aluminum-based MMR welds or composites (MMCs) using the fusion welding method [5]. Fusion welding methods are unable to produce aluminum-based MMC materials with effective welds in terms of strength [6].

Suryanarayanan et al. [7] pointed out that joinability is constrained for Al-MMCs due to non-homogenous reinforcement particles (RPs) distribution and interface of RPs and metal (molten) in the fusion weld zone. According to currently accessible research, it has been determined that combining these materials with the FSW technique, as reported by Starink et al. [8], can lead to fusion welding defects during the process. Numerous research publications based on various aspects of the FSW technique have recently been published. An overview of the most recent advancements in welding procedures is provided by Zhang et al. in their publication [9]. The preliminary studies were viewed as critical assessments of the use of the FSW method for aluminum alloys by Threadgill et al. [10].

A complete and methodical study based on friction stir processing (FSP) and FSW was carried out, according to Mishra and Ma [11]. A detailed study related to particle/particulate-reinforced aluminum particulate-reinforced composites (PRCs) and magnesium matrix composites (MgMCs) is provided by Lloyd et al. [12]. Ibrahim et al. [13] published an investigation of PRCs. Ellis [14] emphasizes the manufacture and assembling of aluminum matrix composites utilizing customary methods.

Rosso [15] provided extensive routes and properties of MMCs and ceramic matrix composites. Tjong [16] evaluated approaches, and architectures, of new nanoparticle-added MMCs. However, little knowledge of the production of MMCs composed of reinforcement of aluminum along weld interface using FSW, according to the literature, is now available on various MMCs. The first three sections of this chapter focus on the overview of MMCs, a brief description of the FSW method, and Al-alloy/FSW-based MMR welds joinability. After that, attention is focused on a thorough analysis of the various crucial problems that can arise when friction stir welding MMR joints that are similar and dissimilar, paying special attention to the macro/micro traits, mechanical characteristics using various parameters, and wear resistance of the MMR welds. Finally, conclusions have been reached based on the critical review, taking future challenges and potential avenues for further research into account.

12.2 Friction Stir Welding (FSW)

Friction stir welding procedure was developed at The Welding Institute (TWI, Cambridge, United Kingdom) in 1991 [17] (Figure 12.1). Because the peak temperature produced by the FSW process is only 80% of the melting point of the base metal, it can be characterized as a hot working technique [19]. The base material (BM) is not macroscopically melted, which prevents the shrinkage, porosity, and cracking phenomena [20, 21]. Over the past ten years, the FSW technique has made great strides in the joining of aluminum alloys, and as a result, it is currently thought to be the most practical way to combine MMR welds with these AA2xxx, AA6xxx, and AA7xxx alloys [22]. Mukhopadhyay [23] claims that the best FSW technology was employed to create inexpensive joints with great mechanical performance. Zhang et al.'s [24] and Chen et al.'s [25] stated that a variety of welds, including straight butt joints, lap joints, T-joints, and 90o corner joints can be produced using FSW, BM, heat-affected zone (HAZ), thermomechanical affected zone (TMAZ), and the nugget zone (NZ) are the primary four zones, first recognized by Mishra and Ma [11], that

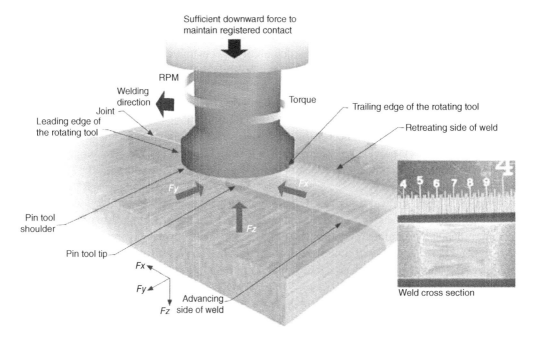

Figure 12.1 FSW process. Source: Lohwasser and Chen [18]/with permission of Elsevier.

are created as a result of the intricate interaction with various thermomechanical-based phenomenon that occurred in FSW. The unaffected material (BM) underwent a heated cycle in the absence of plastic material deformation with roughly the same granular size [18].

The thermomechanically influenced zone, according to Pantelis et al. [26], experiences dynamic recrystallization (DRX) in partial, lower strain per rate and temperature. The majority of the bent and elongated grains were produced as a result of the pattern-based flow regimes in periphery NZ formed. While, the NZ suffered severe plastic deformation (SPD) culminating in the DRX that leads to fine and equiaxed grains according to Cheng et al. [27].

12.3 Reinforcing Materials-Based Fabrication of FSW Welds

Novel structural and functional materials that are intriguing and appealing a lot of interest for a wide range of essential applications in transportation, maritime structures, and vehicle door panels are metal-matrix composites [28, 29]. MMCs are made of Matrix and reinforcement (as depicted in Figure 12.2). Matrix consists of light metal alloys that are strengthened with particulates/precipitates mainly aluminum, magnesium, and titanium and reinforcements [3, 15]. Many researchers have used different approaches to create MMCs by depositing RPs in various reservoirs

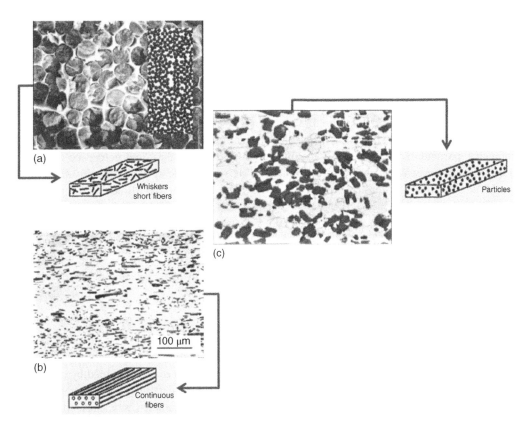

Figure 12.2 Three different types of reinforcement: (a) continuous fibers (also known as long fibers), (b) and (c) particles [3]. Source: Rosso et al. [15]/with permission from Elsevier.

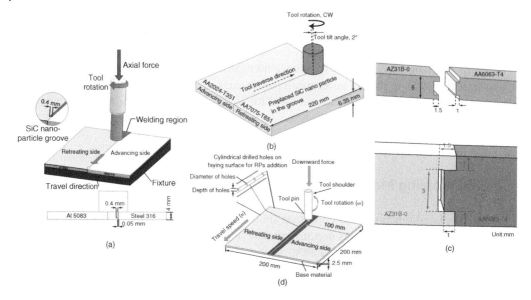

Figure 12.3 Reservoirs made on BM, interface for the addition of RPs, including (a) top-open narrow grooves, Source: Fallahi et al. [30])/with permission of Elsevier, (b) interface lock grooves. Source: Adapted from Asl et al. [31]. (c) profiles (rectangular shape) cut. Source: Adapted from Kumar et al. [32], and (d) holes (drilled). Source: Singh et al. [33, 34].

on the BM interface as illustrated in Figure 12.3a–d. The basic alloy matrix of MMCs can be reinforced with RPs to increase their properties significantly changing their weight, according to Rino et al. [35]. According to Huang et al. [36], the kind and amount of reinforcement employed throughout the manufacturing and production process was the only factor affecting the attributes of MMCs.

In accordance to Kunze and Bamptom [6], high-strength aluminum alloys are increasingly being replaced by Al-MMCs in addition to particles especially ceramic in scale to micrometer and nanometer, particularly in space shuttle tubing, roller coaster brakes, and aerospace engines. According to Evans et al. [37], the US air force uses MMC tubes made up of graphite in addition to Mg which is the most frequent construction. Similarly, to this, Hunt and Miracle [38] noted that using DRA, the B4C combined MMCs can be regarded as weldable MMCs and are used for contamination and frames. Additionally, Prasada and Asthana [39] claimed that the armor is thought to have the largest variety of applications for MMCs due to its combination of static and high ballistic performance.

12.4 Joinability of Reinforced FSW Welds

Aluminum can reduce weight due to its lightweight and pliability, according to Mouritz [2], but it can only be used in structural and engineering applications. According to Mukhopadhyay [23], the diverse and exceptional mechanical properties of aluminum alloys can be obtained by carefully controlling the percentage of alloying components and various heat treatments. According to Mathers [40], precipitation hardening is used to strengthen heat-treatable (2xxx, 6xxx, and 7xxx), while cold working or work hardening is used to strengthen weldable alloys (1xxx, 3xxx, and 5xxx series). To weld aluminum alloys, the traditional fusion welding process, however, requires careful consideration due to this issue, which includes the most frequent defects: loss of strength and the emergence of defects liquidation [30, 41–44]. Ellis [14] has produced Al-based MMC welds using a variety of fusion welding techniques. They asserted that higher thermal cycles (lower heat input) have taken place during the fabrication of SiC-based MMCs with aluminum matrix using electron beam welding (EBW) because SiC particles do not absorb energy, which would otherwise cause uncontrolled heating input and the interface reactions and the formation of more Al4C3 phases at Al/SiC. Besides, Ellis et al. [45] reported substantial segregation of SiC RPs around nugget (melted) during welding (resistance) as this effect is linked to the centrifugal force that arises.

Additionally, Storjohann et al. [46] compared the FSW procedure with the fusion welding process to create AA6061/Al_2O_3/20p and AA2124/SiC/20w. Fusion welding processes include EBM, LBM, and GTAW. They noticed that the strength of the welds decreased during welding (fusion) because of the complete suspension of Al_2O_3 RPs. Whereas the interactions occurred at a fast pace between SiC and aluminum that would cause Al_4C_3 phase production and phase containing Si, weak joints with lower joint characteristics would result, as previously mentioned by Mishra and Ma [11]. However, by carefully regulating the heat index (HI) with optimal specified parameters, the best high-strength welds using all composite materials are formed by employing FSW. The findings of these tests provide clear proof of the applicability of the FSW technique to strengthen various MMR welds using different RPs.

12.5 Metallurgical Characteristics of FSW Reinforced Welds

The quality of the weldment was largely assessed during the FSW of MMR welds by examining the macrostructure and microstructure. The workpiece material in the FSW was put through a lot of heat cycles with SPD because of the peak condition of temperatures produced by the FSW tool rotation [47–51]. The viscous-plastic material flow and RPs distribution supplied to FSW-based welds were the only factors affecting the macro-micro structural behavior [52]. The reinforced FSW butt joints run the risk of developing joint flaws such as tunnels and cavities via insufficient heat in the processed zone [53]. FSW-R joints are thought to require a very crucial optimum selection of the parameters to minimize such flaws [54]. As a result, the following sections have a thorough investigation of the different macro/microstructural properties of FSW-R welds.

12.5.1 Macrostructure of Reinforced FSW Welds

The initial stage in examining the reinforced FSW joints' macrostructure properties is to evaluate their surface and metallurgical quality [55]. As illustrated in Figure 12.4, the macrostructure of

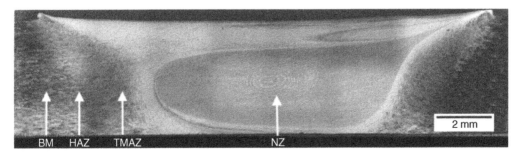

Figure 12.4 Macrostructure of FSW-based AA6082-T6/SiC, O/TiO_2 weld. Source: Salih et al. [56]/with permission from Elsevier.

reinforced FSW welds contains four unique zones: the NZ, TMAZ, HAZ, and the BM. The RPs addition and BM properties are important variables that significantly affect the origin of various macrostructural features [33, 34, 57–60].

12.5.1.1 Morphological Characteristics of the Processed Zone

At various magnifications (low or high), the typical distinctive traits such as the onion ring (OR) pattern, banded structure (BS) creation, shape/size of the NZ, and surface appearance are visually inspected [41, 61–65]. The OR pattern, which was developed in New Zealand, is a significant feature that distinguishes the macrostructure of reinforced FSW welds. Bahrami et al. [66] observed OR structure in NZ during FSW using SiC nanoparticles, threaded tapered, triangular, square, and four flute square pins to weld AA7075-O. They found that the reason for it was either inadequate generation of heat and partial DRX in NZ during FSW, or it was caused by material movement from the top to bottom beneath the NZ.

The individual OR observed in NZ [67] that provides a clear illustration of material transfer processes arises throughout FSW. Sun and Fujii [55] reported a pattern resembling an OR in FSW of pure copper plates in the NZ formed. They claimed that the ORs were created as a result of the parent metal's maelstrom current, which occurred during FSW and allowed SiC nanoparticles to travel to the bottom of AS with the advancement/revolution of the welding tool, creating a swirling OR in NZ. Along with ORs, NZ would also have a feature known as the BS (Figure 12.5a–c).

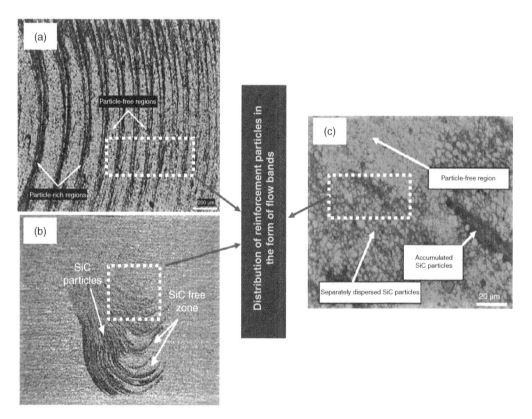

Figure 12.5 (a) AA7075-O/SiC. Source: Hamdollahzadeh et al. [41]/with permission from Elsevier, (b) AA6061-T6/AA2024-T351/SiC. Source: Moradi et al. [68], and (c) AA7075-O/SiC are examples of banded structure creation in FSW welding. Source: Bahrami et al. [66]/with permission from Elsevier.

These macrostructures, according to research, resulted from the mixing of incompatible alloys, which caused grain size changes, distribution of particle/precipitate, the density of dislocation, and chemical composition [11, 31, 56, 69–82]. Hamdollahzadeh et al. [41] found BS (particle-rich/free) zones in NZ. They asserted that this is how ORs grow and the band space is correlated with tool development (Figure 12.5a). Bahrami et al. recently reported [66] particle-rich/free BS in FSW. Besides, Moradi et al. [68] found a similar concentric ring structure (banded) made up of SiC-rich/free structural bands. They concluded that the SiC nanoparticles in NZ produced these zones (Figure 12.5b,c).

The size of NZ was another important characteristic that was found during the macrostructure analysis of reinforced FSW welds, as shown in Figure 12.6. The tool specifications, working temperature, materials' heat conduction, and the presence of reinforcing particles all have a significant impact on the variation in NZ morphology. For AA7075-O/SiC FSW, Hamdollahzadeh et al. [41] detected the basin-shaped NZ in the weld center (Figure 12.6a). They emphasized that the NZ's form was not significantly affected by the tool's rotating orientation being reversed in between passes. Although the morphology of the NZ is unaffected by changes in the tool (geometry) during FSW turns, the insertion of reinforcing particles in the NZ significantly influenced how the NZ appeared overall [66]. Huang et al. [36] recently reported the elliptical-shaped NZ that formed with FSW of 5083Al-H112/Ti and proposed that FSW runs increased as well as concurrent variability in the traveling rotational direction in passes (Figure 12.6b). Nikoo et al. [83] on the other hand thought that the inclusion of RPs, changing traveling speed, and post-weld heat treatments had no effect on the NZ form during FSW (Figure 12.6c) for AA6061-T6/Al_2O_3 welds.

12.5.1.2 Welding Parameters Effect on the Morphology of NZ

Numerous studies examined during FSW the effects of adding RPs in NZ and varying weld parameters on NZ size. The nanoparticle's addition in the NZ produces significant localized strain, which impedes heat input and their flow and causes a change in NZ size [84]. According to Babu et al. [85], a low amount of heat at fast traveling velocity results in a smaller NZ in FSW of AlMg5/Al2O3, while an increment in tool rotational speed (TRS) generates a higher heat intensity and causes a higher localized strain rate in NZ which results in a wider NZ and more plastic deformation. Naghibi et al. [86] have recently described the diverse morphologies as bowl-shaped NZ in FSW of AA6061-T6/Al_2O_3. They discovered reduced plastic flow and decreased heat input

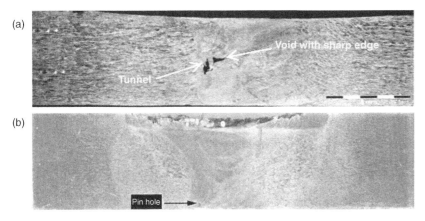

(a)

(b)

Figure 12.6 NZ shapes: (a) basin. Source: Hamdollahzadeh et al. [41]/with permission from Elsevier. (b) elliptical. Source: Huang et al. [36]/with permission from Elsevier.

generation, resulting in nuggets with a bowl-like shape. In contrast, Shahedi et al. [87] concluded NZ size is stretched on the advancing front of BM using TiO_2 nanoparticles which is due to high TRS on AS in comparison to RS.

12.5.2 Microstructural Characteristics of Reinforced FSW Welds

The base material often undergoes significant plastic deformation during FSW in the employment of shoulder surface topologies and tool specifications, according to Thomas et al. [17]. DRX, a process that refines grains in the NZ, is the result of the breaking of coarse particles brought on by the transiently high temperature [41].

12.5.2.1 Reinforcement Particle's Addition Effect on the Evolution of Grain

In addition to helping to supply sites for new grain nucleation during the DRX process, the RPs addition in NZ also functions as a barrier to the migration of grain borders, or what is known as the "pinning effect" [16]. The inclusion of RPs during the FSW of FSW-R welds in the creation of low-angle grain boundaries (LAGBs) or breaking up of initial grains because of SPD of BM [86]. Singh et al. [57, 58] reported that microstructure (finer) in processed NZ was created in FSW-R joints, resulting in the creation of primary DRX granular. Two significant aspects were described by Gallais et al. [88]: the shear (adiabatic) and DRX took place during the FSW. They claimed that during DRX, the fine equiaxed grains are produced by the SPD, and heat input is generated by friction between the tool and the workpiece. Similarly to this, Saeidi et al. [89] found temperature gradients and strain rate were responsible for the development of highly characterized microstructures during FSW.

12.5.2.2 Grain Size Variation in Reinforced FSW Welds

Factors influencing the grain size variation in NZ of MMR welds are DRX, an effect due to annealing, and brought on by heat generated [90]. Fallahi et al. [30] reported that FSW passes' increment causes a decrease in grain size [91] and is responsible for reduction as shown in Figure 12.7.

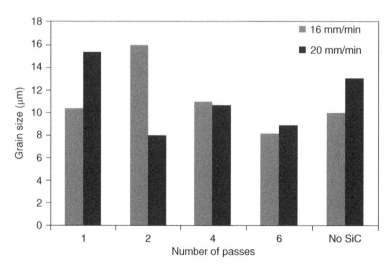

Figure 12.7 Traverse velocities 16, 20 mm/min for FSW passes-based FSW welds. Source: Fallahi et al. [30]/with permission of Elsevier.

Similarly to this, Hamdollahzadeh et al. [41] stated that increasing the number of one to two passes causes grain size reduction. Additionally, they noticed that increasing the number of FSW passes causes the grain size to increase from 4.1 to 4.3 mm resulting in grain coarsening at the highest intensity inputs, where the HI outweighs the effect of pinning. The way that initiates a reduction in grain size without RPs FSW welds, according to Mirjavadi et al. [92], is heat input and temperature. Asl et al. [31] hypothesized that the considerable NZ microstructure refinement-based pinning effect is caused by the addition of reinforcing nanoparticles (Figure 12.8). They also noted that high HI has two different effects on the microstructure: (a) it may reduce the effects of recrystallization and annealing, and (b) nanoparticle agglomeration may lessen the effects of mechanisms for refining, which may cause grain enlargement. According to Shahedi et al. [87], the reinforced particles and higher/lower heat input caused by increasing TRS/TS in FSW of Cu/TiO2 had the opposite influence on the size of grain in NZ.

The reason for NZ's finer grain size in particle-rich, free zones are because of the SiC nanoparticle's presence, which creates an effect (pinning) that is essential to the DRX of the matrix [93]. Pantelis et al.'s [94] observation (12 μm) in particle-poor opposed to the particle-rich region (7 μm) is as a result of the effect (pinning) produced via the presence of SiC that hinders grain growth. While Sun and Fujii [53] reported (2 μm) as opposed to (8 μm) during Pure Cu/SiC. They claimed that this was caused by the inclusion of SiC nanoparticles in NZ prevented the grains' expansion and prevented the average grain size from exceeding the critical maximum radius (Figure 12.9a) according to Bahrami et al. [66], which is to blame for the finer size (Figure 12.9b).

When welding AA5083-H112/Ti using FSW and the underwater (UFSPW), changes in grain size were noted by Huang et al. [36]. They reported average fine equiaxed grain sizes of 3.6 and 0.8 m in both situations. They concluded that the Ti particles and water cooling are the reasons for the fluctuation in NZ grain size in UFSPW, as illustrated in Figure 12.10a–c. However, this results in the creation of dislocation loops around the SiC particles, which refines the size of reinforced NZ grains more than NZ without [47]. While Naghibi et al. [86] discovered fine Al2O3 nanoparticles that create the pinning effect with no difference in granular size for post-weld heat treatment (PWHT) during AA6061-T6/Al$_2$O$_3$.

In contrast, Moradi et al. [68] discovered a larger grain size (of 8.1 ± 0.9 μm) and $(7.1 \pm 0.9$ μm) in the absence of PWHT and SiC nanoparticles than vice versa. The homogeneous distribution in the

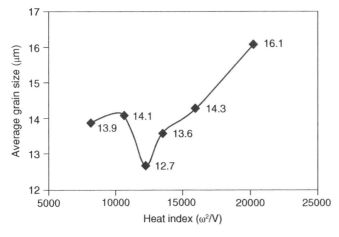

Figure 12.8 Heat index, HI, the average grain size in SZ (ω^2/V). Source: Asl et al. [31]/with permission of Elsevier.

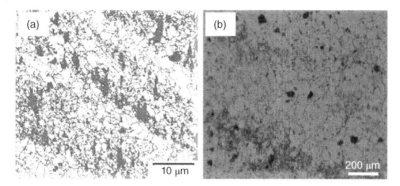

Figure 12.9 Effect of the insertion of RPs: (a) pure Cu with SiC. Source: Sun and Fujii [55]/with permission from Elsevier, and (b) AA7075-O and SiC. Source: Bahrami et al. [66]/with permission from Elsevier.

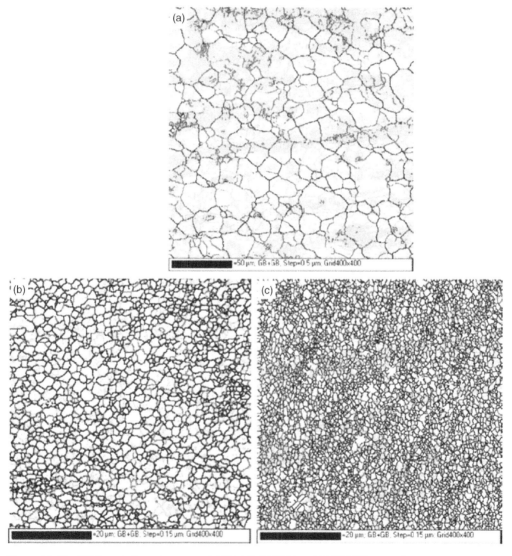

Figure 12.10 EBSD grain boundary maps, (a) BM, NZ, (b) FSW, and (c) UFSPW, green lines represent misorientation (2–5°), red lines (in the range 5–15°), and black lines greater>15°. Source: Huang et al. [36]/with permission from Elsevier.

NZ of SiC nanoparticles that offers sites/locations for recrystallizing grains to form, and the low material's flow stress under annealed conditions, which causes more strain formation in the NZ, are both responsible for this refinement in grain size. In contrast, PWHT causes grain growth in New Zealand, where abnormal grain growth would result in bigger grain sizes. The high heating intensity produced at lower TS accounts for Fallahi et al.'s [30] observation of aberrant grain development in NZ at the second FSW pass after a low travel speed weld. Sato and Kokawa [95] and Ma et al. [96] noticed aberrant grain growth and claimed that grain growth was caused by static recrystallization.

12.5.2.3 Reinforcement Particles Distribution

The revolving FSW tool stir generates 80% heat in NZ that results in RPs refinement and redistribution, and grain growth [11, 67, 83–87, 96–105]. According to Dragatogiannis et al. [106], RPs are either distributed uniformly throughout a matrix or they are clustered. They asserted that the clustered RPs in the NZ break apart due to the stirring action of the rotating FSW tool, resulting in a homogeneous distribution. Uneven distribution aggregated NZ SiC in a single pass is caused by insufficient material flow, which results in SiC particles that were not dispersed in the matrix [93]. According to Mirjavadi et al. [92], the clustered particles ordinarily distributed in NZ during FSW are broken down, resulting in a more uniform distribution of nano-sized TiO_2 particles. This is accomplished by increment in FSW runs to four and the addition of TiO_2 nanoparticles in BM.

Jamalian et al. [52] and Fallahi et al. [30] found that increment in passes from two to six results in homogeneous SiC nanoparticle cluster distribution that increases dislocation. Quantitative analyses and electron dispersive spectroscopy (EDS) were used to determine the RPs distribution. Processing parameters during FSW, as indicated in Table 12.1, also have an impact on RPs distribution in NZ during MMR welds [94]. Nikoo et al. [97] believed that maintaining constant TRS at 1250 rpm leads to more uniform Al_2O_3 nanoparticles.

Fallahi et al. [30] examined more homogenous distribution at a lower 16 mm/min TS that leads to an increase in SiC nanoparticles and fragments percentage, especially steel, in NZ with a lower agglomeration of SiC nanoparticles, while severe SiC nanoparticles agglomeration at 200 mm/min even after the 6 FSW passes were observed in NZ as shown in Figure 12.11a–c. Abdolahzadeh et al. [96] state that for nano-sized SiC particles, either a higher heating intensity or a low heating index would be desirable. But Abbasi et al. [47] reported that low TRS causes inadequate SiC particle distribution instead of greater interparticle space between them. This is attributed to the FSW tool's insufficient stirring motion caused by the lower heating index. Low interparticle spacing and a homogeneous distribution of SiC particles are produced in the NZ via faster tool rotation.

While an increase in TS causes a nonuniform RPs aggregation in the NZ, while uniform RPs distribution is due to an increase in TRS. Numerous articles have also asserted that differences in tool specifications, particularly the FSW pin, have a substantial impact. Saeidi et al. [89] claimed that employing a pin (square) delivers more pulse action for mixing particles and SPD. According to Sun and Fujii [55], the FSW second pass causes the distribution of SiC nanoparticles at Cu grain boundaries with a significant density of dislocation as shown in Figure 12.12a.

However, Jamalian et al. [52] have found that when fabricating AA5086-H34/SiC, switching the tool direction between passes causes a more NZ homogenous distribution of SiC nanoparticles attributed to the direction of material flow in the direction of AS and the action (pulsating) via pin, which improves SiC mixing with BM [66] (Figure 12.12b). The B4C uniform distribution in comparison to Al_2O_3 and SiC nanoparticles according to Abioye et al. [118], can stir action because this action would cause large-size B4C fragmentation into small ones via the occurrence of lower resistance due to these particles for AA6061-T6/B4C/Al_2O_3/SiC.

Table 12.1 Effect of additional reinforcement on the effectiveness of FSW welds.

S/N	Particle type	Particle size	Process parameters	FSW passes	Findings	Ref.
1	Al_2O_3	15 nm	2000 rpm; 70–80 mm/min	Single	The reinforced sample's UTS is 250 MPa while the unreinforced sample's is 200 MPa.	[107]
2	Cu powder	20 μm	660 r/min; 63 mm/min	Double	The highest UTS for Cu-reinforced-based FSW joints was 180 MPa.	[108]
3	SiC	20 μm	1400 4 r.min^{-1}; 50 mm. min^{-1}	Single	The reinforced sample attained a maximum hardness of approximately 116 HV.	[109]
4	SiC, Al_2O_3, B_4C	3–15 μm	850 r.min^{-1}, 45 mm/min	Single	Comparing reinforced joints made of B4C, Al_2O_3, and SiC to unreinforced joints with a hardness value of 65 HV, improvements of 42%, 34%, and 22% have been noted. The wear rate is reduced by a factor of 1.7 and 1.9 for the reinforcement-added welds.	[110]
5	Cu powder	30 μm	450, 600, 750 r/min; 44 mm/min	Single	59.5 HV unreinforced joint hardness and 144.41 MPa UTS, compared to Cu-reinforced joint's 81.68 HV and 179.34 MPa.	[111]
6	B_4C	7 μm	1000 r/min, 50–400 mm/min	Single	The strongest joint efficiency, at almost 70%, is shown in the reinforced sample. Aging helped to increase joint efficiency.	[112]
7	(SiC + Gr) and (SiC + Al_2O_3)	20 μm	900, 1120, 1400 r/min	Single	Higher 219 MPa tensile strength at 900 r/min was reached.	[113]
8	Cu + SiC		710, 1000, 1400 r/min; 50, 63, 80 mm/min	Single	In comparison to an unreinforced welded junction, the reinforced joint has a 96% efficiency at 1000 r/min/80 mm/min and improves in elongation.	[114]
9	SiC	60 nm	550, 600, 650 r/min; 25 and 35 mm/min	Single	With 650 rpm and 35 mm/min as the process conditions, the reinforced sample outperformed the unreinforced joint by 28% in terms of UTS. Additionally, the reinforced sample has a higher mean microhardness than an unreinforced sample.	[115]
10	B_4C	5–7 μm	1200 r/min; 100 mm/min	Multi-pass	Surface welds with reinforcement had 98 HV, which is higher than the sample without reinforcement (53 HV). An increased amount of FSW passes within NZ improved the homogenous dispersion of particles	[116]
11	SiC + Gr	20 μm	650 r/min/3.4 m/s	Single	At ideal conditions, there was a decrease in wear rate. Gr serves as a solid lubricant that reduces wear rate. Under ideal circumstances, the presence of hard SiC particles and their pinning action results in an increase in microhardness.	[117]

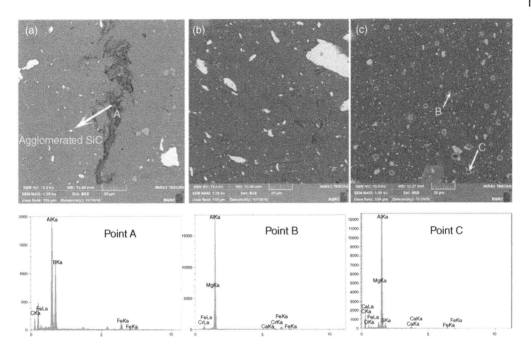

Figure 12.11 At TS of 16 mm/min, FE-SEM pictures together with an EDS examination of the designated sites in dissimilar welds SZ were taken for (a) a single FSW pass, (b) four FSW passes, and (c) six FSW passes. Source: Fallahi et al. [30]/with permission of Elsevier.

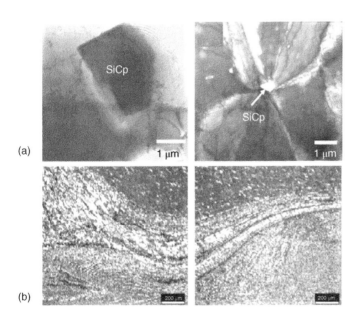

Figure 12.12 (a) TEM pictures. Source: Sun and Fujii's [55]/with permission from Elsevier.
(b) SEM micrographs demonstrating the impact of tool geometries on the distribution of RPs. Source: Bahrami et al. [66]/with permission from Elsevier.

12.5.2.4 Morphology of Reinforcement Particles

The FSW tool and the SPD's vigorous stirring action during the process alter the shape of RPs for reinforced welds. The pinning impact of smaller RPs, particularly nanoparticles in NZ, was greater than that of larger particulates, which aids in microstructure improvement [92]. According to Kumar et al. [32], the force produced by the tool and stirring effect cause nano-sized B4C particles to obstruct the dislocation migration in BM, which causes to decrease [66]. According to Abioye et al. [118], they reported that in the presence of small-size particles, there is an intense B4C nanoparticles fragmentation in NZ due to lower resistance induced by FSW tool stir action in FSW of AA6061-T6/B4C/Al$_2$O$_3$/SiC. Additionally, Huang et al. [36] noted that due to the second-phase RPs fractured strength, there is a breakdown of precipitates of Mg substrate and Ti particles (Figure 12.13a–f).

12.5.2.5 Intermetallic Particles

The qualities of a friction stir welded joint can be significantly influenced by several essential variables, including grain size, dislocation density, and defect development [36, 45, 66, 119]. The intermetallic complexes (IMCs/IMPs) production changes the NZ characteristics that result in brittle fractures of the welded joints, according to Fallahi et al. [30]. The solid-state FSW technique causes IMCs to dissolve and reprecipitate because the heat produced by the FSW revolving tool raises the temperature of NZ near to solidus temperature of BM.

Mishra and Ma [11] reported that it is very challenging to avoid IMC formation during FSW when there is a fluctuation in NZ temperature at both high and low levels of eutectic temperature. The volume and thickness of IMCs formed throughout the process are significantly influenced by changes in parameters (HI), according to Karakizis et al. [120]. Additionally, Fallahi et al. [30] discovered thin IMPs with steel fragment streaks other than IMCs thick at interface Al/Steel for AA5083-H321/A316L/SiC nanoparticles [119, 121].

Fragments of steel are delaminated from steel plates while traveling at a greater speed (20 mm/min), and after six FSW passes, they are mainly removed on RS in NZ. Besides, the good weld quality without defect welds with fine homogenous steel fragments was seen at 16 and 20 mm/min, respectively, after six passes; the top, center, and base weld zones also lost their continuous intermetallic layer as seen in Figure 12.14.

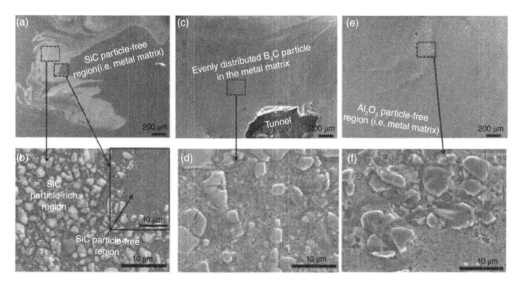

Figure 12.13 SEM pictures from SZ center for reinforced FSWed welds (a, b) SiC, (c, d) B4C, and (e, f) Al$_2$O$_3$. Source: Abioye et al. [118]/with permission from Elsevier.

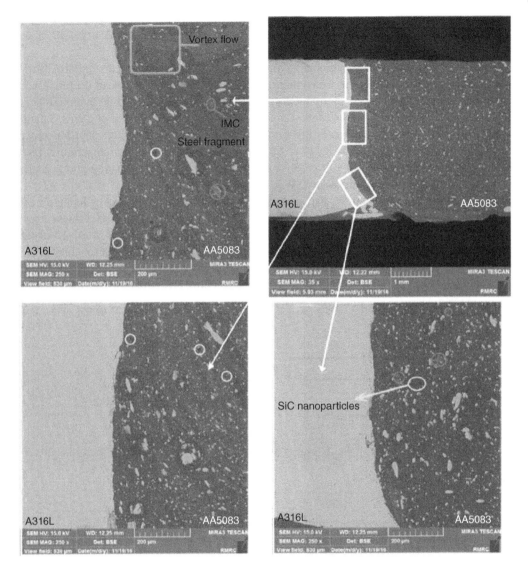

Figure 12.14 FE-SEM images at interface Al–Fe for welds produced at 16 mm/min (W16P6) after six FSW passes (different magnifications). Note: SiC nanoparticle: green circle, IMC: red circle, and steel fragment: blue circle. Source: Fallahi et al. [30].

12.6 Mechanical Behavior of Reinforced FSW Welds

12.6.1 Microhardness Characterization

The variables causing the heterogeneity in FSW welds microhardness are because of microstructural characteristics, grain size, and dislocation differential [10]. These three most significant elements, according to Singh et al. [122–125], cause anisotropy in the microstructure affect NZ center microhardness during FSW of MMR welds. According to Bahrami et al. [66], the homogeneous NZ RPs distribution in FSW causes the dislocations to pinch up and increased dislocation density via CTE difference in RPs/matrix, which generates residual stress during cooling and increases the

NZ microhardness (Hall–Petch relationship). In contrast, Sun and Fujii [55] reported that a more uniform distribution of microhardness in particle-rich is seen compared to the first FSW pass, which results in an uneven distribution (particle-free) (Figure 12.15a).

According to Mirjavadi et al. [92], higher FSW passes lead to TiO_2 nanoparticle's uniform distribution in NZ with grain refinement, which increases NZ microhardness because of the hindrance of grain boundaries migration via pinning effect in FSW of $AA5083/TiO_2$. While microhardness decreased with NZ distance from the center (HAZ) because of the development of grains. Fallahi et al. [30] noted that after six FSW passes, higher microhardness was observed (top (300HV), center (270HV), and bottom (240HV)) due to lower heating index and peak cooling temperature. Like this, Jamalian et al. [52] noted that more FSW passes resulted in uniform SiC nanoparticles distribution and turn higher microhardness in FSW of AA5086-H34/SiC.

Besides, Paidar et al. [54] noted WC nanoparticle inclusion causes an increase in microhardness by inhibiting grain development, which causes grain refining and results in a microhardness increment in FSW of AA5182/WC [48] (Figure 12.15b). Jamalian et al. [52] reported that a 1000–1250 rpm increase in TRS with constant TS leads to higher microhardness on AS than RS. In conclusion, the refinement of grain size, the density of dislocations, the addition of RPs, and their distribution all affect the MMR FSW joint's microhardness, while strengthening precipitates dissolving causes a decrease in microhardness (Figure 12.15c).

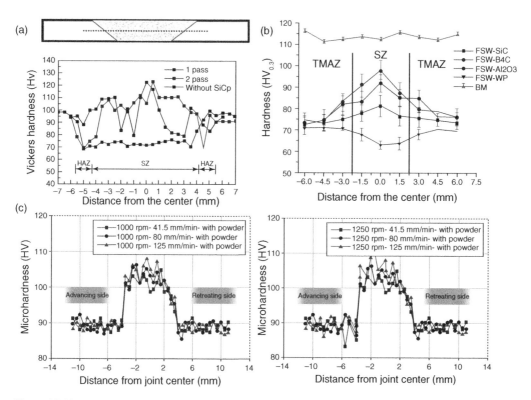

Figure 12.15 (a) Cu/SiC microhardness profile. Source: Sun and Fujii [119]/with permission of Elsevier, (b) microhardness profile: BM, with/without particles reinforcement FSW welds. Source: Abioye et al. [118]/with permission of Elsevier, and (c) single-pass FSW/SiC weld microhardness. Source: Jamalian et al. [52]/with permission of Elsevier.

12.6.1.1 Tensile Behavior

The factor that influences the mechanical characteristics of FSW-based MMR welds is grain size. [31]. In contrast, Abbasi et al. [47] examined the severe NZ RPs. Tensile characteristics of FSW welds were depending on the quality of RPs and matrix bonding according to Abioye et al. [118]. According to Pantelis et al. [99], the heating index is the foremost factor that influences the FSW weld's performance, IMC formation, and FSW passes are other factors affecting the tensile characteristics of reinforced FSW welds [30, 36, 86].

12.6.1.2 Process Parameters Effect on UTS

Fallahi et al. [30] found reduction in TS and increment in TRS lead to get higher travel speeds than higher mechanical strength with percentage elongation of MMR welds. This is due to the high heating index generated at low TS which leads to vigorous material flow mixing maintaining adequate interparticle bonding. Whereas, Paidar et al. [54] thought that the intersection of nanoparticles fragmented in corresponding to dislocations enhanced mechanical properties during FSW of AA5182/WC. Naghibi et al. [86] found that reinforced welds' UTS increased at higher TS compared to lower TS. This is explained by Al_2O_3 uniform nanoparticle distribution in NZ, which prevents dislocations and promotes grain refinement, both of which increase the strength of the weldment in FSW of AA6061-T6/Al_2O_3.

12.6.1.3 FSW Passes and Addition of RPs Effect on UTS

Mirjavadi et al. [92] reported that increased FSW passes, the density of dislocation, rate of hardening, refinement, deformation load, and volume of grain boundaries that act as a hindrance to the dislocation all increased as well. This resulted in reinforced welds having better tensile properties. Sun and Fujii [55] observed higher UTS allegedly seen in six FSW passes. Fallahi et al. [30] reported that because an increase in dislocation density aids in enhancing DRX, cohesion (Orowon strengthening) and strong interface bonding (Al/SiC) resulted in grain refinement.

Jamalian et al. [52] found that a single FSW pass causes a reduction in UTS (Figure 12.16a,b). According to Abdolahzadeh et al. [96], the pinning effect limits the grain growth, and RPs/Al interface reactions resulted in strong Al/SiC bonding via smaller SiC (Figure 12.16c). As a result of homogeneous NZ, Al_2O_3 nanoparticle distribution serves as a nucleator. Naghibi et al. [86] reported UTS improvement and stated that UTS, YS, and %EL can be greatly influenced by several factors, including TS, RPs–base matrix interface bonding, and grain refinement in FSW of AA6061-T6/Al_2O_3.

12.6.1.4 Fracture behavior

The welded samples fracture in a variety of places as a result of the FSW process' diverse processing conditions. BM is fractured at 45° along the slip plane via fracture mechanism leading to decreased hardness for FSW-MMR welds AA5086-H34/SiC [52]. Furthermore, Sun and Fujii [55] observed HAZ fracture in the second FSW pass because of the large Cu matrix grain size, while Jamalian et al. [52] observed BM fracture in multi-pass FSW which is attributed to Al/SiC strong bonding interface that hinders crack nucleation and causes ductile failure. Shahedi et al. [87] observed fracture along HAZ, which is related to the HAZ's high grain size in both reinforced and unreinforced welds in FSW of Cu/TiO_2 as illustrated in Figure 12.17a. While Karakizis et al. [120] thought it was because of the largest thermal cycle encountered. In both the first and second FSW passes, Hamdollahzadeh et al. [56] noted that adequate Al/SiC bonding with fine NZ grain size helps in shifting loads from weak BM to RPs matrix (and vice versa) resulting in ductile failure and fracture outside NZ. Similarly, Bahrami et al. [105] observed fracture in TMAZ due to lower deformation that repletes the density of dislocations resulting in crack nucleation and brittle failure,

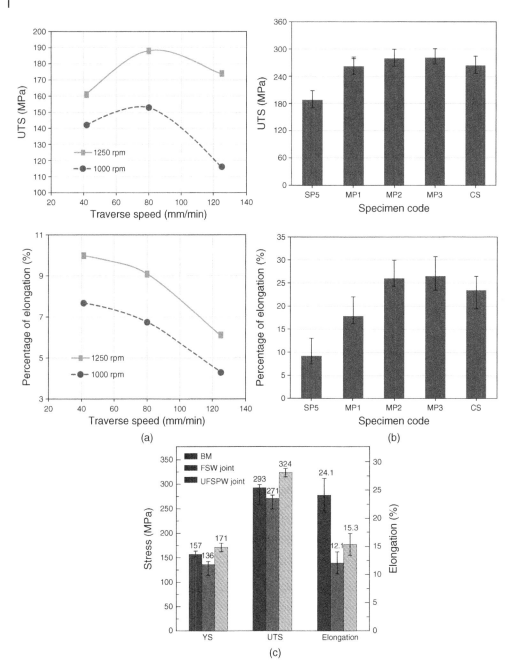

Figure 12.16 (a, b) FSW process parameters (TRS, TS) and SiC reinforcement addition effect on UTS and % El. Source: Jamalian et al. [52]/with permission of Elsevier, and (c) comparison of BM, FSW weld, and UFSPW welds UTS, YS, % El. Source: Huang et al. [36]/with permission of Elsevier.

while reinforced welds fracture outside NZ, i.e. in the BM, indicating the strong Al/SiC bonding. When performing FSW and UFSPW, Huang et al. [36] noticed a fracture in NZ on AS toward BM outside the NZ (Figure 12.17b).

The positions of fracture for MMR welds were altered specifications noted by Abdolahzadeh et al. [96]. According to Naghibi et al.'s [86] observed fracture at lower TS in NZ, the NZ's

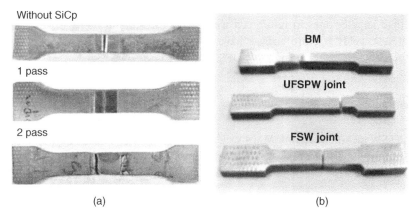

Figure 12.17 Fracture locations in MMR-based FSW welds, (a) NZ. Source: Sun and Fujii [119]/with permission from Elsevier. (b) AS of NZ. Source: Huang et al. [36]/with permission from Elsevier.

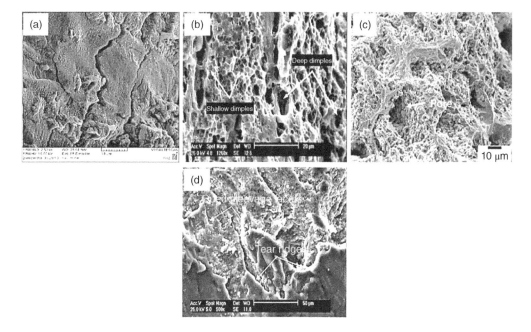

Figure 12.18 Fractured surfaces of reinforced FSW welds: (a) cleavage (Quasi). Source: Jamalian et al. [52]/ with permission from Elsevier. (b) shallow (or deep) dimples Source: Hamdollahzadeh et al. [56]/with permission from Elsevier. (c) formation of pores. Source: Sun and Fujii [55]/with permission from Elsevier. and (d) ridges (tear). Source: Asl et al. [31]/with permission from Elsevier).

heterogeneous dispersion and agglomeration of Al_2O_3 nanoparticles are that cause sites for nucleation to hinder crack growth and concentration of stress. According to Jamalian et al. [52], the dimples (shallow) and tear ridges seen on the fracture surface of SiC nanoparticle-reinforced welds indicate the fracture mode as quasi-cleavage with a ductile failure mechanism. While without SiC nanoparticles welds contain small as well large size dimples resulting in an increment in elongation. While during multiple-FSW pass welds, dimple rapture and ductile failure mechanisms were seen (Figure 12.18a–d). When ductile and brittle fracture modes are mixed, Naghibi et al. [86]

observed on the fracture surface both dimples and ridges at reduced TS with PWHT for MMR welds leading to a reinforcement–matrix interface mechanism. Asl et al. [31] noticed larger and deeper BM dimples than smaller ones in reinforced welds, which results in increased elongation [36]. while the UFSPW welds' fracture surfaces have shallower, smaller dimples, which cause the elongation of the AA5083-H112/Ti during FSW/UFSPW to rise.

12.7 Conclusions

The application of the FSW technique to link MMR welds or joints highlights the current state of the art in this chapter. The topics of applications of FSW process-based MMCs are discussed in detail.

1) The FSW has been shown and is regarded as a well-known method for combining different alloys and composites or joints made from their metal matrix reinforcement.
2) The FSW technique is more appealing and practical for combining aluminum alloy series (AA2xxx, AA6xxx, and AA7xxx) due to potential benefits.
3) By preventing the possibility of unfavorable chemical interactions between the RPs and the BM, the solid-state behavior of the FSW method helps connect MMR welds.
4) According to the literature, using small-scale nanoparticles rather than micro-sized particulates, significantly increased the mechanical strengthening effect of MMR welds.
5) Careful control of temperature variation factors that affect heat input generation improves the quality of FSW-based reinforced joints.
6) During FSW, heating index and control on temperature via welding specifications resulted in grain structure refinement and RPs distribution (uniform) via pinning effect, whereas severe tool force applied to the NZ caused pulsating stir action, which improved RP distribution, shape, and size.
7) The interplay between BM and RPs type or composition, the bonding mechanism between interparticle, RPs, and the BM contact type, and the window of FSW parameters determine the mechanical behavior and characteristics of MMR welds.

12.8 Future Challenges

1) As an illustration, the current need is to carry out a systematic study in which parameters such as reinforcing particle volume fractions, kinds, and impacts on MMR welds properties need to be taken into account.
2) Different alloys or materials other than Al-alloys need to be used to join MMR welds; for example, magnesium alloys are a candidate alloy for aerospace applications.
3) There is very little open literature on the fracture characteristics of Al-alloy and data on MMR welds produced using FSW.
4) More work is needed to fully understand the fracture mechanism employing various RP types and BM to assess their influence on weld characteristics of reinforced FSW welds.
5) Limited research on wear characteristics of reinforced FSW welds that are influenced by RPs, BM contact, microhardness, and process specifications.

References

1 Davis, J.R. (1990). *Properties and Selection: Nonferrous Alloys and Special-Purpose Materials*, 10e. UK: ASM Int.

2 Mouritz, A.P. (2012). *Introduction to Aerospace Materials*, 1e. US: Woodhead Publishing.

3 Chawla, N. and Chawla, K.K. (2006). *Metal Matrix Composites*, 1e, 1–41. US: Springer.

4 Kaczmar, J., Pietrzak, K., and Wlosinski, W. (2000). The production and application of metal matrix composite materials. *J. Mater. Process. Technol.* **106** (1): 58–67.

5 Rawal, S.P. (2001). Metal-matrix composites for space applications. *JOM* **53** (4): 1417.

6 Kunze, J.M. and Bampton, C.C. (2001). Challenges to developing and producing MMCs for space applications. *J. Miner. Met. Mater. Soc.* **53**: 22–25.

7 Suryanarayanan, K., Praveen, R., and Raghuraman, S. (2013). Silicon carbide reinforced aluminium metal matrix composites for aerospace applications: a literature review. *Int. J. Innov. Res. Sci. Eng. Technol.* **2** (11): 6336–6344.

8 Starink, M.J., Deschamps, A., and Wang, S.C. (2008). The strength of friction stir welded and friction stir processed aluminum alloys. *Scr. Mater.* **58** (5): 377–382.

9 Zhang, Y.M., Na, S.J., and Cook, G.E. (2014). Recent developments in welding processes. *J. Manuf. Processes* **16** (1): 1–156.

10 Threadgill, P.L., Leonard, A.J., Shercliff, H.R., and Withers, P.J. (2009). Friction stir welding of aluminum alloys. *Int. Mater. Rev.* **54** (2): 49–93.

11 Mishra, R.S. and Ma, Z.Y. (2005). Friction stir welding and processing. *Mater. Sci. Eng., R* **50**: 1–78.

12 Lloyd, D.J. (1994). Particle reinforced aluminum and magnesium matrix composites. *Int. Mater. Rev.* **39** (1): 1–23.

13 Ibrahim, I.A., Mohamed, F.A., and Lavernia, E.J. (1991). Particulate reinforced metal matrix composites – a review. *J. Mater. Sci.* **26**: 1137–1156.

14 Ellis, M.B.D. (1996). Joining of aluminum-based metal matrix composites. *Int. Mater. Rev.* **41** (2): 41.

15 Rosso, M. (2006). Ceramic and metal matrix composites: routes and properties. *J. Mater. Process. Technol.* **175** (1-3): 364–375.

16 Tjong, S.C. (2007). Novel nanoparticle-reinforced metal matrix composites with enhanced mechanical properties. *Adv. Eng. Mater.* **9** (8): 639–652.

17 Thomas, W.M., Nicholas, E.D., Needham, J.C., et al. (1996). Friction stir welding, UK Patent Application. 2306366.

18 Lohwasser, D. and Chen, Z. (2010). *Friction Stir Welding from Basics to Applications*. US: CRC.

19 Fu, R.D., Sun, R.C., Zhang, F.C., and Liu, H.J. (2012). Improvement of formation quality for friction stir welded joints. *Welding J.* **91**: 169-s-173-s.

20 Rai, N., DebRoy, T., and Bhadeshia, H.K.D.H. (2008). Recent advances in friction stir welding–process, weldment structure, and properties. *Prog. Mater Sci.* **53**: 980–1023.

21 Rai, R., De, A., Bhadeshia, H.K.D.H., and DebRoy, T. (2011). Review: friction stir welding tools. *Sci. Technol. Weld. Joining* **16** (4): 325–342.

22 Cavaliere, P., Santis, A.D., Panella, F., and Squillace, A. (2009). Effect of welding parameters on mechanical and microstructural properties of dissimilar AA6082–AA2024 joints produced by friction stir welding. *Mater. Des.* 609–616.

23 Mukhopadhyay, P. (2012). Alloy designation, processing, and use of AA6XXX series aluminium alloys. *Int. Schol. Res. Network ISRN Metall.* https://doi.org/10.5402/2012/165082.

24 Zhang, Y.N., Cao, X., Larose, S., and Wanjara, P. (2012). Review of tools for friction stir welding and processing. *Can. Metall. Q.* **51** (3): 250–261.

25 Chen, X.G., da Silva, M., Gougeon, P., and St-Georges, L. (2009). Microstructure and mechanical properties of friction stir welded AA6063-B$_4$C metal matrix composites. *Mater. Sci. Eng., A* **518** (1–2): 174–184.

26 Pantelis, D.I., Karakizis, P.N., Dragatogiannis, D.A., and Charitidis, C.A. (2019). Dissimilar friction stir welding of aluminum alloys reinforced with carbon nanotubes. *Phys. Sci. Rev.* **1** (1): 1–20.

27 Cheng, C.I., Lee, C.J., and Huang, J.C. (2004). Relationship between grain size and Zener–Holloman parameter during friction stir processing in AZ31 Mg alloys. *Scr. Mater.* **51**: 509–514.

28 Wan, L., Huang, Y., Wang, Y. et al. (2015). Friction stir welding of aluminum hollow extrusion: weld formation and mechanical properties. *J. Mater. Sci. Technol.* **31**: 12.

29 Fonda, R.W., Bingert, J.F., and Colligan, K.J. (2004). Texture and grain evolutions in a 2195 friction stir weld. *Proceedings 5th Int Conf on Friction stir welding*, TWI, Metz France, September.

30 Fallahi, A.A., Shokuhfar, A., Moghaddam, A.O., and Abdolahzadeh, A. (2017). Analysis of SiC nano-powder effects on friction stir welding of dissimilar Al-Mg alloy to A316L stainless steel. *J. Manuf. Processes* **30**: 418–430.

31 Asl, N.S., Mirsalehi, S.E., and Dehghani, K. (2019). Effect of TiO$_2$ nanoparticles addition on microstructure and mechanical properties of dissimilar friction stir welded AA6063-T4 aluminum alloy and AZ31B-O magnesium alloy. *J. Manuf. Processes* **38**: 338–354.

32 Kumar, K.S.A., Murigendrappa, S.M., Kumar, H., and Shekhar, H. (2018). Effect of tool rotation speed on microstructure and tensile properties of FSW joints of 2024-T351 and 7075-T651 reinforced with SiC nanoparticle: The role of FSW single pass. In: *A.I.P. Conf. Proc.*, 1943.

33 Singh, T., Tiwari, S.K., and Shukla, D.K. (2021). Influence of nanoparticle addition (TiO$_2$) on microstructural evolution and mechanical properties of friction stir welded AA6061-T6 joints. In: *Advances in Production and Industrial Engineering* (ed. P.M. Pandey, P. Kumar, and V. Sharma). Springer, Singapore: Lecture Notes in Mechanical Engineering.

34 Singh, T., Tiwari, S.K., and Shukla, D.K. (2020). Effects of Al$_2$O$_3$ nanoparticles volume fractions on microstructural and mechanical characteristics of friction stir welded nanocomposites. *Nanocomposites* 6 (2): 76–84.

35 Rino, J.J., Chandramohan, D., and Sucitharan, K. (2012). An overview on development of aluminium metal matrix composites with hybrid reinforcement. *Int. J. Sci. Res.* **1** (3): 2319–7064.

36 Huang, G., Wu, J., and Shen, Y. (2018). A strategy for improving the mechanical properties of FSWed joints of non-heat-treatable Al alloys through a combination of water cooling and particle addition. *J. Manuf. Processes* **34**: 667–677.

37 Evans, A., Marchi, C.S., and Mortensen, A. (2003). *Metal Matrix Composites in Industry: An Introduction and a Survey*. Dordrecht Netherlands: Kluwer Academic Publishers.

38 Hunt, W.H. and Miracle, D.B. (2001). Automotive applications of metal matrix composites. In: *ASM Handbook* (ed. D.B. Miracle and S.L. Donaldson), 1029–1032. ASM International.

39 Prasada, S.V. and Asthana, R. (2004). Aluminum metal-matrix composites for automotive applications: tribological considerations. *Tribol. Lett.* **17**: 445–453.

40 Mathers, G. (2002). *The Welding of Aluminum and its Alloys*. Cambridge: Woodhead Publishing Limited.

41 Hamdollahzadeh, A., Bahrami, M., Nikoo, M.F. et al. (2015). Microstructure evolutions and mechanical properties of nano-SiC-fortified AA7075 friction stir weldment: the role of second pass processing. *J. Manuf. Processes* **20**: 367–373.

42 Rani, P., Misra, R.S., and Mehdi, H. Effect of nano-Sized Al$_2$O$_3$ Particles on Microstructure and Mechanical Properties of Aluminum Matrix Composite Fabricated by Multipass FSW. Proceedings of the Institution of Mechanical Engineers, Part C: Journal of Mechanical Engineering Science. 2022;0(0)

43 Sharma, V. and Tripathi, P.K. (2022). Approaches to measure volume fraction of surface composites fabricated by friction stir processing: a review. *Measurement* 193: 110941.

44 Anand, R. and Sridhar, V.G. (2021). Effects of SiC and Al_2O_3 reinforcement of varied volume fractions on mechanical and micro structure properties of interlock FSW dissimilar joints AA7075-T6-AA7475-T7. *Silicon* 13: 3017–3029.

45 Ellis, M.B.D., Gittos, M.F., and Threadgill, P.L. (1994). Joining of aluminum-based metal matrix composites, Initial studies, Members Report 501. Abington, TWI.

46 Storjohann, D., Barabash, O.M., Babu, S.S. et al. (2005). Fusion and friction stir welding of aluminum metal-matrix composites. *Metall. Mater. Trans. A* **36A**: 3237–3247.

47 Abbasi, M., Abdollahzadeh, A., Bagheri, B., and Omidvar, H. (2015). The effect of SiC particle addition during FSW on microstructure and mechanical properties of AZ31 magnesium alloy. *J. Mater. Eng. Perform.* **24** (12): 5037–5045.

48 Abioye, T.E., Zuhailawati, H., Anasyida, A.S. et al. (2021). Enhancing the surface quality and tribomechanical properties of AA 6061-T6 friction stir welded joints reinforced with varying SiC Contents. *J. Mater. Eng. Perform.* 30: 4356–4369.

49 Suresh, S., Venkatesan, K., Natarajan, E. et al. (2021). Performance analysis of nano silicon carbide reinforced swept friction stir spot weld joint in AA6061-T6 alloy. *Silicon* 13: 3399–3412.

50 Singh, T. (2021). Processing of friction stir welded AA6061-T6 joints reinforced with nanoparticle. *Results Mater.* 12: 100210.

51 Abioye, T., Zuhailawati, H., Anasyida, A. et al. (2021). Effects of particulate reinforcements on the hardness, impact and tensile strengths of AA 6061-T6 friction stir weldments. *Proc. Inst. Mech. Eng., Part L: J. Mater.: Des. Appl.* 235 (6): 1500–1506.

52 Jamalian, H.M., Ramezani, H., Ghobadi, H. et al. (2016). Processing-structure-property correlation in nano-SiC-reinforced friction stir welded aluminum joints. *J. Manuf. Processes* **21**: 180–189.

53 Kumar, P.V., Reddy, G.M., and Rao, K.S. (2015). Microstructure and pitting corrosion of armor grade AA7075 aluminum alloy friction stir weld nugget zone – effect of post-weld heat treatment and addition of boron carbide. *Def. Technol.* **11** (2): 166–173.

54 Paidar, M., Asgari, A., Ojo, O.O., and Saberi, A. (2018). Mechanical properties and wear behavior of AA5182/WC nanocomposite fabricated by friction stir welding at different tool traverse speeds. *J. Mater. Eng. Perform.* **27** (4): 1714–1724.

55 Sun, Y.F. and Fujii, H. (2011). The effect of SiC particles on the microstructure and mechanical properties of friction stir welded pure copper joints. *Mater. Sci. Eng., A* **528** (16–17): 5470–5475.

56 Salih, O.S., Neate, N., Ou, H., and Sun, W. (2019). Influence of process parameters on the microstructural evolution and mechanical characterizations of friction stir welded Al-Mg-Si alloy. *J. Mater. Process. Technol.* **275**: 116366.

57 Singh, T., Tiwari, S.K., and Shukla, D.K. (2019). Friction-stir welding of AA6061-T6: The effects of Al_2O_3 nano-particles addition. *Results Mater.* 1: 100005.

58 Singh, T., Tiwari, S.K., and Shukla, D.K. (2020). Novel method of nanoparticle addition for friction stir welding of aluminium alloy. *Adv. Mater. Process. Technol.* 8 (1): 1160–1172.

59 Mehdi, H., Mehmood, A., Chinchkar, A. et al. (2022). Optimization of process parameters on the mechanical properties of AA6061/Al_2O_3 nanocomposites fabricated by multi-pass friction stir processing. *Mater. Today Proc.* 56 (Part 4): 1995–2003.

60 Kesharwani, R., Jha, K.K., Imam, M. et al. (2022). The optimization of the groove depth height in friction stir welding of AA 6061-T6 with Al_2O_3 powder particle reinforcement. *J. Mater. Res.* 37: 3743–3760.

61 Tabasi, M., Farahani, M., Givi, M.K.B. et al. (2016). Dissimilar friction stir welding of 7075 aluminum alloy to AZ31 magnesium alloy using SiC nanoparticles. *Int. J. Adv. Manuf. Technol.* **86** (1–4): 705–715.

62 Varun Kumar, A., Selvakumar, A.S., Balachandar, K. et al. (2021). Correlation between material properties and free vibration characteristics of TIG and laser welded stainless steel 304 reinforced with Al2O3 microparticles. *Eng. Sci. Technol. Int. J.* 24 (5): 1253–1261.

63 Singh, T., Tiwari, S.K., and Shukla, D.K. (2020). Mechanical and microstructural characterization of friction stir welded AA6061-T6 joints reinforced with nano-sized particles. *Mater. Charact.* 159: 110047.

64 El-Sayed Seleman, M.M., Ataya, S., Ahmed, M.M.Z. et al. (2022). The additive manufacturing of aluminum matrix nano Al_2O_3 composites produced via friction stir deposition using different initial material conditions. *Materials* 15: 2926.

65 Kesharwani, R., Jha, K.K., Anshari, A.A. et al. (2022). Comparison of microstructure, texture, and mechanical properties of the SQ and thread pin profile FSW joint of AA6061-T6 with Al_2O_3 particle reinforcement. *Mater. Today Commun.* 33: 104785.

66 Bahrami, M., Dehghani, K., and Givi, M.K.B. (2014). A novel approach to develop aluminum matrix nano-composite employing friction stir welding technique. *Mater. Des.* 53: 217–225.

67 Kartsonakis, I.A., Dragatogiannis, D.A., Koumoulos, E.P. et al. (2016). Corrosion behavior of dissimilar friction stir welded aluminum alloys reinforced with nano additives. *Mater. Des.* **102**: 56–67.

68 Moradi, M.M., Aval, H.J., and Jamaati, R. (2017). Effect of pre and post-welding heat treatment in SiC-fortified dissimilar AA6061-AA2024 FSW butt joint. *J. Manuf. Processes* **30**: 97–105.

69 Jain, S. and Mishra, R.S. (2022). Effect of Al_2O_3 nanoparticles on microstructure and mechanical properties of friction stir-welded dissimilar aluminum alloys AA7075-T6 and AA6061-T6. *Proc. Inst. Mech. Eng., Part E: J. Process Mech. Eng.* 236 (4): 1511–1521.

70 Jain, S. and Mishra, R. (2022). Influence of micro-sized Al_2O_3 particles on microstructure, mechanical, and wear behavior of dissimilar composite joint of AA6061/AA7075 by friction stir processing/welding. *Proc. Inst. Mech. Eng., Part C: J. Mech. Eng. Sci.* 236 (16): 9047–9060.

71 Jain, S. and Mishra, R.S. (2022). Microstructural and mechanical behavior of micro-sized SiC particles reinforced friction stir processed/welded AA7075 and AA6061. *Silicon* 14: 10741–10753.

72 Acharya, U., Yadava, M.K., Banik, A. et al. (2021). Effect of heat input on microstructure and mechanical properties of friction stir welded AA6092/17.5 SiCp-T6. *J. Mater. Eng. Perform.* 30: 8936–8946.

73 Kesharwani, R., Jha, K.K., Sarkar, C., and Imam, M. (2022). Numerical and experimental studies on friction stir welding of 6061-T6 AA with Al_2O_3 powder particle reinforcement. *Mater. Today Proc.* 56 (Part 2): 826–833.

74 Thomas, W.M., Nicholas, E.D., and Smith, S.D. (2001). Friction stir welding-tool developments. In: *Proceedings of Aluminum Automotive and Joining Sessions*, 213. Warrendale, PA: TMS.

75 Ogunsemi, B.T., Eta, O.M., Olanipekun, E. et al. (2022). Tensile strength prediction by regression analysis for pulverized glass waste-reinforced aluminum alloy 6061-T6 friction stir weldments. *Sādhanā* 47: 53. https://doi.org/10.1007/s12046-022-01830-5.

76 Kumar, M., Kumar, R., and Kore, S.D. (2022). Modeling and analysis of effect of tool geometry on temperature distribution and material flow in friction stir welding of AA6061-T6. *J. Braz. Soc. Mech. Sci. Eng.* 44: 153. https://doi.org/10.1007/s40430-022-03456-4.

77 Raja, N.D. and Prakash, D.S. (2020). Experimental investigation of hardness and effect of wear on sintered composites containing AA6061 matrix and tib$_2$/Al_2O_3 reinforcements. *IOP Conf. Ser.: Mater. Sci. Eng.* 988: 012104.

78 Raja, N.D. and Shankar, R.N. (2021). Optimization of friction STIR welded AA6061 + SiCp metal matrix composite to increase joint tensile strength and reduce defects. *Int. J. Simul. Multidisci. Des. Optim.* 12: 28.

79 Radhika Chada, N., Kumar, S., and Reddy, I.R. (2022). Investigation of microstructural characteristics of friction stir welded AA6061 joint with different particulate reinforcements addition. *AIP Conf.erence Proc.* 2418: 050010. https://doi.org/10.1063/5.0081797.

80 Abu-warda, N., López, M.D., González, B. et al. (2021). Precipitation hardening and corrosion behavior of friction stir welded A6005-TiB$_2$ nanocomposite. *Met. Mater. Int.* 27: 2867–2878. https://doi.org/10.1007/s12540-020-00688-8.

81 Mohapatra, D.K. and Mohanty, P.P. (2023). Parametric studies of dissimilar friction stir welded AA2024/AA6082 aluminium alloys. In: *Recent Advances in Mechanical Engineering*, Lecture Notes in Mechanical Engineering (ed. P. Pradhan, B. Pattanayak, H.C. Das, and P. Mahanta). Singapore: Springer https://doi.org/10.1007/978-981-16-9057-0_21.

82 Ali, M.H., Wadallah, H.M., Ibrahim, M.A. et al. (2021). Improving the microstructure and mechanical properties of aluminium alloys joints by adding SiC particles during friction stir welding process. *Metall. Microstruct. Anal.* 10: 302–313. https://doi.org/10.1007/s13632-021-00743-9.

83 Nikoo, M.F., Parvin, N., and Bahrami, M. (2017). Al$_2$O$_3$-fortified AA6061-T6 joint produced via friction stir welding: the effects of traveling speed on microstructure, mechanical, and wear properties. *Proc. Inst. Mech. Eng. Part L. J. Mater. Des. Appl.* **231** (6): 534–543.

84 Bodaghi, M. and Dehghani, K. (2017). Friction stir welding of AA5052: the effects of SiC nano-particles addition. *Int. J. Adv. Manuf. Technol.* **88** (9–12): 2651–2660.

85 Babu, N.K., Kallip, K., Leparoux, M. et al. (2016). Characterization of microstructure and mechanical properties of friction stir welded AlMg5-Al$_2$O$_3$ nanocomposites. *Mater. Sci. Eng., A* **658**: 109–122.

86 Naghibi, H.Y., Omidvar, H., and Nikoo, M.F. (2018). Investigating the effects of traveling speeds and post-weld heat treatment on mechanical properties of nano-Al$_2$O$_3$-fortified AA6061-T6 friction stir welds. *Proc. Inst. Mech. Eng. Part L. J. Mater. Des. Appl.* **232** (10): 816–828.

87 Shahedi, B., Damircheli, M., and Shirazi, A. (2019). Experimental investigation of the effects of welding parameters and TiO$_2$ nanoparticles addition on FSWed copper sheets. *Mater. Res. Express* **6** (2).

88 Gallais, C., Simar, A., Fabregue, D. et al. (2007). Multiscale analysis of the strength and ductility of AA 6056 aluminum friction stir welds. *Metall. Mater. Trans. A* **38A**: 964–981.

89 Saeidi, M., Givi, M.K.B., and Faraji, G. (2016). Study on ultrafine-grained aluminum matrix nanocomposite joint fabricated by friction stir welding. *Indian J. Eng. Mater. Sci.* **23** (2–3): 152–158.

90 Saeidi, M., Behnagh, R.A., Manafi, B. et al. (2016). Study on ultrafine-grained aluminum matrix nanocomposite joint fabricated by friction stir welding. *Proc. Inst. Mech. Eng. Part L. J. Mater. Des. Appl.* **230** (1): 311–318.

91 Miracle, D.B. and Maruyama, B. (2001). Metal matrix composites for space systems: current uses and future opportunities. *JOM* **53** (4): 14–17.

92 Mirjavadi, S.S., Alipour, M., Emamian, S. et al. (2017). Influence of TiO$_2$ nanoparticles incorporation to friction stir welded 5083 aluminum alloy on the microstructure, mechanical properties, and wear resistance. *J. Alloys Compd.* **712**: 795–803.

93 Ahn, B.W., Choi, D.H., Kim, Y.H., and Jung, S.B. (2012). Fabrication of SiCp/AA5083 composite via friction stir welding. *Met. Soc. China.* **22**: s634–s638.

94 Pantelis, D.I., Karakizis, P.N., Daniolos, N.M. et al. (2015). Dissimilar friction stir welding of AA5083 with AA6082 reinforced with SiC particles. *Mater. Manuf. Processes* **31** (3): 150312112220005.

95 Sato, Y.S., Kokawa, H., Enomoto, M., and Jogan, S. (1999). Microstructural evolution of 6063 aluminum during friction-stir welding. *Metall. Mater. Trans. A* **30** (9): 2429–2437.

96 Abdolahzadeh, A., Omidvar, H., Safarkhanian, M.A., and Bahrami, M. (2014). Studying microstructure and mechanical properties of SiC-incorporated AZ31 joints fabricated through FSW: the effects of rotational and traveling speeds. *Int. J. Adv. Manuf. Technol.* **75** (5–8): 1189–1196.

97 Nikoo, M.F., Azizi, H., Parvin, N., and Naghibi, H.Y. (2016). The influence of heat treatment on microstructure and wear properties of friction stir welded AA6061-T6/Al$_2$O$_3$ nanocomposite joint at four different traveling speeds. *J. Manuf. Processes* **22**: 90–98.

98 Liu, S., Paidar, M., Mehrez, S. et al. (2022). Development of AA6061/316 stainless steel surface composites via friction stir processing: effect of tool rotational speed. *Mater. Charact.* 192: 112215.

99 Hassanifard, S., Alipour, H., Ghiasvand, A., and Varvani-Farahani, A. (2021). Fatigue response of friction stir welded joints of Al 6061 in the absence and presence of inserted copper foils in the butt weld. *J. Manuf. Processes* 64: 1–9.

100 Hassanifard, S., Ghiasvand, A., and Varvani-Farahani, A. (2022). Fatigue response of aluminum 7075-T6 joints through inclusion of Al$_2$O$_3$ particles to the weld nugget zone during friction stir spot welding. *J. Mater. Eng. Perform.* 31: 1781–1790.

101 Vuherer, T., Milčić, M., Glodež, S. et al. (2021). Fatigue and fracture behaviour of friction stir welded AA-2024-T351 joints. *Theor. Appl. Fract. Mech.* 114: 103027.

102 Tayebi, P., Fazli, A., Asadi, P. et al. (2021). Formability study and metallurgical properties analysis of FSWed AA 6061 blank by the SPIF process. *SN Appl. Sci.* 3: 367.

103 Zhu, N., Avery, D.Z., Rutherford, B.A. et al. (2021). The effect of anodization on the mechanical properties of AA6061 produced by additive friction stir-deposition. *Metals* 11 (11): 1773.

104 Balajikrishnabharathi, A., Suyamburajan, V., Kaliappan, S. et al. (2022). Study of mechanical properties on Nano metal matrix Composites: Duralcan process. *Mater. Today Proc.* 62 (Part 2): 1282–1287.

105 Bahrami, M., Nikoo, M.F., and Givi, M.K.B. (2015). Microstructural and mechanical behaviors of nano-SiC-reinforced AA7075-O FSW joints prepared through two passes. *Mater. Sci. Eng., A* **626**: 220–228.

106 Dragatogiannis, D.A., Koumoulos, E.P., Kartsonakis, I.A. et al. (2016). Dissimilar friction stir welding between 5083 and 6082 Al alloys reinforced with TiC nanoparticles. *Mater. Manuf. Processes* 31 (16): 2101–2114.

107 Singh, T., Tiwari, S.K., and Shukla, D.K. (2020). Preparation of aluminum alloy-based nanocomposites via friction stir welding. *Mater. Today Proc.* 27 (3): 2562–2568.

108 Garg, A. and Bhattacharya, A. (2019). Influence of Cu powder on strength, failure and metallurgical characterization of single, double pass friction stir welded AA6061-AA7075 joints. *Mater. Sci. Eng., A* 661–679.

109 Fernández, C.M.A., Rey, R.A., Ortega, M.J.C. et al. (2018). Friction stir processing strategies to develop a surface composite layer on AA6061-T6. *Mater. Manuf. Processes* 33 (10): 1–8.

110 Wang, G., Zhao, Y., and Tang, Y. (2020). Research progress of bobbin tool friction stir welding of aluminum alloys: a review. *Acta Metall. Sin.* 33: 13–29. https://doi.org/10.1007/s40195-019-00946-8.

111 Zeng, X.H., Xue, P., Wu, L.H. et al. (2019). Microstructural evolution of aluminum alloy during friction stir welding under different tool rotation rates and cooling conditions. *J. Mater. Sci. Technol.* 35: 972–981.

112 Leon, J.S. and V. (2014). Jayakumar Investigation of mechanical properties of aluminum 6061 alloy friction stir welding. *Int. J. Stud. Res. Technol. Manag.* 2 (4): 140–144.

113 Gomathisankar, M., Gangatharan, M., and Pitchipoo, P. (2018). A novel optimization of friction stir welding process parameters on aluminum alloy 6061-T6. *Mater. Today* 5 (6): 14397–14404.

114 Khojastehnezhad, V.M. and Pourasl, H.H. (2018). Microstructural characterization and mechanical properties of aluminum 6061-T6 plates welded with copper insert plate (Al/Cu/Al) using friction stir welding. *Trans. Nonferrous Met. Soc. China* 28: 415–426.

115 Periyasamy, P., Mohan, B., and V. (2012). Balasubramanian effect of heat input on mechanical and metallurgical properties of friction stir welded AA6061-10% SiCp MMCs. *J. Mater. Eng. Perform.* 21: 2417–2428.

116 Asadollahi, M. and Khalkhali, A. (2018). Optimization of mechanical and microstructural properties of friction stir spot welded AA 6061-T6 reinforced with SiC nanoparticles. *Mater. Res. Express* 5: https://doi.org/10.1088/2053-1591/aadc3a.

117 Asadi, P., Besharati Givi, M.K., Abrinia, K. et al. (2011). Effects of SiC particle size and process parameters on the microstructure and hardness of AZ91/SiC composite layer fabricated by FSP. *J. Mater. Eng. Perform.* 20: 1554–1562.

118 Abioye, T.E., Zuhailawati, H., Anasyida, A.S. et al. (2019). Investigation of the microstructure, mechanical and wear properties of AA6061-T6 friction stir weldments with different particulate reinforcements addition. *J. Mater. Res. Technol.* 8 (5): 3917–3928.

119 Bassani, P., Capello, E., Colombo, D. et al. (2007). Effect of process parameters on bead properties of A359/SiC MMCs welded by laser. *Compos. Part A Appl. Sci. Manuf.* 38 (4): 1089–1098.

120 Karakizis, P.N., Pantelis, D.I., Fourlaris, G., and Tsakiridis, P. (2018). Effect of SiC and TiC nanoparticle reinforcement on the microstructure, microhardness, and tensile performance of AA6082-T6 friction stir welds. *Int. J. Adv. Manuf. Technol.* 95 (9–12): 3823–3837.

121 Karakizis, P.N., Pantelis, D.I., Fourlaris, G., and Tsakiridis, P. (2018). The role of SiC and TiC nanoparticle reinforcement on AA5083-H111 friction stir welds studied by electron microscopy and mechanical testing. *Int. J. Adv. Manuf. Technol.* 94 (9–12): 4159–4176.

122 Singh, T., Tiwari, S.K., and Shukla, D.K. (2019). Effect of nano-sized particles on grain structure and mechanical behavior of friction stir welded Al-nanocomposites. *Proc. Inst. Mech. Eng. Part L. J. Mater. Des. Appl.* 234 (2): 274–290.

123 Singh, T., Tiwari, S.K., and Shukla, D.K. (2019). Processing parameters optimization to produce nanocomposite using friction stir welding. *Eng. Res. Express* 1: 025048.

124 Singh, T., Tiwari, S.K., and Shukla, D.K. (2019). Production of AA6061-T6/Al_2O_3 reinforced nanocomposite using friction stir welding. *Eng. Res. Express* 1: 025052.

125 Singh, T. (2023). Nanoparticle reinforced joints produced using friction stir welding: a review. *Eng. Res. Express* 5 (2): 022001.

13

Summary of Efforts in Manufacturing of Sandwich Sheets by Various Joining Methods Including Solid-State Joining Method

Divya Sachan[1], R. Ganesh Narayanan[1], and Arshad N. Siddiquee[2]

[1] Department of Mechanical Engineering, Indian Institute of Technology Guwahati, Guwahati, India
[2] Department of Mechanical Engineering, Jamia Millia Islamia, New Delhi, India

13.1 Introduction

There is need of advanced materials which facilitate better fuel economy in modern transportation systems without compromising safety and performance. Replacing components which are made from heavier conventional ferrous alloys with components made from high specific strength materials such as high-strength steel, alloys of magnesium, and aluminum- and fiber-based composites enables significant weight reduction in vehicles and thermal management across walls. Reduction in fuel consumption is also feasible [1]. Sandwich sheet structures are found to be a good alternative solution for many engineering applications. Apart from lightweight, sandwich sheet structures also offer excellent insulation for heat and noise as well as excellent energy absorption capacity [2]. Sandwich sheets are widely used in many industries including space, aircraft, land transportation vehicles, ships, wind energy systems, and buildings [3]. They also hold the same stiffness and strength with their lightweight structures. It has been observed that weight savings to the tune of 60–70% can be obtained by employing viscoelastic material core material-based sandwich panels [4]. Solid-state joining process produces lightweight sandwich panels for aerospace applications [5]. Satheeshkumar et al. [6] reviewed several sustainable forming and joining processes such as multitool forming, solid-state joining, and hybrid joining useful for manufacturing sandwich structures of different shapes and sizes.

13.2 Sandwich Sheets

Sandwich sheet structures consist of core material of low density between two skin layers. The excellent bending stiffness-to-weight ratio of sandwich panel is the source of their increasing popularity. The core of the sandwich panels/structure can be made from a number of materials including polymeric or metallic foams, wood, cork agglomerates, and ceramics. In the sandwich structures, the in-service condition mainly induce compressive, bending and shear loading in the core.

Friction Stir Welding and Processing: Fundamentals to Advancements, First Edition.
Edited by Sandeep Rathee, Manu Srivastava, and J. Paulo Davim.
© 2024 John Wiley & Sons, Inc. Published 2024 by John Wiley & Sons, Inc.

The sandwich structures as primarily designed with an objective weight minimization by selecting suitable core material, spatial distribution of matter, or both, and at the same time maintain structural integrity under the applied loads [7]. In Boeing 707, 8% of the wetted surface is honeycomb sandwich, while in Boeing 757/767, 46% of the wetted surface is honeycomb sandwich structures. Fuselage shell of Boeing 747 comprises of Nomex honeycomb sandwich; its floors/side panels/overhead bins/ceiling are all made from sandwich panels. Sandwich structures also have a wide range of applications in satellites, automobile, aircraft, cars, porta cabins, wind energy systems, etc. [8].

13.3 Classification of Sandwich Sheet Structures

Sandwich structure consists of two high stiffness face sheets capable of taking axial and compressive loads, bending moments, in-plane shears, and in between a low-density core capable of bearing flexural shears [1]. On the basis of face sheet material, the sandwich sheet can be classified as composite and non-composite. The sandwich panels with various core designs comprise of honeycomb, folded, corrugated, egg box, or lattice cores, etc. The sandwich structures having cellular foam cores are classified as stochastic and periodic structures, which are further categorized as foam cores, 2D cores, and 3D cores [9].

Composite sandwich sheet structures include mainly glass fiber reinforced composite (GFRP) and carbon fiber reinforced composite (CFRP). Fiber reinforced polymer composite possess better strength-to-weight ratio as compared to metallic face sheet. Due to high strength-to-weight ratio of GFRP sandwich panels, these find large applications in aerospace, marine, wind turbines, and bridges; other characteristics of these panels include very high resistance to corrosion and flexibility of customized designs. Figure 13.1 shows the GFRP face sheets and paulownia-web core (GFPW) panel used as floor panels during construction of building [10].

Non-composite sandwich structure includes aluminum, metal, plastic, etc., as the face sheet. Aluminum foam sandwich (AFS) sheets have a wide range of applications. Instead of using conventional structure for the want of higher strength, it may be better to increase the strength of Al/Polypropylene/Al sandwich structure by increasing the thickness of polypropylene sheet [11].

Figure 13.1 Nanjing Foshou lake building with GFPW panels as floor panels during construction. Source: Used with permission from Zhu et al. [10]/Elsevier.

13.4 Applications

13.4.1 High-Speed Train Nose

Finite element analysis was conducted using LS-DYNA for numerical simulation of frontal collision of the nose of high-speed train by varying the layer thickness of honeycomb and nose lengths. It was concluded that the AA5052-H38 alloys offered better performance during simulated frontal collision of these trains, as it absorbed 35% higher impact energy to an empty shell [12]. Investigation of four types of core materials, namely polyethylene terephthalate (PET), poly vinyl chloride (PVC), styrene acrylonitrile (SAN), polyurethane (PUR); skins made from two self-extinguishing polyester resins as matrix, and two types of glass reinforcement were employed to develop sandwich structures for use in rapid transport trains. In the peel test, about 92% of maximum strength of PET core made by vacuum bagging process in comparison core comprising of SAN and PET prepared by prepreg technology was observed. PET foam Aires T90 is found to be ideal for concrete transport applications [13].

13.4.2 Marine

Helsinki University of Technology conducted 10 years research on sandwich structures and found that laser welded sandwich panels offer 30–50% weight reduction as compared to the conventional steel structure. They also provide resistance to fire and blast. Improved fatigue strength and impact strength were observed after redesigning of hoistable car deck by introducing sandwich sheet structures instead of traditional paneled structure [14]. In marine industry, sandwich panels are widely used in the construction of primary hull shell, appendages, and deck housings. Sandwich panel with PVC as core material and glass fiber/epoxy or carbon fiber/epoxy as face sheets were subjected to water impact slamming and different pressures and moments were observed. Therefore, applied loads should be treated as nonuniform while designing the sandwich sheet structures [15].

13.4.3 Wind Turbine Blades

It is reported that at present mainly GFRP composites are used. In future, the CFRPs can be used for very large blades of wind turbines as sandwich structures will provide better lightweight structures with additional buckling capacity. This will also reduce the buckling of spar flange in compression during extreme wind conditions [16]. Siemens wind power manufactured wind turbine blades of sandwich sheet structure with face sheets of GFRP laminates and core of balsa wood by unique manufacturing technique called "integral blade." Wrinkle defects were tested under in-plane compression loading during which the structure was found to fail under three different modes. Inter fiber/Inter laminar failure criterion is capable of predicting initiation of layer-wise delamination [17].

13.4.4 Aerospace

The sandwich panels are largely employed in the aviation section, specifically in the fairings and floor panels of passenger compartment. The honeycomb composite sandwich are largely employed in Airbus A380 in floor panels, nose gear doors, flap track fairings, and landing gear doors with carbon/glass epoxy as skin sheets [18]. To fulfil the requirements at various locations of aircraft, a variety of face sheet materials and the core materials are utilized. For face sheet, mostly glass and carbon fiber prepregs with epoxy resin matrix are used with core material NOMEX honeycomb. The ongoing research is more focused on the application of CFRP sandwich structures [19]. The aircraft interior is mostly made up of composite sandwich structure and with the growing demand of air travel growth by 4.4% by 2036, there will be need of partial replacement on yearly basis and overall refurbishments in every eight years of sandwich panels. Hence, there is a need of automated production technology for the manufacturing of sandwich panels. A technology for small batch-type production of honeycomb sandwich panels suitable for aircraft interiors has been proposed. This concept reduces cost, time, and improves quality [20].

13.4.5 Ship Building

Ship performance can be increased by reducing ship weight and increasing corrosion and fatigue resistance. This can be achieved by implementing sandwich panels in the ship structures. IHC Holland dredgers started applying steel sandwich panels and it resulted in the reduction of structural weight by 39%. In ship structure, sandwich panels are mainly applied in the trailing suction hopper dredgers [21]. Sandwich plate system has been widely used for ship building and ship repairing. It can be used as an alternative to the conventional stiffened plate construction with a rough estimation of overall weight reduction from 10% to 70%. SPS face plate includes steel or aluminum and core is mainly of PUR elastomer, which makes the structure less complex and flush [22].

13.5 Fabrication Methods

The sandwich structures are conventionally and most commonly fabricated via adhesive bonding, layup, autoclave, and liquid molding routes (resin transfer molding, i.e. RTM). Conventional welding techniques using heat sources like arc, laser, and sheet resistance cannot be used to join sandwich sheets of steel and polymer due to their peculiar nature [23]. Welding aluminum foam sandwich by TIG, MIG, and laser welding is not suitable as metal foam core is sensitive to increase in temperature and providing suitable process parameters in these cases is practically difficult [24]. Solid-state welding methods like friction stir welding and ultrasonic welding are better alternatives for fabrication of sandwich sheet structures.

13.5.1 Lay-Up Method

This method is one of the simplest, oldest, and cheapest methods for manufacturing of sandwich sheet structures. Sandwich panels with core made from PUR foam glass epoxy skin as face sheets are prepared by hand lay-up technique. A mixture of epoxy and hardener is mixed in a ratio of 100 : 54 for about two minutes and after that it is spread with the help of wide brush and hand roller. Large variations in strength are observed due to surface irregularities, contaminants, and entrapment of air bubbles [25]. Sandwich structure employing core made of medium or low-density fiber board, fiber composite laminate employed in the face sheet can be produced by hand lay-up technique. The performance of these sandwich structure is better because debonding strength laminated/MDF core sandwich structure is found to be more than that of the core itself [26]. The hand lay-up method, however, suffers from several limitations including low production volume, labor-intensive and consequently employed for the fabrication of large components, such as swimming pool, playhouse structures, and boat hulls.

13.5.2 Prepreg Method

Prepegs are used to construct sandwich structures meant for applications in yachts and in aerospace sector. Merits of prepegs include well-impregnated reinforcement and use of resins with better properties as compared to wet layup. Layers of preimpregnated (prepreg) carbon fibers and Nomex honeycomb are laid up by hand, and after that, cured in an autoclave. To reduce the overall capital cost, autoclave is replaced with vacuum-only oven cure, which will help in elimination of batch components for cure resulting in improved work flow and efficiency [27].

13.5.3 Adhesive Bonding

The aforementioned methods of manufacturing sandwich sheets are one step processes; however, not reliable always. In several situations, the skin and core are bonded in separate manufacturing step like in adhesive bonding. Adhesive layers are applied between the face sheets and the core, and the assembly is treated at elevated temperature and pressure. After that, the sandwich is cooled. To improve the structural performance, the bonding takes place in a vacuum or in an autoclave. This fabrication method is mostly used in the aircraft and spacecraft industry where high stiffness is required with minimum weight [28]. In this method, an adhesive is placed between the face sheet and the core, mainly porous cellular core, pressure is applied on the assembly of face sheet and core, and subsequently the adhesive is cured. Wastage of adhesive is minimized by the development of film adhesives. Ceramic and steel sheets were bonded to honeycomb-core structures using an adhesive film of polyolefin with a slit thermally sensitive pattern to increase the bonding strength. No debonding was observed during three-point bend tests [29]. This method is time consuming and costly due to large curing time and extensive surface preparation [30].

13.5.4 Solid-State Joining Methods

Fabrication of sandwich sheet structures by direct fusion welding is not suitable because the face sheet and core materials are chemically, structurally, and physically incompatible. CO_2 laser welding of viscoelastic core material and steel face sheet caused evaporation of the core layer [31]. In pulsed laser welding due to high cooling rate, cracks were observed in the joint during joining of polymeric core sandwich sheet structure [32]. Diffusion bonding has been employed for the bonding of titanium alloy used in light sandwich panels for high-temperature applications [33]. Due to the numerous disadvantages in the conventional welding for the fabrication of sandwich sheet structures, solid-state welding methods are good alternatives.

13.5.4.1 Friction Stir Welding (FSW)/Friction Stir Spot Welding (FSSW)

Feasibility study on FSW of two skin steel and polymer core layer by using pin and pin-less tool was done by Buffa et al. [2]. Polymer burning was observed due to the heat and complex plastic flow when FSW/FSSW is performed by tool with pin. Weld defects were observed and optimized process parameters are required to avoid weld defects. Figure 13.2 shows the stress–strain behavior of

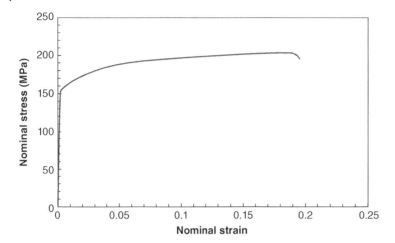

Figure 13.2 Stress–strain plot of metal–polymer–metal sandwich sheet structures. Source: Reproduced with permission from Buffa et al. [2]/Elsevier.

metal–polymer–metal sandwich composite, which is characterized by increasing stress with strain until failure followed by slight drop was seen [2].

AFS prepared through FSW are found to possess several advantages such as high flexural and impact strengths and better acoustic damping in comparison to those fabricated via adhesive or prepreg routes. Joining of face sheet to aluminum foam by FSW/FSSW mainly occurs due to plasticized material flow. Figure 13.3 depicts transverse section detailing morphology of AFS and it demonstrates that such panels have good formability at interface which is free from holes and cracks. FSW led to the grain refinement of face sheets, and therefore, improves crucial mechanical properties [34].

FSSW of AA6082-T6/HDPE/AA6082-T6/HDPE/AA6082-T6 sandwich sheets with different pin profiles and shoulder surface was studied by Ravi et al. [35]. Tool with square pin profile result in higher interface bond area, significantly lower the size of hook defect and enhance the joint strength among all the pin profiles. During lap shear test, nugget-pull out failure was observed [35]. When FSSW is employed for face sheets made from AA5052-H32/HDPE/AA5052-H32, a tool rpm of 1800 rpm and above is found to produce sound sandwich sheet structures [36]. The morphology of the hooks, which are formed in pairs, in the sandwich leads to improvement in mechanical performance when the welding is performed at higher rotational speeds. The sandwich sheet failure was found to occur in two modes: partial nugget failure and nugget pull-out failure. Figure 13.4 shows that sandwich sheet consists of fine grains as compared to bimetallic sheets when formed at high speed in the range of 1200–1800 rpm, which is attributed to relatively lower peak temperature in the former case [36]. The study of FSW of AFS combines the welding of precursor skin sheets

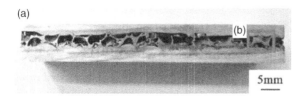

Figure 13.3 Morphology of AFS [(a) macrograph, (b) region of importance]. Source: Used with permission from Peng et.al. [34].

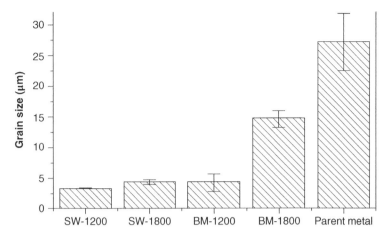

Figure 13.4 Grain size measurement of sandwich sheet and bimetallic sheets [SW: sandwich sheet; BM: Bimetallic sheet]. Source: Reproduced with permission from Rana et al. [36]/Elsevier.

by FSW and the precursor foaming together for better interface characteristics in the welded AFS [37]. FSW comes out to be an effective and economical method for producing large-size AFS in comparison to limitation of size of AFS by powder metallurgy method. Core shear is the mode of failure. FSW results in metallurgical bonding and three-point bending shows that FSW does not affect the bending strength of welded AFS.

13.5.4.2 Accumulative Roll Bonding
Accumulated roll bonding (ARB) is a severe plastic deformation process, under which the stacked sheets are subjected to multiple rolling passes operations (Figure 13.5). The plastic deformation during rolling develops progressing interface bonding. When the ARB is performed at room temperature, it is known as cold ARB and that at above recrystallization, it is termed as hot ARB. During ARB, every pass is aimed at predetermined level of reduction in the thickness reduction. Sandwich structures and multicomponent materials with graded and tailored properties can be made by ARB. Formation of ultrarefined grains with strength improvement is one of the merits of the process [38]. Production of aluminum alloy sandwich structures at room temperature and at higher temperatures are possible as done by Hausol et al. [39, 40]. Fabrication of multicomponent structures with fibers, particles, and foils is also possible [41].

13.5.4.3 Diffusion Bonding
In AFS panels manufactured by diffusion bonding, aluminized steel sheets facilitate the bonding process and metallurgical bonding is observed between the aluminum foam core and steel sheets. Hot pressing temperatures, holding time, and applied pressure are the important process parameters in this technique [42]. AFS panel was successfully fabricated by diffusion welding method by Lin et al. [42]. The comparison between the diffusion welding and adhesive-bonded AFS panels reveals that the metallurgical interface joints provide better fatigue strength and peel strength leading to a longer fatigue life. Moreover, these AFS panels possess good mechanical properties because the interface of core and face sheet are compact and continuous. It was observed that under the same load, fatigue life of AFS with metallurgical interface joints is significantly higher than that of adhesive bonded sample [43]. Intermetallic compounds, $Al_{20}CaTi_2$ and Al_2Ti, were formed during fabrication of titanium sheet enhanced AFS by transient liquid-phase diffusion bonding.

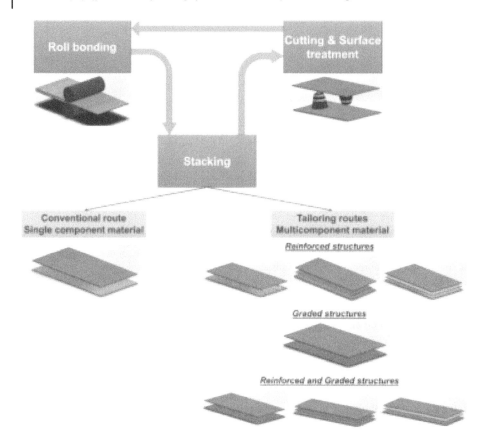

Figure 13.5 Schematic of ARB process. Source: Reproduce with permission from Ebrahimi and Wang [38]/ Elsevier.

Thickness of the interface layer increases as the holding time is increased leading to excellent properties such as low density, better thermal insulation, and high energy absorption and specific strength [44].

13.6 Summary

The present article attempts to summarize the efforts in manufacturing sandwich structures in applications such as high-speed train nose, marine, aerospace, and ship building.. Depending upon the core and face sheets structure, the manufacturing technique of sandwich structure is decided. Joining of AFS by TIG, MIG, or by laser welding is difficult and to identify suitable process parameters is practically impossible as core layer is sensitive to the increasing temperature. In brazing, brittle intermetallic compounds are formed due to the interfacial reaction between AF and face sheets. Also, the sandwich panels formed by adhesive cannot be used at higher tempera-ture condition. The conventional manufacturing method of sandwich structure such as lay-up and adhesive bonding can be replaced with the solid-state joining techniques like FSW/FSSW and diffusion bonding. FSW fabricates the joint in solid state which has a number of advantages such as low residual strains, energy efficient, and environmentally friendly.

AFS prepared by FSW interface joining has better flexural and impact strengths and higher acoustic damping in comparison with those prepared via conventional routes such as adhesive or brazing. FSW led to the grain refinement and improves mechanical properties. AFS panels produced by solid-state joining method results in metallurgical bonding with improvised flexural and fatigue strength. Joining of polymers is a difficult process due to different rheological and physical differences in the polymer. Rolling direction of skin sheets, double-stage FSW, hybrid manufacturing methods such as adhesive bonding + FSW, are some ways to improve the feasibility of solid-state joining of sandwich sheets.

References

1 Patel, M., Pardhi, B., Chopara, S., Pal, M (2018). Lightweight composite materials for automotive – a review. *Int. Res. J. Eng. Technol.* 5: 41–47.

2 Buffa, G, Campanella, D, Forcellese, A, Fratini, L, Simoncini, M 2020, 'Solid state joining of thin hybrid sandwiches made of steel and polymer: a feasibility study', *Procedia Manuf.*, vol. 47, pp. 400–405, DOI: https://doi.org/10.1016/j.promfg.2020.04.315.

3 Karlsson, KF, Tomas Astrom, B 1997, 'Manufacturing and applications of structural sandwich components', *Composites, Part A*, vol. 28, pp. 97-111, DOI:https://doi.org/10.1016/S1359-835X(96)00098-X.

4 Hara, D, Ozgen, G 2016, 'Investigation of weight reduction of automotive body structures with the use of sandwich materials', *Transp. Res. Procedia*, vol. 14, pp. 1013 -1020, DOI: https://doi.org/10.1016/j.trpro.2016.05.081.

5 Lee, H.S., Yoon, J.H., and Yoo, J.T. (2020). Application of solid state joining technologies in aerospace parts. *Key Eng. Mater.* 837: 69–73. https://doi.org/10.4028/www.scientific.net/kem.837.69.

6 Satheeshkumar, V., Narayanan, R.G., and Gunasekera, J.S. (2023). Sustainable manufacturing: material forming and joining. In: *Sustainable Manufacturing Processes*, 53–112. Academic Press. ISBN: 978-0-323-99990-8.

7 Xiong, J, Du, Y, Mousanezhad, D, Asl, M, Narato, J, Vaziri, A 2018, 'Sandwich structures with prismatic and foam cores – a Review', *Adv. Eng. Mater.*, vol. 21, pp. 1-19, DOI:https://doi.org/10.1002/adem.201800036.

8 Vinson, J 2005'Sandwich structures: past, present, and future', *Adv. Sandwich Struct. Mater.*, vol. 7, pp. 3-12, DOI: https://doi.org/10.1007/1-4020-3848-8_1.

9 Sivaram, A, Manikandan, N, Krishnakumar, S, Rajavel, R, Krishnamohan, S, Vijayaganth, G 2020, 'Experimental study on aluminium based sandwich composite with polypropylene foam sheet', *Mater. Today Proc.*, vol. 24, pp. 746–753, DOI:https://doi.org/10.1016/j.matpr.2020.04.331.

10 Zhu, D, Shi, H, Fang, H, Liu, W, Qi, Y, Bai, Y 2018, 'Fiber reinforced composites sandwich panels with web reinforced wood core for building floor applications', *Composites Part B*, vol. 150, pp. 196-211, DOI:https://doi.org/10.1016/j.compositesb.2018.05.048.

11 Ma, Q, Rejab, M, Siregar, J, Guan, Z 2021, 'A review of the recent trends on core structures and impact response of sandwich panels', *J. Compos. Mater.*, vol. 55, pp. 1-43, DOI:https://doi.org/10.1177/0021998321990734.

12 Amraei, M, Shahravi, M, Noori, Z, Lenjani, A 2013, 'Application of aluminium honeycomb sandwich panel as an energy absorber of high-speed train nose', *J. Compos. Mater.*, vol.48, pp. 1–11, DOI: https://doi.org/10.1177/0021998313482019.

13 Rusnakova, S, Zaludek, M, Fojtl, L, Rusnak, V 2014, 'Design and verification of sandwich structures for high speed trains', *Key Eng. Mater.*, vol. 586, pp. 72-75, DOI: https://doi.org/10.4028/www.scientific.net/KEM.586.72.

14 Kujala, P. and Klanac, A. (2005). Steel sandwich panels in marine applications. *Brodogranja* 56: 305–314. https://hrcak.srce.hr/457.

15 Battley, M.A. and Allen, T. (2011). Dynamic performance of marine sandwich panel structures. In: *18th International Conference on Composite Materials (ICCM-18)*. Korea: Jeju.

16 Thomsen, O 2009, 'Sandwich materials for wind turbine blades – present and future', *J. Sandwich Struct. Mater.*, vol.11, pp. 7-26, DOI: https://doi.org/10.1177/1099636208099710.

17 Leong, M, Lars, C.T, Overgaard, Thomson, et al. 2012, 'Investigation of failure mode mechanisms in GFRP sandwich structures with face sheet wrinkle defects used for wind turbine blades', *Compos. Struct.*, vol. 94, pp. 768-778, DOI:https://doi.org/10.1016/j.compstruct.2011.09.012.

18 Castanie, B, Bouvet, C, Ginot, M 2020, 'Review of composite sandwich structure in aeronautic applications', *Composites, Part C: Open Access*, vol. 1, pp. 1-23, DOI: https://doi.org/10.1016/j.jcomc.2020.100004.

19 Herrmann, A, Zahlen, P, Zuardy, I 2005, 'Sandwich structures technology in commercial aviation – present applications and future trends', *Adv. Sandwich Struct. Mater.*, vol. 7, pp. 13-26, DOI:https://doi.org/10.1007/1-4020-3848-8_2.

20 Eschen, H, Harnisch, M, Schuppstuhl, T 2018, 'Flexible and automated production of sandwich panels for aircraft interiors', *Procedia Manuf.*, vol. 18, pp. 35-42, DOI:https://doi.org/10.1016/j.promfg.2018.11.005.

21 Kortenoeven, J, Boon, B, Bruijn, A 2008, 'Application of sandwich panels in design and building of dredging ships', *J. Ship Prod.*, vol. 24, pp. 125-134, DOI: https://doi.org/10.5957/jsp.2008.24.3.125.

22 Ramakrishnan, K.V. and Sunil, P. (2016). Applications of sandwich plate system for ship structures. *IOSR J. Mechan. Civil Eng.* 83–90.

23 Rana, P.K., Narayanan, R.G., and Kailas, S.V. (2021). Assessing the dwell time effect during friction stir spot welding of aluminum polyethylene multilayer sheets by experiments and numerical simulations. *Int. J. Adv. Manuf. Technol.* 114: 1953–1973. https://doi.org/10.1007/s00170-021-06910-0.

24 Urso, D, Maccarini G 2012, 'The formability of aluminium foam sandwich panels', *Int. J. Mater. Form.*, vol. 5, pp. 243-257, DOI:https://doi.org/10.1007/s12289-011-1036-9.

25 Krzyzak, A, Mazur, M, Gajewski, M, Drozd, K, Komorek, A, Przybylek, P 2016, 'Sandwich structured composites for aeronautics: methods of manufacturing affecting some mechanical properties', *Int. J. Aerosp. Eng.*, vol. 2016, pp. 1-10, DOI: https://doi.org/10.1155/2016/7816912.

26 Hassan, M 2017, 'Characterization of face sheet/core debonding strength in sandwiched medium density fiberboard', *Mater. Sci. Appl.*, vol. 8, pp. 673-684, DOI: https://doi.org/10.4236/msa.2017.89048.

27 Crump, DA, Dulieu-Barton, JM 2010, 'The manufacturing procedure for aerospace secondary sandwich structure panels', *J. Sandwich Struct. Mater.*, vol. 12, pp. 421-447, DOI: https://doi.org/10.1177/1099636209104531.

28 Zukas, F.J., Stafford, J.P., and Condon, N.J. (1996). Preparation of adhesively bonded sandwich structures. U.S. Patent 5,944,935.

29 Pereira, AB, Fernandes, FAO 2019, 'Sandwich panels bond with advanced adhesive films', *J. Compos. Sci.*, vol. 3, pp. 1-15, DOI:https://doi.org/10.3390/jcs3030079.

30 Amancio-Filho, ST, Santos, JF 2009, 'Joining of polymers and polymer–metal hybrid structures: recent developments and trends', *Polym. Eng. Sci.*, vol. 49, pp. 1461–1476, DOI: https://doi.org/10.1002/pen.21424.

31 Salonitis, K, Stavropoulos, P, Fysikopoulos, A, Chryssolouris, G 2013, 'CO$_2$ laser butt-welding of steel sandwich sheet composites', *Int. J. Adv. Manuf. Technol.*, vol. 69, pp. 245–256, DOI:https://doi.org/10.1007/s00170-013-5025-7.

32 Gower, HL, Pieters, RRGM, Richardson, IM 2006, 'Pulsed laser welding of metal polymer sandwich materials using pulse shaping', *J. Laser Appl.*, vol. 18, pp. 35-41, DOI: https://doi.org/10.2351/1.2080307.

33 Lee, H.S., Yoon, J.H., and Yoo, J.T. (2018). A study on solid state welding of aerospace materials. *Key Eng. Mater.* 762: 343–347.

34 Peng, P, Wang, K, Wang, W, Huang, L, Qiao, K, Che, Q, Xi, X, Zhang, B, Cai, J 2019, 'High performance aluminium foam sandwich prepared through friction stir welding', *Mater. Lett.*, vol. 236, pp. 295-298, DOI:https://doi.org/10.1016/j.matlet.2018.10.125.

35 Ravi, KK, Narayanan, RG, Rana, PK 2019, 'Friction Stir Spot Welding of Al6082-T6/HDPE/Al6082-T6/HDPE/Al6082-T6 sandwich sheets: hook formation and lap shear test performance', *J. Mater. Res. Technol.*, vol. 8, pp. 615-622, DOI: https://doi.org/10.1016/j.jmrt.2018.05.011.

36 Rana, PK, Narayanan, RG, Kailas, SV 2017, 'Effect of rotational speed on friction stir spot welding of AA5052-H32/HDPE/AA5052-H32 sandwich sheets', *J. Mater. Process. Technol.*, vol. 252, pp. 511-523, DOI: https://doi.org/10.1016/j.jmatprotec.2017.10.016.

37 Su, X, Huang, P, Feng, Z, Gao, Q, Wei, Z, Sun, X, Zu, G, Mu, Y 2021,'Study on aluminium foam sandwich welding by friction stir welding', *Mater. Lett.*, vol. 304, pp. 1-4, DOI:https://doi.org/10.1016/j.matlet.2021.130605.

38 Ebrahimi, M. and Wang, Q. (2022). Accumulative roll-bonding of aluminum alloys and composites: an overview of properties and performance. *J. Mater. Res. Technol.* 19: 4381–4403.

39 Hausol, T., Hoppel, H.W., and Goken, M. (2010a). Tailoring materials properties of UFG aluminium alloys by accumulative roll bonded sandwich-like sheets. *J. Mater. Sci.* 45: 4733–4738.

40 Hausol, T., Maier, V., Schmidt, C.W. et al. (2010b). Tailoring materials properties by accumulative roll bonding. *Adv. Eng. Mater.* 12: 740–746.

41 Narayanan, R.G. (2019). Sustainability in joining. In: *Sustainable Material Forming and Joining*, 39–57. CRC Press.

42 Lin, H, Luo, H, Huang, W, Zhang, X, Yao, G 2016, 'Diffusion bonding in fabrication of aluminium foam sandwich panels', *J. Mater. Process. Technol.*, vol. 230, pp. 35-41, DOI:https://doi.org/10.1016/j.jmatprotec.2015.10.034.

43 Yao, C, Hu, Z, Mo, F, Wang, Y 2019, 'Fabrication and fatigue behaviour of aluminium foam sandwich panel via liquid diffusion welding method', *Metals*, vol. 9, pp. 1-11, DOI:https://doi.org/10.3390/met9050582.

44 An, Y, Yang, S, Zhao, E, Huang, X 2019, 'Diffusion bonding in fabrication TA2 sheets enhanced aluminum foam sandwich', *Mater. Sci. Forum*, vol. 898, pp. 950-956, DOI: https://doi.org/10.4028/www.scientific.net/MSF.898.950.

14

Defects in Friction Stir Welding and its Variant Processes

Vinayak Malik[1,2] and Satish V. Kailas[1]

[1] *Department of Mechanical Engineering, IISc, Bangalore, Karnataka, India*
[2] *Department of Mechanical Engineering, KLS Gogte Institute of Technology, Visvesvaraya Technological University, Belagavi, Karnataka, India*

14.1 Introduction

Friction stir welding (FSW) has evolved as a revolutionary method of joining in modern manufacturing. This solid-state welding process offers numerous benefits compared to traditional fusion welding methods, such as superior mechanical properties, enhanced weld integrity, and minimal distortion. FSW has found applications in various industries, which include aerospace, automotive, shipbuilding, and construction, owing to its ability to join dissimilar materials and produce high-quality welds.

However, like any manufacturing process, FSW is not immune to imperfections. Flaws in friction stir process can arise due to a combination of factors, including material properties, process parameters, and tool design [1]. Understanding and mitigating these defects is crucial for ensuring the reliability and longevity of FSW joints.

14.2 General Defects in FSW

This section discusses a few of the general defects observed in normal FSW. One of the primary defects encountered in FSW is the presence of voids within the weld. These voids can compromise the mechanical strength and fatigue resistance of the joint. They are typically caused by air entrapment during this solid-state welding process. Factors such as improper tool design and inadequate movement of plasticized material can contribute to the formation of these voids [2].

Another common defect is the occurrence of micro-cracks in the weld zone. Cracks can be developed due to a variety of reasons, such as high residual stresses, improper heat input, or distorted movement of plasticized material beneath the tool [3]. Cracks weaken the joint and provide pathways for the initiation and propagation of fatigue cracks under cyclic loading conditions. Therefore, their prevention is vital for ensuring the structural integrity of the welded components.

In addition to voids and cracks, defects such as lack of fusion and incomplete penetration can also occur in FSW. Lack of fusion refers to the incomplete bonding between the abutting workpiece materials, resulting in a weak joint interface. Incomplete penetration, on the other hand, signifies insufficient weld depth, resulting in a reduced cross-sectional area and compromised joint strength. These defects can be attributed to factors such as improper tool design [4], inappropriate process parameters, or inadequate clamping of the workpieces [5].

The subsequent sections provide readers with foundational information on types of defects in the normal friction stir process followed by distinctive defects in a few of its variant processes. The later section discusses possible measures to avoid or eliminate these flaws and undesirable features in the joint region.

14.2.1 Defect Classification

In general, defects in friction stir-based processes can be categorized into two classes: (i) geometric related and (ii) concerned with material flow. Examples of geometric-type defects are lack of fusion and lack of penetration (LOP). They particularly originate due to irregularities in the adjoining geometry of weld pieces and/or weld tool, and anomalies in the initial layup. For instance, when the tool pin on the weld is too short to fully encompass the material at the weld's root, LOP faults occur. This results in a small area of the weld seam remaining unconsolidated. Similarly, when the weld tool is not aligned with the weld seam and the initial weld interface is not entirely consumed during welding, lack of fusion flaws may develop [6]. The structural integrity of the weld can be compromised by both types of flaws; however, flow-related defects and their interrelation with the process dynamics are more important to study because geometric-related faults can usually be reduced during the weld fit-up and design. These friction stir defects can be further classified as external and internal defects (Tables 14.1 and 14.2).

These defects are easy to identify as they are present on the outside of the weld. A mere visual inspection would serve the purpose of figuring out their severity. A few external defects are shown in Figure 14.1.

Table 14.1 External defects.

Sl. no.	Defect name	Description of defect	Possible reasons for their occurrence
1	Flash	A strip of expelled material is generally seen on the retreating side	Excessive plunge depth or axial load, thickness mismatch in plates of advancing and retreating side, excessive temperature leading to overheated weld
2	Lack of fill at the surface	A continuous or intermittent groove at the weld surface. Sometimes it might be in the form of a crack. It is normally located on advancing portion of the cross-sectioned weld	A large gap between faying plates during welding, insufficient forge pressure at the posterior of the tool, inadequate plunge depth
3	Galling at surface	Galling and tearing of material at weld surface. It imparts poor finish to the weld	Excessive sticking of work material to tool due to high temperatures

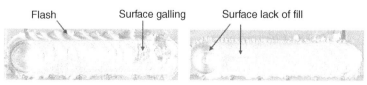

Figure 14.1 Weld photographs showing (a) flash and surface galling, (b) surface lack of fill.

Table 14.2 Internal defects.

Sl. no.	Defect name	Description of defect	Possible reasons for their occurrence
1	Void	A continuous channel-like hole runs in the longitudinal direction of weld. When it is seen on the advancing side, it is called a wormhole. Many a times number of small holes are observed randomly distributed over the weld nugget	Inadequately consolidated material due to lack of forge pressure behind the tool, insufficient heat input leading to cold welding conditions, i.e. excessive tool traverse speed for a given tool rotation speed
2	Joint line remnant (JLR)	It is a discontinuous and wavy surface left in the weld because of the natural oxide layer present on the original faying surfaces. It is also referred as a lazy S or zigzag curve. If JLR is connected to the small undisrupted original faying line at the bottom of the weld, it is called a kissing bond (KB), weak bond, or root flaw	Large tool offset from original faying surfaces, excessive heat input leading to deflection of the original faying line away from the pin and stir zone, insufficient tool pin length, improper tip radius of tool pin

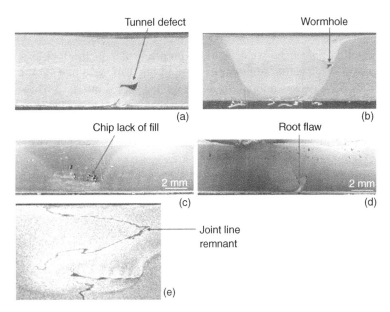

Figure 14.2 Weld microstructures showing (a) tunnel defect, (b) wormhole, (c) chip lack of fill, (d) root flaw, and (e) joint line remnant. Source: Dialami et al. [7]/with permission from Elsevier.

These defects are of more concern as they are not easily discernible. They are concealed within the welds. Few of the defects are identifiable with specific nondestructive tests. A few internal defects are shown in Figure 14.2.

Table 14.3 Characteristic difference in butt and lap welds.

Characteristic	Butt FSW	Lap FSW
Joint configuration	Face-to-face	Overlapping
Metal flow	More even	More uneven
Weld quality	Generally better	Generally worse
Thickness range	Thinner materials	Thicker materials

14.3 Characteristic Defects in Friction Stir Butt and Lap Joints

14.3.1 Friction Stir Butt joints

Table 14.3 summarizes the key differences between butt FSW and lap FSW.

LOP is significant to the butt type of joints in FSW. It largely arises from inadequate downward force and/or insufficient tool penetration. It is primarily the tip of the unwelded butting edge on the flip side of the weld surface (Figure 14.3).

14.3.2 Friction Stir Lap Joints

The overlapping joint configuration aligns the weld interface perpendicular to the tool axis. Therefore, the joining mechanism is altogether different in Friction Stir Lap Welding (FSLW). Such welding is characterized by hook defects originating at the welded sheets' interface. These defects generally appear during the assurgent movement of bottom sheet material in the lap joint, mainly during tool pin penetration in bottom sheets. It is a defect of concern as the metallurgical joints around these regions are incomplete, weakening the entire weld. Figure 14.4 shows the

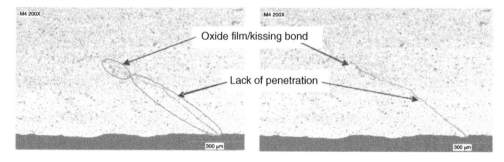

Figure 14.3 Microstructures of weld cross-sections showing LOP and kissing bond. Source: Mandache et al. [8]/Taylor & Francis Group.

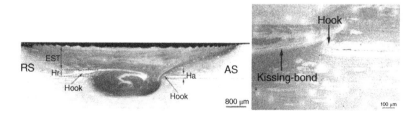

Figure 14.4 Hook defects observed in a lap weld cross sections: (a) overall image, (b) zoomed image of a hook defect. Source: Chen et al. [9]/with permission from Elsevier.

locations and appearance of hook defects in a typical lap joint configuration. The morphology of the hook defect primarily changes with varying tool pin lengths and can significantly affect the lap shear failure loads of the joints [10].

14.4 Distinctive Defects in Major Friction Stir Variants

14.4.1 Bobbin Stir Welding

The notable defects distinctive in Bobbin Stir Welding (BSW) primarily emerge during the weld's beginning and end. They are termed entry and exit defects. Figure 14.5 shows these types of defects. A typical feature called ejected tail (region 1) is characteristic of entry defect essentially seen on the retreating side accompanying a visible discontinuity (region 2) on the advancing side. At the weld exit, a similarly enlarged discontinuity is observed on the advancing side (region 3). The free tool trailing edge without any blocking mass for flowing plasticized material leads to material ejection in the entry zone. Further, this lost ejected material finally results in discontinuity on the advancing side. It is reported that the severity of the defects mentioned above is directly linked to the ratio (ω/v) of tool rotational speed(ω)/tool traverse speed(v) which is expressed in revolutions per mm [12].

The free surface at the tool's trailing edge and the absence of a solidified mass at the posterior obstructing the flow are the primary reasons for the material ejection in the entry zone.

The above defects can be minimized by adopting the following suggested solutions:

i) use of predrilled pilot holes and
ii) use of dual-rotation bobbin tools where top and bottom shoulders have the provision to change their speeds independently [13].

The other characteristic defects in BSW are bobbin marks and drag defects. The bobbin mark defect in BSW refers to the formation of visible marks or grooves on the surface of the weld joint. These marks originate from the interrupted interaction between the rotating bobbin tool and the workpiece and can impact the weld's aesthetic appearance and may act as stress concentration points. The bobbin drag defect refers to the unwanted dragging or pulling of the workpiece material by the rotating bobbin tool. It can occur when the frictional forces at the tool/workpiece interface are excessive or when the tool encounters resistance during welding. This defect can cause distortion, irregular material flow, and compromised joint quality.

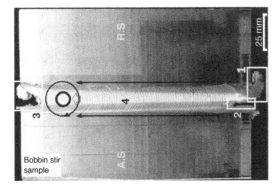

Figure 14.5 A photograph depicting the entry and exit defects in a typical bobbin friction stir welded aluminum plate with marked regions. Source: Tamadon et.al. [11]/MDPI.

14.4.2 Counter-rotating Twin-Tool Friction Stir

A common observation reported by many researchers is that re-welding helps minimize defects and improve mechanical properties. It is referred to as multi-pass in friction stirring community. To achieve two passes in a single attempt, twin-tool technique was devised [14, 15]. It assists in reducing the clamping force as the resultant forces from both tools counter each other. The second pass takes place simultaneously with the first pass without any time delays (Figure 14.6). This reduces the torque requirement in the second pass as the material is sufficiently soft from the continuing initial pass. The defects observed here are similar to conventional friction stir. However, its occurrence is relatively less in counter rotating twin-tool friction stir (CRTTFS).

Further, asymmetric recess defects can be observed in CRTTFS when one side of the weld joint exhibits a more extensive and deeper recess than the other. This defect is a fallout of the unbalanced heat generation and material flow caused by the contrarotation of tools. The differences in the thermal profile and material flow behavior on each side of the weld can result in asymmetric recess defects.

14.4.3 Refill Friction Stir Spot Welding

Refill friction stir spot welding (RFSSW), or Friction Spot Welding (FSpW), is a preferred method to join materials possessing low melting point, for instance, Al and Mg alloys. It is also a preferred choice for dissimilar combinations of Al/Ti, Al/Cu, Al/Steel, and Al/Mg. The key attractive feature of this process is obtainment of keyhole-free surface friction stirred spot weld. Typical defect-like geometric features in this process are hooking, partial bonding, and bonding ligament. Hooking is produced by the upward orientation of the initial sheet interface produced during the penetration and retraction of the sleeve. The mechanical performance of the joint is significantly hampered by its presence [16]. Generally, deeper plunge depths result in severe hooking flaws. Bonding ligament refers to a zone where successful joining is expected. Partial bonding is a transitional region between bonding ligament and base material. A higher degree of intermixing avoids these types of undesirable features and can be achieved through a suitable tool design. Figure 14.7a shows these typical features occurring in FSpW. Further, incomplete refill defects can show up in RSSSW. It is failure of the material to completely fill the void created during the plunging and retraction cycles [19]. The complex nature of the RFSSW process, involving multiple plunges and retracts, introduces additional challenges in achieving a complete refill, leading to this defect. Figure 14.7b shows a typical incomplete refill defect.

14.4.4 Friction Stir Additive Manufacturing

Friction Stir Additive Manufacturing (FSAM) is derived from FSLW, wherein multiple laps are added in a stepwise fashion to the former until the necessary job thickness is attained. However,

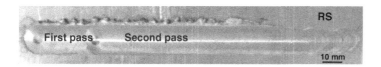

Figure 14.6 Photograph showing the top-view of the weld made from counter-rotating twin friction stir method. Source: Kumari et al. [14]/with permission from Elsevier.

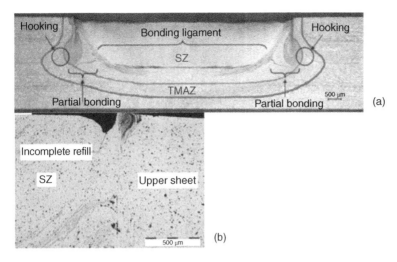

Figure 14.7 Cross section of FSpW joint indicating (a) hooking flaw, partial bonding, and bonding ligament. Source: Rosendo et al. [17]/John Wiley & Sons, Inc. and (b) incomplete refill. Source: Kubit and Trzepiecinski [18]/Springer Nature.

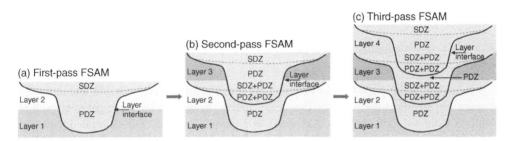

Figure 14.8 Schematic showing the emergence of complex microstructure in FSAM due to overlapping of different zones. Source: Hassan et al. [20]/MDPI/CC BY 4.0.

the FSAM exhibits complex microstructural zones due to the overlap of multiple processed sheets. The microstructural evolution in FSLW is characterized by two zones: (i) shoulder driven zone (SDZ) and (ii) pin driven zone (PDZ). There is a difference in these microstructural zones and corresponding mechanical properties primarily due to strain, strain rate, and temperature gradients arising from tool shoulder and pin geometries. The complexity augments with added layers in FSAM, as shown in Figure 14.8.

Kissing defects appear in FSAM layers due to tenacious oxidized contents on lapping areas of stacking sheets (Figure 14.9). They can be minimized by suitably increasing the plunging force during the process or pretreatment of sheets to reduce oxidation. Other common lap configuration defect like Hook is noticed, which was discussed in prior sections of the chapter. The other anomaly seen in FSAM is layer inconsistencies. They occur when there are variations in layer height, width, or material distribution within the built part. These inconsistencies can result from irregular material flow, tool misalignment, or fluctuations in process parameters. Layer inconsistencies can lead to dimensional inaccuracies and surface roughness variations that compromise the overall part quality.

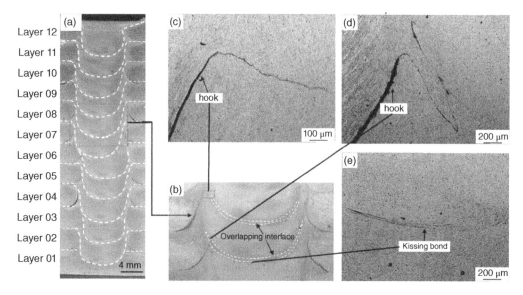

Figure 14.9 FSAM macrostructure with hooking and kissing bond defects captured at layers 7 and 8. Processed material is 7N01-T4 aluminum alloy. Source: Hassan et al. [20]/MDPI.

14.5 Solutions to Avoid Defects in Friction Stir-Based Processes

The following section highlights a few measures that can be incorporated to prevent defects in friction stir processes.

Optimize the process parameters: Process variables like the axial force, traverse speed, and tool rotation speed can significantly influence the weld's quality. By optimizing these parameters, it is possible to achieve a stable process that produces high-quality welds. For example, increasing the rotational speed can increase the quantum of heat input and improve the weld quality. Similarly, decreasing the traverse speed can help to ensure adequate material flow and prevent voids from forming.

Use appropriate tool materials: The material of the tool performs a crucial role in the FSW process, generating frictional heat and facilitating material flow. The tool should be made from a material that can endure the elevated temperatures and stresses encountered during welding. For example, tools made from polycrystalline cubic boron nitride (PCBN) or tungsten-rhenium (W-Re) alloys have been shown to perform well in FSW applications.

Maintain proper clamping: Proper clamping is essential to ensure that the workpiece is securely held during welding. Inadequate clamping can lead to tool deflection, material movement, and joint misalignment, resulting in defects such as underfill and flash. By maintaining proper clamping, it is possible to make certain that the material is securely held in its intended position and that the weld is formed correctly.

Provide adequate cooling: Cooling is an important factor to be considered during FSW as it can help control the weld's temperature and prevent defects such as local melting, chip lack of fill, and micro-cracking. Using liquid cooling or air-cooled backing plates can help maintain the weld's temperature below critical levels.

Use appropriate pre-weld preparation techniques: The quality of the weld can be significantly impacted by the condition of the workpiece prior to welding. Adequate preparation techniques such as cleaning, de-burring, and surface treatment can help to ensure that the material is free from contaminants and that the joint is properly aligned.

14.6   Summary and Concluding Remarks

To summarize, though FSW offers numerous advantages, defects can still occur during the process. Voids, micro-cracks, lack of weld consolidation, incomplete penetration, and joint line remnant/kissing bonds are among the most common defects encountered in friction stir processes [21]. Understanding the causes and implementing appropriate measures to minimize these defects is essential for achieving high-quality friction stir joints with superior mechanical properties and structural integrity. By continually improving process control, optimizing parameters, refining tooling techniques, appropriate tool materials, proper clamping, adequate cooling, and appropriate pre-weld preparation techniques, the industry can further enhance the reliability and acceptance of FSW as a robust joining method. Adopting these guidelines makes it possible to achieve high-quality welds that meet the requirements of various applications.

Acknowledgment

The authors would like to thank Indian Institute of Science, Bangalore, and KLS Gogte Institute of Technology for extending support in research activities and providing computing facilities needed for data analysis and writing scientific reports. The expertise and understanding acquired from the Institute has made a substantial contribution to the quality and depth of the reports.

References

1 Malik, V. and Kailas, S.V. (2020). Understanding the effect of tool geometrical aspects on intensity of mixing and void formation in friction stir process. *Proc. Inst. Mech. Eng., Part C: J. Mech. Eng. Sci.* 095440622093841. https://doi.org/10.1177/0954406220938410.

2 Malik, V. and Saxena, K. (2022). Understanding tool–workpiece interfacial friction in friction stir welding/processing and its effect on weld formation. *Adv. Mater. Process. Technol.* 8 (sup4): 2156–2172. https://doi.org/10.1080/2374068X.2022.2036042.

3 Al-Moussawi, M. and Smith, A.J. (2018). Defects in friction stir welding of steel. *Metall. Microstruct. Anal.* 7 (2): 194–202. https://doi.org/10.1007/s13632-018-0438-1.

4 Malik, V., Sanjeev, N.K., Hebbar, H.S., and Kailas, S.V. (2014). Investigations on the effect of various tool pin profiles in friction stir welding using finite element simulations. *Procedia Eng.* https://doi.org/10.1016/j.proeng.2014.12.384.

5 Podržaj, P., Jerman, B., and Klobčar, D. (2015). Welding defects at friction stir welding. *Metalurgija* 54 (2): 387–389. Available at: https://hrcak.srce.hr/128969.

6 Arbegast, W.J. (2008). A flow-partitioned deformation zone model for defect formation during friction stir welding. *Scr. Mater.* 58 (5): 372–376. https://doi.org/10.1016/J.SCRIPTAMAT.2007.10.031.

7 Dialami, N., Cervera, M., Chiumenti, M. et al. (2019). Prediction of joint line remnant defect in friction stir welding. *Int. J. Mech. Sci.* 151: 61–69. https://doi.org/10.1016/j.ijmecsci.2018.11.012.

8 Mandache, C., Levesque, D., Dubourg, L., and Gougeon, P. (2012). Non-destructive detection of lack of penetration defects in friction stir welds. *Sci. Technol. Weld. Joining* 17 (4): 295–303. https://doi.org/10.1179/1362171812Y.0000000007.

9 Chen, H., Fu, L., Liang, P., and Liu, F. (2017). Defect features, texture and mechanical properties of friction stir welded lap joints of 2A97 Al-Li alloy thin sheets. *Mater. Charact.* 125: 160–173. https://doi.org/10.1016/j.matchar.2017.01.038.

10 Aldanondo, E., Vivas, J., Álvarez, P., and Hurtado, I. (2020). Effect of tool geometry and welding parameters on friction stir welded lap joint formation with AA2099-T83 and AA2060-T8E30 aluminium alloys. *Metals* 10 (7): 872. https://doi.org/10.3390/met10070872.

11 Tamadon, A., Pons, D.J., Sued, K., and Clucas, D. (2018). Formation mechanisms for entry and exit defects in bobbin friction stir welding. *Metals* 8 (1): 33. https://doi.org/10.3390/met8010033.

12 Tamadon, A., Pons, D., and Clucas, D. (2019). Structural anatomy of tunnel void defect in bobbin friction stir welding, elucidated by the analogue modelling. *Appl. Sys. Innov.* 3 (1): 2. https://doi .org/10.3390/asi3010002.

13 Wang, G.-Q., Zhao, Y.-H., and Tang, Y.-Y. (2020). Research progress of bobbin tool friction stir welding of aluminum alloys: a review. *Acta Metall. Sin. Engl. Lett.* 33 (1): 13–29. https://doi.org/ 10.1007/s40195-019-00946-8.

14 Kumari, K., Pal, S.K., and Singh, S.B. (2015). Friction stir welding by using counter-rotating twin tool. *J. Mater. Process. Technol.* 215: 132–141. https://doi.org/10.1016/j.jmatprotec.2014.07.031.

15 Thomas, W.M., Staines, D.J., Watts, E.R. et al. (2005). The simultaneous use of two or more friction stir welding tools. *Int. J. Join. Mater.* 17 (1): 1–5. Available at: https://www.twi-global.com/ technical-knowledge/published-papers/the-simultaneous-use-of-two-or-more-friction-stir-welding-tools-january-2005.

16 de Sousa Santos, P., McAndrew, A.R., Gandra, J., and Zhang, X. (2021). Refill friction stir spot welding of aerospace alloys in the presence of interfacial sealant. *Weld. World* 65 (8): 1451–1471. https://doi.org/10.1007/s40194-021-01113-3.

17 Rosendo, T., Tier, M., Mazzaferro, J., and Mazzaferro, C. (2015). Mechanical performance of AA6181 refill friction spot welds under lap shear tensile loading. *Fatigue Fract. Eng. Mater. Struct.* 38 (12): 1443–1455. https://doi.org/10.1111/ffe.12312.

18 Kubit, A. and Trzepiecinski, T. (2020). A fully coupled thermo-mechanical numerical modelling of the refill friction stir spot welding process in Alclad 7075-T6 aluminium alloy sheets. *Arch. Civ. Mech. Eng.* 20 (4): 117. https://doi.org/10.1007/s43452-020-00127-w.

19 Malik, V. NK Sanjeev, HS Hebbar, SV Kailas (2014) 'Finite element simulation of exit hole filling for friction stir spot welding – a modified technique to apply practically', in *Procedia Engineering*. https://doi.org/10.1016/j.proeng.2014.12.405.

20 Hassan, A., Pedapati, S.R., Awang, M., and Soomro, I.A. (2023). A comprehensive review of friction stir additive manufacturing (FSAM) of non-ferrous alloys. *Materials* 16 (7): 2723. https:// doi.org/10.3390/ma16072723.

21 Bhardwaj, N., Narayanan, R.G., Dixit, U.S., and Hashmi, M.S.J. (2019). Recent developments in friction stir welding and resulting industrial practices. *Adv. Mater. Process. Technol.* 5 (3): 461–496. https://doi.org/10.1080/2374068X.2019.1631065.

15

Nondestructive Ultrasonic Inspections, Evaluations, and Monitoring in FSW/FSP

Yuqi Jin[1,2], Teng Yang[1,3], Narendra B. Dahotre[1,3], and Tianhao Wang[4]

[1] *Center for Agile and Adaptive Additive Manufacturing, University of North Texas, Denton, TX, USA*
[2] *Department of Physics, University of North Texas, Denton, TX, USA*
[3] *Department of Materials Science and Engineering, University of North Texas, Denton, TX, USA*
[4] *Energy and Environment Directorate, Pacific Northwest National Laboratory, Richland, WA, USA*

15.1   Introduction

Ultrasonic inspections and evaluations, as one of the most common approaches in nondestructive testing/nondestructive evaluation (NDT/NDE), can provide substantial information for both academia and industries. In principle, ultrasound waves are high-frequency vibrational oscillations in which the wave propagation behaviors are tightly related to the physical properties of the medium, including mechanical properties, uniformity, anisotropy, microstructure, and stress conditions. The deviations in wave propagation behaviors between the tested sample and the reference ideal case can be applied to evaluate the physical property differences between the tested sample and reference sample via reverse engineering, which is the foundational principle of ultrasonic inspections and evaluations. Consequently, to understand the tests and the potential information from the ultrasound NDT/NDE on FSWed/FSPed samples, the wave propagation behaviors need to be briefly introduced along with the consideration of the influence of discontinuities, microstructures, and residual stresses, which are all unique in FSWed/FSPed samples compared to other industrial products. The present chapter explains the basic ultrasound wave propagation behaviors associated with the macrostructure, microstructure, and residual stresses (Section 15.2). Then, the explanations of applying those principles to ultrasonic inspections, evaluations, and monitoring are given in Section 15.3. The last section introduces and discusses some case studies regarding the recent ultrasonic NDT/Es in friction stir-based manufacturing processes (Section 15.4).

15.2   Ultrasonic Wave Behaviors in FSWed/FSPed Samples

Before applying the ultrasonic wave to testing, the basic understanding of the ultrasound waves needs to be reviewed briefly. First, in relatively large-scale ultrasonic inspections and evaluations, such as FSWed/FSPed samples, time-domain measurements are usually selected based on the

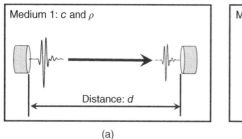 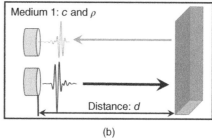

(a)   (b)

Figure 15.1   (a) Bistatic time-of-flight measurement. (b) Monostatic time-of-flight measurement.

principle of time-of-flight (Figure 15.1) to enable the observation of spatial information. According to the sound velocity in the medium c and temporal delay t, the distance of the ultrasonic wave source can be estimated as $d = c\,t$. The same expression can also be used to estimate the thickness T of the ultrasonically tested sample with either transmission configurations ($T = c\,t$, Figure 15.1a) or reflection configurations ($2T = c\,t$, Figure 15.1b). In principle, such measurements can be performed by any waveform as long as the vibrational mode is supported in the propagation medium. In practical nondestructive inspection cases, the preference for a short ultrasound pulse exists due to the convenience of temporal delay estimation.

Unlike infinitely long continuous waves, such as sine waves, a short pulse has a finite length in the time domain. Analytically, based on Fourier transformation, a short length pulse can be understood as a superposition product over infinite numbers of frequency components (sine waves) within a certain frequency bandwidth. The bandwidth can be found via locating the lower frequency band edge f_1 and the higher frequency band edge f_2 in the frequency domain, which can be obtained by performing Fourier transformation on the entire short pulse. With the known band edges, the length of the pulse can be estimated as $\Delta t_w \propto \left(f_1^{-1} - f_2^{-1} \right)$. In other words, a wider pulse bandwidth in the frequency domain refers to a narrower temporal length of the pulse in the time domain. Generally, an ultrasonic pulse with a narrower temporal length provides higher precision in conducting time-of-flight measurements. Furthermore, in practical case, it is worth noting that the ultrasonically tested sample should have a thickness clearly greater than the ultrasound wavelength of at least the lower band edge frequency component as $\lambda_1 = c\,f_1^{-1}$. From the point of view of the pulse, it is preferable that the sample thickness is much larger than the spatial pulse width ($T >> \Delta d_w = c\Delta t_w$). When the operating pulse spatial width or the wavelength of the ultrasound is shorter than the sample thickness, the propagated pulse envelope or the waveform can be destroyed by the presence of the complex transmission and reflection coefficients which leads to strong uncertainty on the time-of-flight measurements.

During the ultrasonic nondestructive tests, many wave behaviors can occur due to the presence of complexity in the tested samples. Note that transmission, reflection, and absorption are the most common wave behaviors. In FSWed/FSPed samples, since the most common samples are metals, the absorption can be generally negligible. When the wave travels from medium (1) to medium (2) with normal incidence, the source pulse is ideally separated into two parts as transmitted pulse and reflection pulse. The transmission coefficient describes the fractions of the potential on each part, $tr = (2Z_2)(Z_1 + Z_2)^{-1}$, and reflection coefficient, $re = (Z_2 - Z_1)(Z_1 + Z_2)^{-1}$, where the acoustic impedance is obtained from the product of the speed of sound and density of the medium as $Z = c\,\rho$. When the source wave propagates into the second medium with a non-perpendicular incidence angle, refraction can occur with an angle dependent on the ratio between the speed of sound

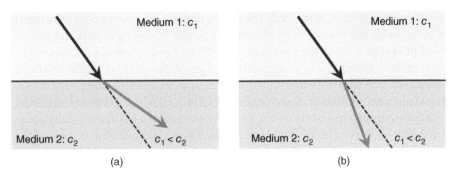

Figure 15.2 (a) Refraction in a medium with higher sound velocity. (b) Refraction in a medium with lower sound velocity.

values in 2nd and 1st mediums as Figures 15.2a,b shows, which is needed to be taken into account during dissimilar metal FSW joint inspections.

Besides the basic transmission, reflection, and refraction wave behaviors, scattering and dispersion are also important in ultrasonic nondestructive tests for analyzing advanced information, including microstructure and residual stresses in FSW and FSP. During testing, both scattering and dispersion [1–3] are associated with the nonuniformity in the tested samples, including not limited to defects, cracks, grain sizes, and residual stresses. In detail, scattering states the wave path deviation and wave energy separation due to the presence of nonuniformity, as Figure 15.3a shows. During the ultrasonic tests, due to the limitation of the locations and the number of detectors, the wave path deviation-induced temporal delay is hardly measured unless the scattering waves are propagated in the source wave direction. Hence, the major contribution of scattering effects can be observed as the change in the attenuation of the wave. Different from scattering, dispersion states the frequency-dependent wave propagation behaviors with the presence of nonuniformity. Note that the dispersion in ultrasonic tests commonly refers to the frequency-dependent speed of sound occurring (Figure 15.3b), with the size of disorder or nonuniformity being much smaller than the operating wavelength. In a broadband ultrasonic pulse, the dispersion effect-induced frequency-dependent speed of sound can introduce out-of-phase destructive interference. Alternatively, the propagated pulse is temporally wide with lower amplitude with respect to the source pulse.

Due to the unique processing procedures (shear-induced severe plastic deformation under the melting point of the workpiece), FSWed/FSPed samples have various regions as the signature of the processes, including base material regions, heat-affected zone (HAZ), thermomechanical affected zone (TMAZ), and stir zone (SZ). The distinguishable topology of microstructures,

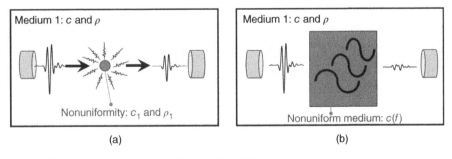

Figure 15.3 (a) Scattering effect. (b) Dispersion effect.

presence of defects, and distribution of residual stresses in these zones provide highly different ultrasonic wave propagation behaviors. Firstly, the discontinuities, including defects and cracks, are the majority of ultrasonic NDE in the conventional FSW/FSP quality control processes. The presence of defects and cracks introduces sharp mechanical properties difference-induced strong acoustic impedance mismatch with respect to the surrounding materials, which leads to the excitation of an additional ultrasonic reflection. Based on the temporal delay of the additional reflection, the location of the discontinuities can be estimated. However, the feasibility of discontinuity detection is highly dependent on the size of the defects or cracks, which are required to be beyond the operating ultrasound wavelength. In FSWed/FSPed samples, sub-surface cracks and lack-of-penetration can commonly be located by the additional echo determinations. In the case of determination of small voids, instead of the additional echo occurring, low amplitude backscattering is possible to present by using advanced ultrasonic equipment and signal processing techniques, which is hard to analyze the information or even the existence of the small discontinuities.

15.2.1 Influence From Defects on Ultrasound Wave Traveling

Cracks or defects commonly present in FSWed/FSPed samples with the processing under improper parameters or tool conditions. In the common flaws in FSW and FSP, flash and surface tunnel defects are optically visible and do not require an ultrasonic inspection to determine. On the contrary, lack-of-penetration flaws and internal cavities due to the parameters and tools errors are generally not visible, which are the majority of the ultrasonic nondestructive inspection targeting at. Unlike the defects and cavities in powder metallurgy manufacturing or laser fusion-based AM processes, the cavities in FSWed/FSPed samples are usually air-infilled instead of powder-infilled. The infilled air can introduce strong enough mismatched acoustic impedance, more than 3 orders of magnitude, with respect to the common metals. Hence, during the ultrasonic inspection, air-infilled cavities and defects should provide sufficient acoustic energy reflected ($\geq 90\%$) in the ideal case. However, in the practical inspection case, physical limitations still commonly exist that weaken the detection performance of ultrasonic inspections due to the operating wavelength and geometry of the defects. As a result, the typical defects are grouped into two categories, including single defects and accumulated small enough voids group.

With a single air-infilled defect in the metal material, when the diameter of the defect is much larger than the operating wavelength $d_d \gg \lambda_{f_1}$, a clear reflected pulse occurs. The reflected pulse is expected to have a comparable waveform with respect to the original source pulse, or at least, the reflected pulse should have comparable bandwidth and distribution of amplitude in the frequency domain after the Fourier transformation. When the diameter of the single defect approaches the operating wavelength ($\lambda_{f_1} \gg d_d \gg \lambda_{f_2}$) of the ultrasonic wave during the inspection, within the frequency bandwidth of the pulse, art of the frequency components with smaller wavelength excites reflection waves at the defect. Instead of reflection, only minor scattering waves are generated with other frequency components with longer wavelengths. Hence, the reflected pulse is not expected to have a comparable waveform and distribution on the frequency spectrum after the Fourier transformation. In addition, in this case, the size of the wave source and receiver start to have a role in the reflection generation. On the receiver, the device usually translates the 2D acoustic pressure distribution into a 1D time-dependent profile. Hence, on the area of the receiver, the defect-affected area of the acoustic pressure needs to be sufficient to deviate the self-averaged acoustic pressure determination. Similarly, the source pulse excited from the emitter usually spatially follows a Gaussian-like distribution of acoustic pressure. Only when the defect is located around the center of the acoustic beam, the reflected signal can be ensured as high enough. Hence, instead of single location measurement, in practical inspection, linear or even 2D raster scans are preferable to increase the possibility of the determination of defects, especially when the defect has a comparable size with respect to the operating wavelength of the sound wave. When the defect has a size clearly smaller than the operating wavelength, $d_d \ll \lambda_{f_2}$, no reflection signal would be generated following the approximation of the diffraction limit. Very minor scattering radiations can be excited, which is experimentally hard to measure. However, in principle, the ultrasonic pulse after the scattering generation has a smaller total acoustic pressure after passing through the small defect. However, such small signal amplitude deviation can be caused by many other physical conditions in the tested samples, such as the deviation in the local grain sizes. Hence, by simply measuring the amplitude change, it is not reliable to lead to the conclusion on the existence of a small defect in practical ultrasonic inspection. In Figure 15.4a–c, the ultrasound wave behavior illustration is posted, corresponding to the varying ratio between the single defect size and the operating wavelength.

In advance, when the defect has a non-equal axial geometry, anisotropic acoustic wave behaviors are introduced in the inspections. In this case, the size of the defect does not follow a simple diameter value of the defect. The length of the defect along the wave oscillation direction is considered the effective length or size of the defect. Hence, using either longitudinal or transversal waves, an identical cylindrical defect can provide a significantly different acoustic pressure decrease due to the different wave oscillation direction. In the extreme case, perhaps, with a long cylindrical defect, the length is greater than the wavelength with orients along the length perpendicular to the wave propagation direction. By using a transversal ultrasonic wave, the attenuation on the transmission wave can be much stronger than using the longitudinal wave. On the contrary, the reflection from the transversal ultrasonic wave is also expected to be stronger with respect to the test with a longitudinal wave.

Lack of penetration is a typical flaw existing in FSWed samples. From the view of ultrasound, lack of penetration can be understood as a plane-like defect with sufficient height and length but narrow width. In general, the height and length are greater than the typical wavelength of the ultrasonic wave since the air-infilled gap carries strong enough acoustic impedance mismatch conditions. Furthermore, the potential location of lack of penetration is predictable, always at the joint interface between the two workpieces. Hence, the incidence wave direction is essential to clearly detect the lack-of-penetration flaws. Without predicting the height of the lack-of-penetration defect, a diagonal incidence ultrasonic wave with respect to the joint interface can be considered the first step to determining the existence of a lack-of-penetration flaw.

When the presented voids are small enough concerning the operating wavelength, the scattering effect induced by ultrasonic radiation is highly challenging to detect due to the low amplitude. However, the transmitted pulse after the scattering equips a low pulse amplitude which can

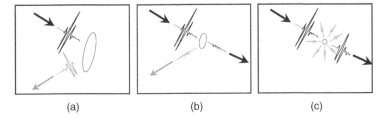

(a)   (b)   (c)

Figure 15.4 (a) Acoustic wave behavior with a defect with a much larger size with respect to sound wavelength. (b) Acoustic wave behavior with a defect with comparable size with respect to operating wavelength. (c) Acoustic wave behavior with a defect with smaller size with respect to sound wavelength.

be measured. As an alternative approach, the metal sample with a small void can be effectively considered as a uniform material with slightly lower physical properties with respect to the same material with void-free conditions, following the homogenization theory. As the wave equation described, the slightly lower effective density and effective elastic constant introduce a lower transmitted acoustic amplitude. Reversely, by measuring the transmission pulse amplitude in a comparison configuration on an ultrasonically tested sample, especially in the same zone (SZ), the small void is possible to be determined at a lower transmission location, referring to a lower effective density. The small void in metal-induced lower ultrasound transmission amplitude and effective density were verified by finite element analysis-based acoustic simulation in Figure 15.5a,b with additional parametric studies on the volume fraction of the voids regarding effective operating wavelength and effective density.

The estimated effective density values from the ultrasound reflections are compared to the theoretical values and showed a good agreement. The theoretical values were obtained from the effective medium approximation theory [4], which is expressed as $\rho_{eff} = \rho_{metal}\sqrt{1-F}$. In the ideal cases, such as the simulation environment, the ultrasound-evaluated effective density values are highly mismatched with the effective medium theory. The slight differences were contributed from the presence of the additional reflection from the pores as illustrated in the previous section. As the size-dependent study demonstrated, with the operating wavelength approaching the size of the pore, higher amplitude additional reflections are expected from the pores, which leads to a more considerable uncertainty to the second major reflection for the calculation by the effective medium theory. However, the operating wavelength can be considered as a variable dependent on the volume fraction of the porosity as expression $1/\lambda_{eff} = F/\lambda_{air} + (1-F)/\lambda_{metal}$ described in Krokhin et al. [4]. In Figure 15.5a, the behavior of the effective operating wavelength was calculated with respect to the porosity fraction. With the increase of the porosity fraction, the decreased effective wavelength approached the size of the porosity, which induced stronger additional scattering from the pores. Hence, from the comparison shown in Figure 15.5b, the difference between two lines increases along the increased porosity fraction. Note that the approach stated here is only suitable for the voids that are much smaller than the operating wavelength and also much smaller than the size of the wave source of the ultrasound during the inspections and evaluations.

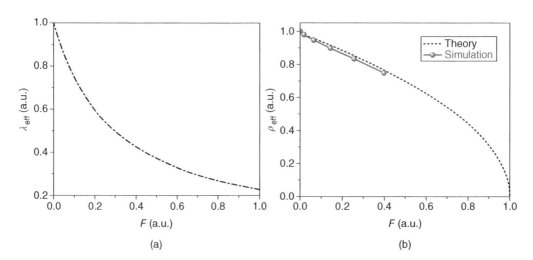

Figure 15.5 (a) Void fraction-dependent effective wavelength. (b) Void fraction-dependent normalized effective density.

Hence, with immersion ultrasonic testing, with no additional internal refection or strong back-scattering signal presented, the effective density of the tested metal plate can be calculated using the following equation [3, 5]:

$$\rho_{\text{eff}} = c^{-1} Z_0 \left(\frac{-1 - \dfrac{p_1}{p_e - p_0} - \sqrt{4 \dfrac{p_1}{p_e - p_0} + 1}}{\dfrac{p_1}{p_e - p_0} - 2} \right) \tag{15.1}$$

where p_e is the peak magnitude of the source pulse from the transducer, p_0 and p_1 are the peak magnitude of reflection signals obtained from the front and back surfaces of the workpiece, t_f and t_i are the time points estimated by the temporal length of the pulse [6]. c is the speed of sound in the sample which is defined as $c = 2d/(t_{i,1} - t_{i,0})$. Z is the acoustic impedance of the scanned sample, and Z_0 is the acoustic impedance of the surrounding ambient material (any fluids with low viscosity).

15.2.2 Influence From Microstructures on Ultrasound Wave Traveling

Due to the unique forged-like behavior in FSW/FSP, the processes revised the base material to refined and equiaxed-grained microstructures in the SZ, which provides limited impacts on the ultrasonic wave propagation behaviors including Rayleigh scattering, frequency dispersion, and anisotropy. On the contrary, in the base material regions and HAZs, the original size or enlarged size grains commonly have a strong anisotropic scattering and frequency dispersion.

In the conventional tensile or indentation tests, the elastic constant of void-free metal can be barely sensed to have strong or even clear differences caused by the difference in the microstructure. Compared to elasticity, plastic properties can more easily show the difference. The requirement for determining the slight difference is very high, especially in metals, within the small elastic deformation range. From the point of view of sound or vibrational waves, the observation scale needs to be small enough to sense the contribution of microstructures clearly. With low strain rate mechanical tests, the effective operating wavelength can be considered infinitely long, beyond the scale of the microstructure and even beyond the sample size. Hence, the elasticity difference contributed by the local microstructure is homogenized. With finite-length operating wavelength ultrasound, the observation scale is generally still large but is closer to the scale of the microstructure with respect to the low strain rate mechanical tests. Additional scattering and dispersion effects are present in the propagation of ultrasound waves, which are both frequency dependent. The frequency dependence deviates from the effective elastic modulus in such a high frequency and introduces the frequency dependence on the elastic constant. With the low-strain mechanical tests, the elastic modulus is approximately constant, also called static elastic modulus. With high-frequency ultrasonic observation, the frequency-dependent elastic modulus is known as dynamic elastic modulus or effective elastic modulus.

As stated, the microstructure-contributed frequency-dependent ultrasonic wave propagation behaviors are introduced by the existence of both grains and grain boundaries in polycrystal metals. In a microscopic view, the difference in crystal types and grain orientation offer local elastic and density deviation along the ultrasound oscillation direction, where the mismatched impedance condition-induced scattering effect occurs at grain boundaries. In a broader view, the strong dispersion effect is present when the disorder of internal structure increases. In polycrystal metals, along the ultrasonic wave propagation direction, the more significant deviation of grain sizes enlarges the dispersion effect at a certain observation scale. In conventionally applied acoustics and NDT fields, the scattering effect-induced acoustic intensity (p_a^2) is concluded as a part of attenuation effects together with other sources of intensity loss, including beam spreading, dispersion-induced out-of-phase interference, and even absorptions. Hence, in the following paragraph, the different scattering effects are analytically summarized in terms of attenuation coefficients.

In detail, the microstructure-induced ultrasonic wave scattering effects are highly frequency-dependent [7, 8]. Similar to the case of defect, the scattering behaviors that occurred during the ultrasonic wave propagations vary according to the relative comparison between the size of grains and operating wavelength. Firstly, similar to elastic scattering (no generation or absence of particles) in light, transversal ultrasonic scattering can be very complex due to abnormal diffraction when the operating wavelength approaches the size of the microstructure, which needs careful analysis case-by-case. At comparable operating frequency f, longitudinal ultrasonic waves have much longer wavelengths and simpler oscillations, the scattering effect is relatively simpler to analyze. When the operating wavelength is clearly larger than the grain size, satisfying $\lambda > 2\pi D$, the ultrasound scattering effect follows Rayleigh scattering approximation, where the ultrasound attenuation coefficient can be expressed as $\alpha \approx K V_{gr}^3 f^4$, where K is bulk modulus and V_{gr} is the volume of the grain. When the operating wavelength is comparable to the grain size, satisfying $\lambda \approx 2\pi D$, the ultrasound scattering effect follows the Stochastic scattering approximation, where the ultrasound attenuation coefficient can be expressed as $\alpha \approx K D f^2$.

Compared with both expressions, due to the contribution of a small number V_{gr}^3, the attenuation effect from Rayleigh scattering is generally weaker than Stochastic scattering with the assumptions of common ultrasonic NDT frequency and common grain size in rolling produced or FSW/FSP metal plates. However, it is noted that the frequency dependences on the attenuation between such two scattering effects are highly different, f^4 in Rayleigh scattering and f^2 in Stochastic scattering. With a proper ultrasonic transmission test, after translating the signal to the frequency domain, the frequency-dependent attenuation can be found to determine whether the averaged grain size is much larger or comparable to the operating wavelength on the ultrasound wave propagation directions. Such experimental approximation in the relative scale can be nondestructively obtained to roughly understand the averaged grain size on the tested samples during the ultrasonic inspection of the SZs on FSWed or FSPed samples. Further conclusions, such as the absolute values of the averaged grain sizes, are still challenging to obtain with the current background of ultrasonic non-destructive tests.

As mentioned previously, in the ultrasonic inspections of FSWed or FSPed samples, the unique refined and equiaxed grains in the SZ do not introduce too much challenge to the analysis of microstructure-induced scattering effects on the ultrasonic wave propagation (Figure 15.6a). However, in the surrounding materials, including HAZ, TMAZ, and base material regions, the grain structures are under enlarged or as-received conditions based on the manufacturing processes and post-treatment of the workpieces, which are generally rolled and annealed. The grain structures in such rolling plates are not equiaxed-grained, usually column-like geometries. Hence, additional anisotropic scattering behaviors are expected (Figure 15.6b). When the ultrasound propagates perpendicular to the length of the grains, less total attenuation results due to more scattering effects are contributed by Rayleigh scattering. On the contrary, with the ultrasound wave propagating along the length direction of the grains, the contribution from Stochastic scattering increases, which leads to increased ultrasonic attenuation. Furthermore, on the path along the length of the column-like grains, the ultrasound-observed grain size (length) deviation is expected to be larger, which enormously increased the disorder in the system leading to a significant dispersion effect-induced pulse attenuation by the out-of-phase interference.

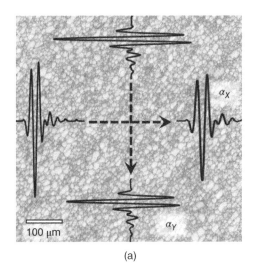 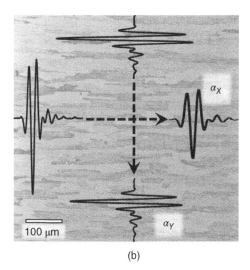

(a)   (b)

Figure 15.6 (a) Attenuation effects in an isotropic microstructure metal. (b) Attenuation effects in an anisotropic microstructure metal.

15.2.3 Influence From Residual Stresses On Ultrasound Wave Traveling

With respect to the laser fusion-based processes, FSW/FSP processes have unique residual stress distributions due to the contribution of mechanical frictions and stir. In bulk mechanical responses, residual stresses affect the plasticity of the workpieces by direct superpositions. In the view of ultrasonic waves, the existence of residual stresses [9, 10] deviates the effective elasticity of the workpiece, which introduces strong modification on the wave propagation behavior. In addition, the nonuniform distribution of residual stresses amplifies frequency dispersion effects on the propagation of ultrasound waves. With the listed dependence between the physical properties and the corresponding ultrasound wave propagation behaviors, a deeper understanding and more information can be obtained from the various nondestructive ultrasonic inspections and evaluations.

Residual stress is the mechanical potential induced by the unrecovered elastic deformation that is stopped by the surrounding plastic deformation. The primary sources of residual stresses are mechanical processing, thermal gradient, and crystal phase transformation. In conventional laser metal processing or laser-fused additive manufacturing, although the residual stresses are strong in terms of magnitude, the residual stress distributions are not as complex as the friction stir-based processes in a bulk view. By involving the linear motional and rotational motional stresses, FSWed or FSPed samples commonly equip strong asymmetric residual stress distribution in terms of compression and tensile with respect to the advancing and retreating side. During ultrasound inspections, when the ultrasound beam has a spatial width clearly smaller than the size of half of width of the SZ, the ultrasonic wave can sense the difference between AS and RS in terms of temporal delay and pulse amplitude if the microstructure on both sides is comparable. However, when the spatial size of the ultrasonic beam is wide enough to cover both sides, commonly existing in ultrasonic immersion scans, the contribution from both tensile and compression residual stresses results in the dispersion effect, where the dispersion effect is proportional to the difference between the tensile and compression residual stresses. Hence, ultrasonic evaluations can map the residual stress distribution with carefully designed experimental configurations. Furthermore, to observe the distribution of the residual stresses, selecting a lower operating frequency of ultrasound is

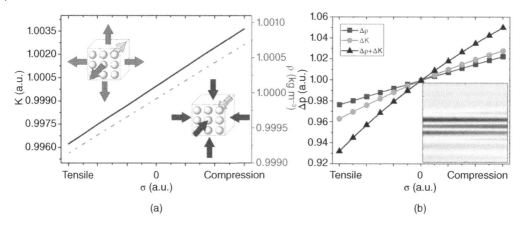

Figure 15.7 (a) Residual stress-dependent effective bulk modulus. (b) Residual stress-dependent acoustic pressure in the medium.

applicable to weaken the deviation of scattering effects due to microstructure differences over the SZ and other regions.

The principle of neutron [11] or X-ray [12] measurements on residual stress or strain were based on the observation of the atomic distance by the diffraction of the wave following the Bragg law. The residual stress-induced atomic distance variation, residual strain, also slightly deviates the density of medium, which was similar to thermal expansions. In the fundamental relation, in a medium with known initial atomic distance a_0 and interatomic potential U, bulk modulus follows the relation: $K = a_0 \left(\dfrac{\partial^2}{\partial r^2} U \right)$. Hence, the bulk modulus is inversely proportional to the interatomic distance r, where the external force with low strain rate on the medium can be balanced out by the changes in atomic force $F \propto -\dfrac{\partial U}{\partial r}$ associated with the change of atomic distance. Overall, the effective bulk modulus can be stated as $K \propto a_0 \dfrac{\partial F}{\partial r}$, that served as the principal observation performance in the atomic distance measurements. With the same expression, in the bulk-scale volumetric form, $K \propto A_0 \dfrac{\partial F}{\partial A}$, where A_0 and ∂A refer to the volume and the change on volume per number of atomic which are inversely proportional to the deviation on the effective density (Figure 15.7a). Such strong changes in the ultrasound wave propagation behaviors deviated speed of sound and amplitude in acoustic simulation models (Figure 15.7b).

15.3 Common Ultrasonic Inspection and Evaluation Methods for FSWed/FSPed Samples

Common nondestructive ultrasonic inspections and evaluations are performed using short and broadband ultrasonic pulses in the time domain. Different from frequency domain measurements, the additional temporal information provides valuable information on FSW/FSP inspections, such as the locations/sizes of defects and speed of sound values of the measured locations. In principle, ultrasonic waves can be grouped into two major groups: longitudinal and transversal waves.

The propagation behaviors are tightly dependent on longitudinal sound velocity/bulk modulus and shear sound velocity/shear modulus. The inspections have commonly relied on phased array imaging systems. Similar to the ultrasonic imaging systems in the biomedical field, the industrial phased array detectors have periodically arranged piezoelectrical elements performing excitations and detections of time-of-flight measurements on the ultrasonic wave arrays, which can be arranged back to images showing the size, location, and even geometry of the discontinuities. Such instant imaging systems are commonly applied to inspect the SZ in FSW/FSP workpieces for estimating the existence of lack-of-penetration defects and large-size cracks. Besides phased array systems, pure longitudinal or transversal mode ultrasound was barely applied on defect/flaw inspections in FSW/FSP. The combined mode ultrasound, such as Lamb and Rayleigh waves, is commonly used in FSW/FSP inspections due to the small workpiece thickness. Depending on the wave generation sources, the observation wavelength of Lamb waves and Rayleigh waves can be tuned from centimeter to micrometer scales by using either piezoelectric transducers or photoacoustic lasers. With the high-frequency ultrasonic waves offered by photoacoustic inspections, some minor size defects or voids can be detected in the thin FSW/FSP workpieces.

15.3.1 Ultrasound Inspections on Defects and Flaws for FSWed/FSPed Samples

A typical pulse-reflection inspection system has three basic components: pulser/receiver, transducer, and oscilloscope. The pulser/receiver provides an electrical pulse to drive transducers for emitting high-frequency (0.5–20 MHz) ultrasonic waves. During the inspection, the acoustic broadband pulse travels through the sample. Some of the power can be reflected when a discontinuity exists in the sample on the wave propagation direction, and the discontinuity size is clearly larger than the acoustic wavelength of the short pulse. The transducer transforms the collected reflections into electrical signals and displays them on an oscilloscope. From the reflection signals' number, amplitude, and time delay, discontinuity location, size, and other information can be further estimated.

Piezoelectric transducers are commonly used to convert electrical analog signals into corresponding mechanical vibrations, thereby generating ultrasound. The energy transformation between an electrical and vibration (strain) wave is reversible. Therefore, in general, a piezoelectric transducer can be performed as an emission source or (and) receiver. Due to the composite materials on the piezoelectric transducers, coupling materials are needed to perform ultrasound inspection in either contacting or immersion tests. During immersion tests, the ambient fluid can behave as ultrasound coupling material; hence, no additional coupling material was needed. But in contacting tests, due to the existence of surface roughness and acoustic impedance mismatching, additional coupling materials are required, which usually are fluids, including water, cutting oil, honey, or aloe vera gel. Note that the applied thickness of the coupling material is better to be thin, under ¼ of the operating wavelength (calculate using the speed of sound in coupling fluid). Cutting oil is a good solution as a coupling material to perform ultrasound tests on water-active metals. Ultrasonic transducers are generally grouped as planar transfers or focused transducers. At the same operating frequency, planar transducers emit plane wave-like ultrasound beam, which is suitable for performing transmission tests or time-of-flight measurements for estimating the elasticity of the tested samples. On the contrary, focused transducers are more appropriate for defecting scans due to the small spatial focal point with relatively high acoustic intensity. Note that the focal length announced on the datasheet of the focus transducers usually refers to the focal distance in the time domain with a broadband pulse. With the application using single frequency component continuous waves, the frequency-dependent

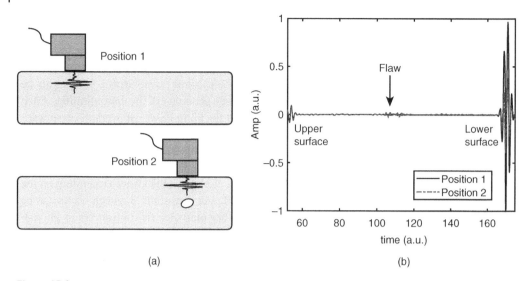

(a)   (b)

Figure 15.8 (a) Acoustic inspection at a location without defect (upper) and a location with defect (lower). (b) Typical reflection signals from the numerical simulations with the conditions in (a).

focal lengths can be present in experiments. Figure 15.8a,b demonstrates the configuration example and expected signals from the flaw detection inspection.

Instead of piezoelectric transducers, laser-induced photoacoustic effects can also be involved in ultrasonic inspections. The laser-exposed area can be heated up with a high-intensity laser focused on the tested sample. The laser focal point behaves in periodic heating and cooling cycles with periodical laser pulses. The heating and cooling provide local tensile and compression cycles following the heat expansion (Figure 15.9a). The tensile and compression cycles can be applied as an ultrasound source. Note that point source is barely existing in high-frequency ultrasound experiments. The photoacoustic source can be considered a point-like source that also provides isotropic waves. Direction-dependent elasticity isotropy tests are feasible to perform using a photoacoustic setup. Moreover, the common photoacoustic tests operate in high ultrasound or even hyper-sound frequency receivers for a high detection resolution. Therefore, the signal can have significant attenuation in large-scale samples. Furthermore, a low-power continuous laser can also be used as an ultrasound receiver on metal samples, known as laser vibration meters (Figure 15.9b). The ultrasound wave can be detected with high resolution by detecting the strain-induced reflection phase shift in the time domain. On the contrary, such setups are highly environmentally vibration sensitive.

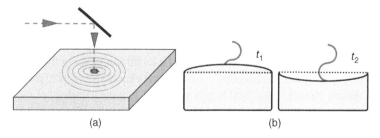

(a)   (b)

Figure 15.9 (a) Photoacoustic emission. (b) Laser vibration meter.

In acoustics and NDTs, the basic experimental configurations can be grouped into either monostatic setups or bistatic setups. Monostatic setup refers to the configuration using one wave device as the emission source and receiver simultaneously. Piezoelectrical transducers can be used in monostatic setups. In general, during one pulse repetition cycle, the piezoelectrical transducer excites a short pulse at the beginning time period in the cycle. Once the ultrasound pulse is emitted, the transducer serves as a receiver for detecting the reflections for the rest of the time. On the contrary, the bistatic configurations separate the emission source and signal receiver as two devices, usually two identical transducers, for diffraction or transmission measurements. In advanced setups, the receiver side in a bistatic setup is not necessary to involve the same transducer when the inspections need a smaller area of interest. Piezoelectrical needle hydrophones or laser vibration meters can also be the proper detectors for the inspection in water or air ambient.

Phased array ultrasonic inspection is an instant ultrasonic imaging or scan system commonly used in industrial and biomedical areas. Multiple piezoelectric elements are integrated into the phased array probe to excite and receive ultrasound waves (Figure 15.10a). By adjusting the phase of excited ultrasound waves from each piezoelectric element, varying wave interferences can offer flexible ultrasonic beamforming in the tested sample. The reflected ultrasonic waves or backscattering from the flaws received at the phased array elements can be constructed back into imaging by computational approaches. The multiple piezoelectric elements on the probe are arranged into either a 1D array or a 2D array. With 1D phased array probes, instant ultrasonic 2D imaging can be conducted with the axes on array direction and sample depth direction (temporal scale). The imaging resolution is generally proportional to the phased array frequency and the number of piezoelectric elements (Figure 15.10a). In the industrial field, ultrasonic phased array systems are commonly applied to determine flaws in welded samples, especially on the joint regions. In FSW/FSP, phased array ultrasonic systems are frequently used to detect the lack of penetration flaws. However, as a contacting ultrasonic testing method, the flashing materials and the thickness variations around the SZs introduce difficulties in the phased array scans. Soft and flexible polymers

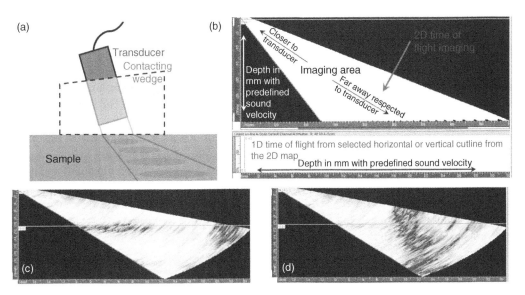

Figure 15.10 (a) Illustration of typical ultrasonic phased array inspection. (b) Illustration of typical interface of phased array inspection system. (c) Illustration of typical defect in FSW workpiece in perpendicular orientation. (d) Illustration of typical defect in FSW workpiece in longitudinal orientation.

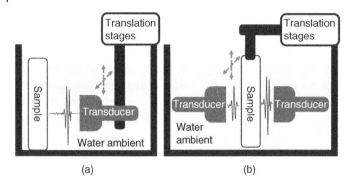

Figure 15.11 (a) Monostatic immersion ultrasonic scan. (b) Bistatic immersion ultrasonic scan.

are involved in making widgets and serve as coupling materials to overcome limitations. Figure 15.10c,d demonstrates the imaging results on a typical defect in FSW workpiece in perpendicular and longitudinal orientation, respectively.

Ultrasonic immersion scans or mapping are usually performed in water ambient with immersion ultrasonic transducers. The water serves as flexible ultrasonic wave coupling material to overcome the surface roughness limitation on the contacting ultrasonic testing. The scans are performed with both monostatic (Figure 15.11a) and bistatic (Figure 15.11b) configurations. With enough apart distance between the ultrasonic probe and tested sample, on the temporal scale, the immersion tests can provide an additional reflection signal occurring at the front surface of the tested samples with respect to the contacting ultrasonic tests. The surface reflection can be used to map distributions of some additional information of the samples, such as relative acoustic impedance and relative surface roughness. On the contrary, due to the front surface reflection, the ultrasonic wave transmitted into the tested sample becomes relatively less, which affects the resolution of the internal flaws or defects detection. Due to the relatively thin sample thickness, the immersion tests are suitable for inspecting FSWed or FSPed samples.

Sensitivity and resolution are two commonly used terminology to illustrate a technical ability to detect defects in ultrasonic inspections. The capacity to detect minor discontinuities is sensitivity referring to the lowest signal amplitude that can be detected. In general, higher frequencies (shorter wavelengths) increase sensitivity. Resolution is the system's capacity, within or close to the surface of the component, to locate discontinuities. As the frequency increases, the resolution also generally increases in principle. On the contrary, as mentioned in the previous section, the increased operating frequency simultaneously increases the signal destructive effects, including scattering and dispersion. The low amplitude or short temporal width information can be hidden in the time-of-flight measurements. Hence, the frequency selections and inspection setup selections are important. Based on the thickness of the sample (along the wave propagation direction) and estimated sound speed values, a short enough ultrasonic pulse needs to be selected. With the basic assumption on the grain sizes over the different zones on the samples, the shortest wavelength in the broadband pulse still needs to be much larger to avoid strong scattering effects. The selected wave propagation direction is better to have a lower deviation in terms of grain size to avoid strong dispersion effects.

15.3.2 Ultrasound Evaluations on Mechanical Properties for FSWed/FSPed Samples

In defect-free or low void fraction FSWed/FSPed samples, with the measurements on longitudinal sound velocity and shear sound velocity, the elastic modulus can be further evaluated in terms of shear modulus, bulk modulus, and Young's modulus. Without additional machining or polishing, ultrasonic evaluations can compare the estimated elastic modulus between different workpiece zones or samples with various processing parameters as a part of quality control or parametric optimization.

Tensile tests and nano-indentation are the two common destructive methods to test the elasticity properties of the samples. Ultrasonic evaluation is the most common nondestructive method to determine the elasticity by measuring the longitudinal and transversal sound speeds. Based on Hooke's Law and Newton Laplace Law, speed of sound values in the media can be used to calculate elasticity based on the known density values, which is stated as $c = (E_{ij}/\rho)^{1/2}$, where c is the speed of sound, ρ is the pre-estimated density the material, and E_{ij} is the elasticity component along the wave oscillating direction (Figure 15.12a).

In metallic materials, the sound velocity of the transversal (shear) wave is slower than the longitudinal sound wave in general. Hence, transversal ultrasound equips a shorter operating wavelength at a certain selected frequency. Additional scattering and dispersion effects can be introduced on the transversal ultrasound wave. Therefore, a slightly lower operating frequency selection on the transversal ultrasound with respect to the longitudinal ultrasound frequency was suggested in such elasticity measurement. In the conventional sound velocity-based elasticity measurements, density ρ and sample thickness d values are pre-estimated as known values. By obtaining the temporal delays of longitudinal and transversal ultrasound waves in the sample with two parallel surfaces using monostatic testing configuration (Figure 15.12a), speed of longitudinal and transversal mode ultrasound waves can be obtained as $c_L = 2d\,t_L^{-1}$ and $c_T = 2d\,t_T^{-1}$, where c_L, c_T are the speed of longitudinal and transversal mode ultrasound waves, and t_L, t_T are the time-of-flight values of the longitudinal and transversal mode ultrasound waves in the sample. With the speed of transversal

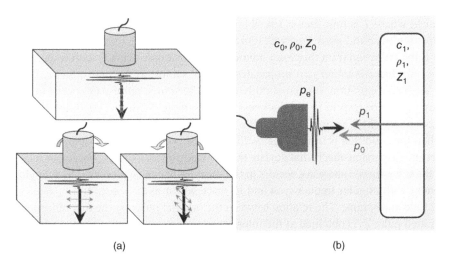

(a)   (b)

Figure 15.12 (a) Wave oscillation is independent with the rotation angle of the longitudinal transducer (topr). Wave oscillation is dependent with the rotation angle of the shear transducer (bottom). With both measurements, the elasticity can be estimated. (b) Immersion ultrasonic test for the estimation of elasticity.

wave cT, the shear modulus G can be calculated as $G = \rho c_T^2$. And, Young's modulus E and Poisson's ratio v of the tested sample are computed by: $c_L = \left(\dfrac{E(1-v)}{\rho(1+v)(1-2v)}\right)^{1/2}$ and $G = \dfrac{E}{2(1+v)}$.

In advance, oscillation angle-dependent transmission tests can be feasible using the transversal ultrasonic speed of sound measurements [13]. As discussed in the previous section, longitudinal ultrasound has an oscillation direction and wave propagation direction on the same axis. However, transversal ultrasonic waves have an oscillation direction perpendicular to the wave propagation direction. Perhaps, the transversal ultrasonic wave propagates along Z-axis, and the wave can have an oscillation direction along X or Y axis or even with any orientations in the XY plane. Hence, by changing the orientation of the shear wave transducers, the angle-dependent anisotropic elasticities (shear modulus and Young's modulus) can be estimated, which is important in metallic samples that equip grain orientations-dependent anisotropic mechanical properties. Such information is hard to obtain with other conventional mechanical properties without advanced cutting and preparations on the samples (Figure 15.12a).

The introduced longitudinal and transversal ultrasound velocity-based measurements are heavily used and studied in academia and industries. However, as mentioned, contacting ultrasonic measurements have limitations in the cases of FSWed and FSPed samples, which have flashing materials and uneven thickness. The contacting measurements can perform well when measuring the elasticity values in only the SZ and base materials. However, when the interests are focused on the elasticity transitions from zone to zone, the contacting measurements can only be conducted appropriately with additional polishing processes. Hence, in this case, another alternative approach for elasticity mapping, determining effective bulk modulus using immersion setup (Figure 15.12b), can overcome the limitations.

In homogeneous fluid ambient, the velocity vector can be described by the wave potential as $\vec{V} = \nabla\varphi$. The potential φ is the factor in the governing wave equation $c^{-2}\ddot{\varphi} + \nabla^2\varphi = 0$, where c is the effective speed of sound. The pressure oscillation of the sound wave is also stated by the potential following $p = -\rho\dot{\varphi}$, where ρ is the density of the fluid. Since the immersion elastography [14] involves short ultrasonic pulse instead of continuous wave. Perhaps a short pulse $\varphi_e(T)$ sent by an ultrasonic transducer, where T is time factor describing the wave traveling in temporal domain. The entire envelope of ultrasound wave separates into two pulses including reflection and transmission. One part is the reflection from the closer boundary of the sample as $\varphi_0(T)$ with respect to the location of the transducer. Another part propagates into the sample as the transmission. It is noted that, under such configuration, the temporal length of the source pulse $\varphi_e(T)$ needs to be short enough comparing with respect to the thickness of the sample. With this condition being satisfied, the first boundary of the sample can be stable when the roundtrip reflection travels back there. Perhaps, when the condition is invalid, the oscillating first boundary offers complex transmission coefficient and dispersion effects that deviate the acoustic pressure values. The linear relationship between the wave speeds and wave vectors in the short pulse can be presented by a single frequency component with angular frequency ω and its wave vector as $k_0 = \omega/c_0$ and $k_1 = \omega/c_1$ stated for ambient fluid and sample. The relation between the original pulse φ_e, the first reflection φ_0, and the transmitted pulse φ_t is obtained at the interface between the sample and the ambient fluid. The acoustic pressure and sound velocity are $p_e + p_0 = p_t$, $V_e + V_0 = V_t$. In terms of potential, the original pulse and two separated pulses are expressed as $\varphi_e(x,T) = e^{ik_1x - i\omega t}$, $\varphi_0(x,T) = r_{0,1}e^{-ik_0x - i\omega t}$, $\varphi_t(x,T) = t_{0,1}e^{ik_1x - i\omega t}$, where $r_{0,1}$ and $t_{0,1}$ are the reflection coefficient and transmission coefficient following $t_{0,1} = 2Z_1/(Z_0 + Z_1)$ and $r_{0,1} = \dfrac{Z_1 - Z_0}{Z_1 + Z_0}$. The acoustic pressure is $p_t = \dfrac{Z_1}{Z_0}\left(p_e - |p_0|\right)$ and $p_t = \dfrac{Z_1}{Z_0}\left(p_e + |p_0|\right)$ with the conditions satisfied $Z_1 > Z_0$ and $Z_1 < Z_0$, respectively.

In experimental cases, p_e can be obtained using a known impedance reference sample in the same ambient fluid. The impedance of the sample Z_1 can be estimated using the relative amplitude of the reflection p_1. Transmission propagating through the first boundary of the sample and left part of the energy inside the sample, the relation between the roundtrip reflection and the first reflection can be expressed using the amplitude of the original pulse following $p_1 = t_{1,0}\, r_{1,0},\ p_t = \dfrac{2Z_0|Z_1 - Z_0|}{\left(Z_1 + Z_0\right)^2}\left[p_e - \mathrm{sgn}\left(Z_1 - Z_0\right)|p_0|\right]$. In practice, the amplitudes of all three signals,

p_e, p_0, and p_1 can be collected as constant to form a factor $\alpha = \dfrac{p_1}{p_e - \mathrm{sig}\left(Z_1 - Z_0\right)|p_0|}$. The acoustic

impedance of the tested sample Z_1 can be further calculated as $\dfrac{Z_1}{Z_0} = \dfrac{-1 - \alpha - \sqrt{4\alpha + 1}}{\alpha - 2}$. In the

experimental data, p_0 and p_1 are taken as the corresponding peak values of the pulses. The speed of sound value c in the target sample can be determined using the method introduced in the previous section with the known sample thickness value. Based on the estimated acoustic impedance and speed of sound of the tested sample, the dynamic bulk modulus and effective density of the scanned sample can be obtained as $K = Zc$ and $\rho = \dfrac{Z}{c}$.

15.3.3 Ultrasound Monitoring for FSW/FSP

Currently, during the FSW and FSP processes, basic monitoring can be conducted via the real-time transition on the machine feedback data, including the normal force, torque, velocity, and displacements of the tool. In advance, introducing a thermal camera and thermal couples in the setup can provide the surface or subsurface temperature distribution on the processing workpiece. Although such monitoring data are helpful in back-track for determining the processing behaviors around the error that occurred, none of the monitored parameters can solely refer to the processing quality. Therefore, studies on ultrasound monitoring for FSW and FSP are conducted to have real-time data regarding the flaw formation and transient mechanical properties of the workpiece during the processing. Ultrasonic monitoring can serve as an in situ observation method collaborating with machine feedback data by applying the concepts mentioned in the previous sections to understand or even predict the formation of welding or processing flaws, which has attracted more attention in academia. The methodologies are various at the current initial study stage of ultrasonic in situ monitoring. However, all provide unique understanding and observations during the FSW/FSP processes, which are worth discussing in detail.

Ultrasonic monitoring systems [15, 16] are usually applied to perform real-time determination of the health of the workpieces in terms of fatigue, corrosion, and defect initiations which do not involve significant displacement or motion on the workpiece. In manufacturing processes, ultrasonic monitoring systems have recently been widely developed for additive manufacturing processes that use relatively static experimental setups. Due to the nature of FSW and FSP, relatively large displacements of the processing tool introduce challenges to adding an ultrasonic sensor

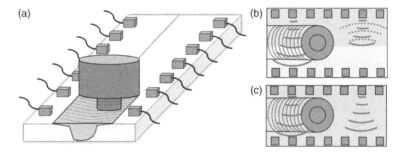

Figure 15.13 (a) Ultrasonic sensors array monitoring in FSW. (b) Top view of ultrasonic sensors array monitoring in FSW. (c) Top view of ultrasonic sensors array monitoring in FSP.

that can follow the motion of the tool using any contacting method. Hence, the alternative solution from the literature points out that the in situ monitoring configurations can be designed using an array of ultrasonic sensors on the workpiece or air-coupled measurement methods.

Welding flaws can be found during the FSW processes, including the bonding quality, lack of penetration flaws, and significant defects. In a linear path FSW or FSP process, the array of ultrasonic sensors can be placed on two sides of the base material regions (Figure 15.13a). The configuration can offer ultrasonic waves propagating in perpendicular directions with respect to the welding path. Alternatively, a single array of sensors can also perform the monitoring by recording the reflections from the welding interface. In the initial state of the FSW cases, the two separated raw metal plates cannot conduct ultrasonic waves to pass through due to a lack of bonding. During the FSW processing, from the receiver closer to the starting location, the ultrasonic waves can pass through the welding interface with the presence of sufficient bonding (Figure 15.13b). To properly detect the bonding condition over the entire depth, Lamb waves and Rayleigh waves can be applied for various thicknesses of the workpieces. With the monitoring of the temporal delay, pulse amplitude, and even minor scattering, lack of penetration flaws and significant defects can be determined by the presence of additional attenuation in transmission, additional reflection pulses, or considerable scattering. Such ultrasonic in situ monitoring method is efficient and has been experimentally verified in literature.

However, there are still several disadvantages and limitations. By using an array of ultrasonic sensors, the locations in the intervals of the sensors cannot be directly monitored. The sensors cannot excite ultrasonic waves simultaneously due to the existence of interference. The monitoring behavior needs to be clarified in the FSP cases (Figure 15.13c). Furthermore, to evaluate the bonding condition of the welded interface, the proposed method has an issue with lack-of-reference, especially in the cases of dissimilar metal welding, which has significant acoustic impedance mismatching conditions even under well-welded workpieces. Overall, ultrasonic array monitoring is still a decent method that can be applied in future industries. To overcome the limitations of the proposed, advanced signal processing techniques, artificial intelligence can be involved to overcome the listed challenges.

Elastic modulus monitoring during FSW and FSP processes is under development in academia as a potential direction in the future. Besides the defects and flaws detection, the transient effective elastic modulus monitoring can indicate excess heating-induced over-softening effects and even predict the potential residual stress behaviors on the final products. With such a monitoring

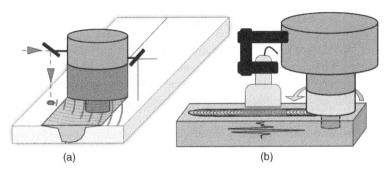

(a)   (b)

Figure 15.14 (a) Air coupled monostatic ultrasonic monitoring in FSW/FSP by photoacoustic emission and laser vibration meter. (b) Air coupled monostatic ultrasonic monitoring in FSW/FSP.

system, the in situ control loop can adjust the rotational speed or linear motion velocity accordingly to retain the desired region's mechanical properties. With the currently proposed ex situ set-ups, the non-contacting nondestructive evaluation methods are possible to translate into in situ monitoring methods for real-time elastic modulus measurements. Firstly, photoacoustic emission and laser vibration meter configuration can excite and receive the ultrasonic wave time-of-flight information following the linear motion of the FSW/FSP tool (Figure 15.14a). The change in the elastic modulus can be obtained by tracking the speed of sound information along the FSW/FSP path. Alternatively, an air-coupled transducer can be involved to collect temporal reflection signals following the tool on the SZ (Figure 15.14b). On the thin samples, effective elastic modulus or the effective acoustic impedance transient behaviors can be captured to analyze the heat and stress evolutions during the FSW or FSP processes. Similarly, in the underwater FSW or FSP, the immersion ultrasonic transducers can enable in situ observation with higher sensitivity.

In addition, there is another in situ monitoring approach needs to be mentioned, which applies an ultrasound emission test during the friction stir welding and processing [17]. Acoustic emission tests are broadly used in plasticity studies for composite materials. The approach has been used for metals to find relative ductile to brittle transition behaviors caused by corrosions or embrittlement during conventional tensile tests. The principle of acoustic emission tests is based on the frequency analysis of the sound emission from a sample when it breaks or forces sharp plastic deformation to determine the ductility of the samples. Imagine that the identical tensile samples are under testing with identical conditions. The harder material, or higher elastic modulus, would emit a high-frequency sound when the sample breaks. In friction stir-based manufacturing techniques, acoustic emission monitoring is focused on the behavioral analysis of the tool–workpiece interactions. It is well known that the friction stir-induced softening effect can transfer the solid metal to softened material with viscoplastic behaviors. Especially, aluminum can have shear rate-dependent viscoplastic behaviors in the deformation field and strain rate-dependent viscoelastic behaviors on high-frequency ultrasound. The softened metals' effective viscosity can achieve zero at a specific temperature and shear rate produced by the rotational tool (shoulder and pin), leading to the sticking-to-slipping transition in FSW and FSP. Such behaviors are represented frequently by numerical simulation with viscoplastic fluidic models but are barely verified in experiments. Hence, the scholars in the mentioned reference cleverly applied acoustic emission monitoring to capture the transition in experimental conditions. The transition between sticking and slipping conditions can be found by tracking the fundamental frequency shifting and further monitoring the stability and void formation during the processes. The work demonstrates that acoustic emission monitoring is a great tool for experimental parametric studies, which is highly valued for parameter-sensitive friction stir-based processes. In this literature work, the works were performed under FSP-like configuration for demonstrations. In the future study, the potential works can be considered to translate the proposed method to FSW with two acoustic emission sensors on both raw workpieces to observe the sticking to slipping transition via the advancing side and retreating side separately. Especially in dissimilar metal welding configurations, two sensors would be beneficial to track the sticking-to-slipping transition on the softer metal side and metal segment formation on the harder metal side. This would be a powerful tool in both practical applications and fundamental studies.

15.4 Case Studies on Recent Novel Ultrasound Evaluation and Monitoring in FSW/FSP

After the explanation and introduction of the ultrasonic wave and tests for FSWed/FSPed samples, the case studies were listed to show the recent novel ultrasonic evaluation works with experimental approaches on FSWed/FSPed samples which provide the signatures of FSWed/FSPed samples and the linkages to the understanding of the FSW/FSP processes, including the elasticity mismatch-induced stress concentration, asymmetric elasticity distribution in SZs, elasticity deviation around the starting/ending locations, and residual stresses distribution over entire FSW/FSP workpiece. In addition, an artificial intelligence-assisted ultrasonic monitoring on the FSW/FSP tool wearing work is also important to discuss a critical component in FSW/FSP quality controls.

15.4.1 Ultrasonic Elastographic Evaluated Dissimilar FSWed Samples

In friction stir welding of dissimilar metals, elasticity mismatching and the formation of intermetallic compounds (IMCs) are the critical factors leading to premature failure under uniform transversal loading. In dissimilar metal combinations with IMCs, the brittle interfacial product connects two elasticity mismatching and plasticity mismatching base materials. When the cross section is under transverse tensile loading, elasticity mismatching location concentrates the stress-induced large magnitude of elasticity strain (stress concentration). The high local stress and large strain introduce an earlier time point of shifting the elastic strain to elastoplastic strain at the softer side material around the IMCs, which leads to stress concertation, therefore, causing the initiation of a crack at the welded interface around the IMC or IMC.

In dissimilar metal combinations without the presence of IMC, the elasticity mismatching and stress concertation behaviors are barely studied in the literature. This section presents an ultrasonic elastograph that predicted failure location work using a dissimilar copper and steel friction stir welded (FSWed) sample without forming IMC. As stated, besides the presence of IMC, the remaining majority of crack initiation is the stress concentration condition under loading caused by the sharp elasticity mismatching. However, by using the conventional elasticity measuring techniques, the elastic modulus distribution is hardly mapped continuously, which would introduce uncertainties in determining the mismatching location. For example, indentation-based mapping methods are commonly used to map the elasticity or even plasticity distributions on the samples. However, certain spatial gaps between the step intervals during the mapping cannot be avoided. The immersion ultrasound raster scan-based measurement was selected to map the elasticity distribution on the dissimilar metal FSWed sample to overcome the limitation in this work.

Ultrasonic wave raster scans are commonly performed in ultrasonic microscopic works. However, ultrasonic microscopic imaging still shows the feature discontinuously in general, because the existence of the spatial internal induces observation gaps. Hence, in this ultrasonic elastograph work, the raster scan spatial interval is selected to be smaller than the ultrasonic beam width from the ultrasonic transducer to avoid discontinuity in the mapping. Therefore, at each scan location, the measured elasticity was the effective value self-averaged from a cylinder volume on the sample, which describes as $V_{\mathrm{eff}} = \pi \left(\dfrac{W_{\mathrm{FWHM}}}{2} \right)^{2} t$, where t is the sample thickness and W_{FWHM} is width of acoustic beam at full width at half max. As long as the relative position of the sample and transducer changed along any axis, the measured dynamic bulk modulus values were the effective values from two different volumes. With a smaller scan step than the beam waist, the measured effective volume at neighbor locations partially overlaps. The non-overlapped volume contributed to a difference in measured elasticity distribution. In such an approach, effectively, the entire area of interest can be adequately scanned without the existence of observation gaps.

Figure 15.15a shows the FSWed copper and steel joint in which the steel and copper were arranged on the advancing and retreating sides, respectively. The welding path has a slight offset toward the copper side. No significant flawing material and surface crack can be found on the sample without any preparation or pre-processing. The central dash box indicates the ultrasonic

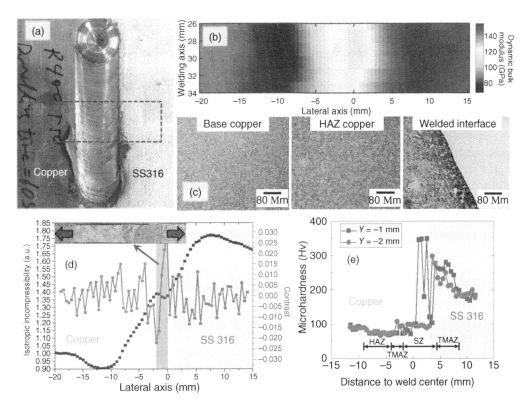

Figure 15.15 (a) Dissimilar copper and steel FSW workpiece. (b) Ultrasonic scanned elastograph on the workpiece. (c) Optical microscopy on the workpiece. (d) Elasticity analysis based on the ultrasonic elastograph, and digital imaging tensile test result on the cross section. (e) Hardness profiles on the cross section. Source: Jin et al. [18]/with permission from Elsevier.

elastograph region covering base material copper, SZ, and base material steel around the middle zone along the longitudinal direction. The corresponding elastograph is posted in Figure 15.15b. Physically, the elastograph presents the averaged effective bulk modulus along the sample thickness direction over the horizontal distribution. In the elastograph, it is clear to observe the copper and steel are located on the left and right sides correspondingly. The SZ in the central region is about 10 mm wide, equipping effective bulk modulus increase and decrease behaviors from the transition between the copper and steel sides. The existence of the HAZ is clearly presented on the copper side at a lateral axis of about-10 mm. A TMAZ is located between the HAZ and SZ on the copper side around the lateral axis-6 mm. On the steel side, no clear HAZ was found. However, a wide TMAZ can be observed between the lateral axis of 5 to 10 mm on the steel side.

Figure 15.15c shows the microscopies of the FSWed sample in copper side base material, copper side HAZ, and the interface between the copper and steel. Besides the commonly existing tensile residual stress in HAZ, the microscopic images show the microstructure influence on the ultrasound measured lower effective elasticity in HAZ copper due to the large grain size with respect to the base material copper. At the interface between the copper and steel, the local microscopy shows a sharp material transition was observed regarding the mismatching condition in microstructure and materials, which was commonly predicted as the failure location during the tensile test.

The ultrasonic elastograph results indicated the mismatched location at different positions. In Figure 15.15d, the elastograph was summarized into a 1D profile, as the black line indicated. Based on the profile, the contrast profile was calculated based on the changing slope of the effective elasticity $\frac{dK}{dx}$ and presented as the red line on the plot. The strongest mismatching location on the effective elasticity profile was observed in the SZ on the copper side, which is about 1–2 mm away from the interface between the copper and steel. Such mismatching conditions can be observed or missed in the conventional microhardness mapping, as Figure 15.15e indicates.

To validate the observations and predictions made by the ultrasonic elastograph, a cross section was cut out from the FSWed sample around the ultrasound-mapped area. The cross-section sample was prepared to perform digital imaging assisting tensile test. As the inserted sub Figure 15.15d shows, the crack initiation location and failure position were not at the interface between the copper and steel. Instead of the interface, the crack initiated and propagated around the position 2–3 mm away from the material interface on the SZ copper side. During the recording of the digital imaging assisting tensile test, the completed strain distribution behaviors agreed with the stress concentration hypnosis. In the initial elastic deformation region, the largest strain was found in the HAZ of the copper side due to the lowest elastic modulus with respect to the other regions over the entire sample. In the elastoplastic deformation region, the strain started to concentrate at the elasticity mismatched location, which is indicated by the ultrasonic elastograph. The concentrated stress induced more considerable strain. Along with the increasing stress and strain, the elasticity mismatched location initiated a failure.

In the case study, a FSWed metallurgically immiscible copper–steel structure was evaluated by the recently developed noncontracting ultrasonic elastomapping [18]. The elasticity map clearly presented the signature zones on the FSWed sample, such as base material, HAZs, and TMAZs. By analyzing the elasticity mismatch conditions, the accurate spatial location of the crack initiation around the welding path was nondestructively predicted. The conventional tensile tests and digital image correlation strain mapping further corroborated this. The proposed methodology properly correlated the elasticity mismatch with crack initiation using the continuous dynamic bulk modulus mapping without the presence of a brittle intermetallic compound. In summary, two main

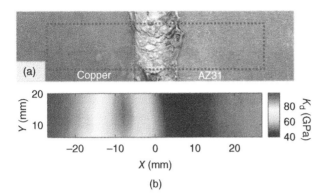

Figure 15.16 Dissimilar copper and AZ31 FSW workpiece and ultrasonic scanned elastograph on the workpiece.

advantages of the proposed technique were claimed. Firstly, the method can obtain the elasticity distribution of dissimilar joints nondestructively without any sample preparation. Consequently, the evaluation of the mechanical process using this technique is much faster than conventional measurement methods. Furthermore, the present technique can be applied to various welded structures to detect any unexpected elasticity drop, which can become a hazardous location while the structure is under mechanical strain.

The second case that needs to be introduced here is another dissimilar metal FSWed sample with copper and magnesium AZ31, shown in Figure 15.16a. The combination of metals is not commonly welded in conventional industries due to the significant corrosion issue. However, the FSWed sample demonstrates performance in terms of ultrasonic evaluations. On the FSWed sample, copper and AZ31 were placed on the advancing and retreating sides, respectively. The elastograph (Figure 15.16b) shows the SZ was from $X = -5$ mm to $X = 5$ mm. A sharp transition was observed on the advancing side (left) of SZ, whereas the retreating side presented a faded-out transition from the SZ to HAZ. The behavior agrees with the common microscopic observation about AS and RS in FSWs. Furthermore, on both copper and AZ31 sides, HAZs were located around $X = -8$ mm and $X = 8$ mm. Due to the high thermal conductivity and thermal diffusion, the softened regions were broad on both sides. Generally, the FSWed copper and AZ31 joint equip all the signatures in dissimilar metal welding but also differ from the typical FSWed workpiece in the bulk view. In dissimilar metal friction stir welding, when the softening temperature or the mechanical properties of the two candidates have substantial differences, the welding parameters are usually selected to fit the metal with a lower softening temperature. The harder metal on the advanced side usually equips a clear TMAZ without forming a significant HAZ due to the relatively low temperature for grain growth and free to stress releasing condition to the softer metal. The observation of this copper and AZ31 joint clearly showed that the HAZ could also be significant with the harder metal on the advancing side. The ultrasonic elastograph captured the unique elasticity distribution with the suitable observation scale.

15.4.2 Ultrasonic Elastographic Evaluated FSPed Sample

The third case introduced in this section is a friction stir processed aluminum plate. As a local and continuous mechanical properties modification method, the mechanical properties distribution on the FSP products are piece-by-piece unique and highly processing parameter dependent.

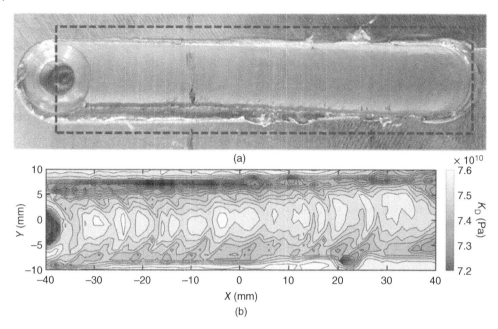

Figure 15.17 (a) Aluminum alloy FSP workpiece. (b) Ultrasonic scanned elastograph on the workpiece.

Figure 15.17a shows a FSPed AA 2024 sample with the advancing side and retreating side close to the upper and lower edges of SZ, respectively. In Figure 15.17b, the ultrasound elastograph is plotted. The map clearly shows the asymmetric elasticity distribution due to the modification of the FSP. Between $Y = -7$ mm and $Y = 6$ mm, the SZ presents properly with slightly higher effective elasticity on advancing side and lower effective elasticity on retreating side, respectively. Moreover, the transition from the advancing side SZ ($Y = 5$ mm) to HAZ was sharper with respect to the transition on the retreating side ($Y = -7$ mm). From the starting point to the end point on the processed path, the highest effective bulk modulus along transversal direction slightly increases. The width of the HAZ on retreating side (lower side) decreases with the increases on the width of the HAZ on the advancing side. Such behaviors can hardly be observed by conventional testing techniques, but the distribution is important for the parametric optimizations in terms of quality control. Hence, the ultrasonic elastograph is considered as a suitable tool for observation and understanding of the mechanical properties' distribution in the bulk view.

15.4.3 Ultrasonic Effective Residual Stress Mapping on FSPed Sample

In the previous section, the methodology of ultrasonic residual stress mapping was introduced based on the approach of estimation of the effective bulk modulus deviation. In this case study, the introduced method was applied via numerical simulation and experiments on a FSPed aluminum plate. Firstly, using molecular dynamics simulation, the stress-dependent effective density and effective bulk modulus on aluminum were obtained and input into bulk-scale finite element analysis-based ultrasonic wave propagation simulations shown in Figure 15.18a. With immersion testing configurations (simulation), the aluminum sample was separated into three regions with the compressional residual stress, stress-free, and tensile residual stress. As the left column figures

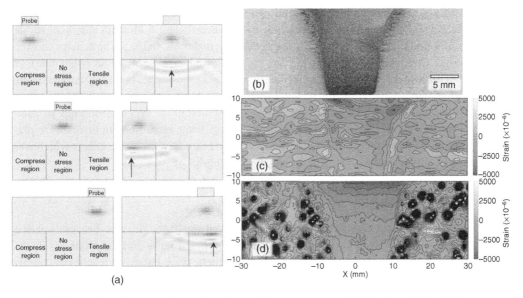

Figure 15.18 (a) Residual stress-dependent ultrasonic wave propagation behaviors obtained from numerical simulations. (b) Cross-section sample from a aluminum alloy FSP workpiece. (c) Ultrasonic immersion test mapped residual strain/residual stress map on the cross-section sample. (d) Synchrotron X-ray mapped residual strain/residual stress map on the cross-section sample.

illustrate, the ultrasonic transducer excites identical ultrasonic waves propagating to the different regions on the aluminum sample (upper, middle, and lower). On the right-side column, the ultrasonic waves have traveled approaching to finish the roundtrip in the aluminum sample. The stress-dependent ultrasonic wave propagation behaviors can be observed, even only ±20 MPa residual stresses were assigned on the regions equip residual stresses. Very slight temporal delays exist between the different regions which can be negligible. However, the amplitude difference between the ultrasonic waves propagated in different regions can be clearly found as the black arrows indicated. Due to the stress-dependent effective bulk modulus and effective density, the residual stresses provide clear acoustic impedance difference. In the case of sample with compressional residual stresses, the higher acoustic impedance region provides stronger reflection amplitude at the upper surface of the sample; hence, less acoustic energy propagates into the sample. The second reflection completed the roundtrip inside the sample provide further strong reflection amplitude difference in comparison. Based on the temporal information from the first and second reflections, the acoustic impedance, effective density, and effective bulk modulus can be estimated using Equation 15.1 (Section 15.2.1) stated in the previous section.

After the verification with nano-to-bulk scale numerical simulations, the ultrasonic residual stress estimation method was applied to cross-section sample from a FSPed aluminum sample (Figure 15.18b) for proof-of-concept. Both ultrasonic wave-based mapping (Figure 15.18c) and standard synchrotron X-ray based mapping (Figure 15.18d) were conducted for comparison. In the plots, both figures were presented in terms of residual strain, which can be translated into residual stress range ±300 MPa (on the limits of the scalebar). Note that the ultrasonic residual stress map presents the distribution averaged residual stress values over the depth (Z-axis): however, the synchrotron map presents the residual stress distribution on the central plane of the sample

(*XY*-plane). Hence, compared with ultrasound and synchrotron, the mapped-out residual stress distributions were not expected to be perfectly identical. Nevertheless, both maps clearly illustrate the SZ and the boundaries between the SZ and surrounding materials. In the SZ, both maps showed slightly stronger compressive residual stresses on the advancing side of the SZ (right side). In addition, both maps also present agreement that the SZ closer to the upper boundary equips strong compressive stress. In the ultrasonic residual stress map, a HAZ with tensile residual stresses is present on the advancing side of the SZ which is not presented in the synchrotron map. In addition, on the base material regions on the synchrotron map, some measurement errors were presented due to the highly changed grain size induced grain boundary scattering effects which was not affecting the ultrasound with long observation wavelength.

15.4.4 Ultrasonic Monitoring on FSW/FSP Tool Wearing

Tool condition is critically related to the quality of the FSWed and FSPed products because wearing of the tools, especially the pin length, can introduce significant flaws and uncertainties in the FSWed and FSPed products. Optical and other electromagnetic wave-based approaches are highly accurate in wearing measurements or monitoring but easily getting influenced by the sticking materials on the tool. Hence, ultrasonic measurement could be a suitable method for monitoring tool wearing by using penetrating waves to measure the pin length through inside from the shoulder [18]. With the wearing on pin length, the decrease in the temporal delay was expected to be observed, as the simulation results in Figure 15.19a illustrate. The time-of-flight behavior can be obtained using numerical simulation of the wave propagation behaviors, as Figure 15.19b shows. With placing the ultrasound transducer on the shoulder of the FSW/FSP tool, the two major internal reflections can be obtained from the lower boundary of the tool shoulder (the interface between shoulder and tool pin) and the pin tip of the tool as the indications marked in Figure 15.19b. With a parametric study on the pin length with fixed pin width, a representative signal bank was obtained describing the wearing process on a tool pin. The summary of the signal bank is posted in Figure 15.19b. The numerical simulation presented no wearing condition on the tool shoulder. Hence, the area of interest on the signal bank was focused on the second reflection, which occurred at the pin tip. However, as the contour shows, due to the finite transducer size and operating wavelength with respect to the size of the tool pin, the wearing-dependent reflection is not linear in terms of time-to-distance relation, reflection amplitude, and waveform. Therefore, the prediction of the tool pin's wearing condition is challenging based on the reflection waveform.

Since the simulation produced signal bank is obtained, a machine learning program can help study the signal bank and predict the pin length. Using the signal bank obtained by numerical simulations in Figure 15.19c, one machine learning program was selected and performed the training and testing, showing the exceptional performance (Figure 15.19d). To test the practical performance of the trained machine learning program, the experimental time-of-flight data bank was measured on the FSW tools with different lengths (Figure 15.19e). The simulation data trained machine learning program was applied to predict the pin length from the experimental obtained ultrasonic wave data. The results show that the numerically trained machine learning AI can perform tool wearing measurements in monitoring configuration on the experimental data, as Figure 15.19f illustrates. One more requirement needs to be fulfilled to achieve the tool wear monitoring in this study for more general cases. The acoustic impedance difference between tool and workpiece materials must be large enough to provide a reflection signal with exceptional amplitude in experiments. In Figure 15.19g, the impedance mismatching-induced reflection energy

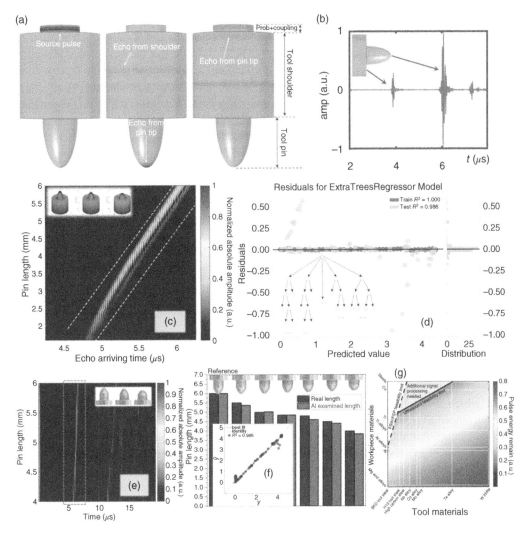

Figure 15.19 (a) Illustration of the ultrasonic time-of-flight measurement on an FSW/FSP tool for pin length detection. (b) Typical reflection signal obtained from the ultrasonic time-of-flight measurement obtained from numerical simulation. (c) Summary of reflection signals with varying pin length obtained from numerical simulation. (d) Training on the machine learning program by the numerical data bank for pin length estimation. (e) Experimentally measured reflection signals with different pins with varying pin length. (f) Numerical data bank trained machine learning program estimated pin length values and accuracy on the experimentally obtained reflection signals. (g) Tool and workpiece materials combination map on the feasibility of the proposed methodology. Source: Jin et.al. [18]/with permission from Elsevier.

remaining amount was plotted with the combination of the broadly applied workpiece and tool materials.

The research in FSW field is growing in academia; however, broadly applying FSW in industries is still challenging due to the lack of properly quality-controlled inspections or monitoring methods. Some proposed solutions were focused on the continuous capturing on the temperature profiles [1], machine feedback (torque and normal force) [19], and processed surface conditions [20]. All the factors were necessary to be involved in the monitoring loop to ensure better stability of the production quality. However, the limitation of instant estimation of tool wearing has not been overcome yet in the existing literature, in which the wear or degradation of the friction stir tool is also a tightly related factor resulting in flaws and even potential failure of the workpieces. Machine learning technique has been applied in friction stir welding field for monitoring the macrostructure profile of the welding joint [21] or the unexpected machine feedback deviation [19].

However, such methodologies sense the flaw formation only if it happened which do not provide much information in terms of flaw prevention. Moreover, properly obtaining a decent training bank based on experimental approaches is also challenging due to the existing human/equipment errors and high degree-of-freedom interference of the practical factors. The experimental human/equipment errors in the training data bank significantly deviate the accuracy of machine learning prediction. Furthermore, in practice, multi-factor machine learning required correspondingly increased amount of training data with the increased number of involved factors. Hence, the prefect condition training data bank using numerical simulation was realized. With intentionally designed numerical models and real experiments, the condition of simulation and experiments can be approximately merged to provide comparable output data as AI training bank. Following the proposed approach, an accurate and cost-friendly machine learning program can be trained with numerical simulated data bank to properly recognize practical data from real experiments [22].

15.5 Roles and Potentials of Ultrasound in Future FSW/FSP

Unlike the conventional material microstructural characterization techniques, ultrasound-based techniques are nondestructive, consistent, have deeper penetration depth, and can provide instant results, which facilitates the application of ultrasonic inspection and evaluation techniques. However, since FSW/P has been increasingly employed in joining and processing aerospace and vehicle structures that require extremely high-quality joining and stable processing qualities, new challenges have been presented to traditional ultrasound testing methods. The industrial application of FSWed/FSPed components is also hindered by needing proper NDT testing. Nevertheless, as a result of the recent progress in the advanced ultrasonic sensors as well as the understanding of the correlation between ultrasonic signals (waves) and macro-microstructures, ultrasonic inspection and evaluation techniques can accelerate the transition progress of FSW/P from laboratory to production lines with proper and cost-effective optimizations for broader impacts and applications. These advanced ultrasonic techniques can be applied as a brand-new control strategy. The current control strategy of FSW/P is mainly based on the machine feedback data (axial forces, torque, etc.) as input. Workpiece and tool temperatures can be detected via thermal cameras or thermal couples, which can be applied as input for various control strategies. All these inputs need to be combined to predict the welding or processing quality since none of them can solely reflect macrostructure or microstructure features. Advanced ultrasonic inspection techniques are based on analyzing the reflected signals (waves), which are continuously traveling through the whole workpiece on the depth. Therefore, ultrasonic techniques can provide real-time data regarding the flaw formation and transient mechanical properties of the workpiece during the processing, which can be applied as new input for further advanced control strategy.

On the other hand, with the recent presentations of advanced ultrasonic technologies, the novel in situ and ex situ ultrasonic technology can be a potential tool for understanding the fundamentals and transient behaviors of the FSW/FSP from a different point of view with respect to the conventional mechanical or electromagnetic tests and characterizations. The understanding and efforts on the ultrasonic techniques studied in FSW/FSP samples and processing can continuously benefit the industries in FSW, FSP, and even friction stir-based additive manufacturing [23]. In the future, the advances in AI industrial applications have opened more opportunities for ultrasound testing applications. With a well-trained AI program, the accuracy, speed, and amount of information for ultrasonic inspections, evaluations, and monitoring can be significantly enhanced, which can be an appropriate approach for industrial piece-by-piece in-line measurements of process and quality control, as well as the monitoring system for the FSW/P process, leading the application of ultrasound technology to the next level.

15.6 Conclusion

In this chapter, the ultrasonic nondestructive inspections, evaluations, and monitoring for friction stir welding and processing are introduced and discussed in detail in terms of principles, methodologies, and practical applications. Firstly, the acoustic propagation behaviors associated with the macrostructure, microstructure, and residual stresses was explained in Section 15.2. In specific, the relationship between the defect size and acoustic wave behaviors was stated. The effective medium theory was also involved to state the potential detection of minor defects which are smaller than the operating wavelength. The microstructure and residual stresses that induced various types of scattering and dispersion effects are explained in detail. The uniformity and anisotropy dependence on the ultrasonic wave propagation behaviors were discussed. Secondly, in Section 15.3, the commonly applied ultrasonic inspections, elasticity evaluations, and in situ monitoring processes which applied the principles are stated in Section 15.2. Each experimental equipment is discussed in Section 15.3 which categorized testing configurations. Especially for in situ ultrasonic monitoring, the in situ observations can provide unique understanding which is hardly obtained from ex situ testing. However, such monitoring methodologies are barely proposed and demonstrated in the existing literature. The discussion provides some potential approaches as future study directions. In Section 15.4, some recent and novel experimental approach-based case studies are introduced and discussed including ultrasonic elastography, ultrasound residual stresses mapping, and ultrasound in situ monitoring on the FSWed and FSPed workpieces and tools.

Acknowledgment

This work is supported from the infrastructure and support of Center for Agile & Adaptive and Additive Manufacturing (CAAAM) funded through State of Texas Appropriation #190405-105-805008-220.

References

1 Dehelean, D., Palumbo, D., et al. (2008). Monitoring the quality of friction stir welded joints by infrared thermography. 52: 621–626.
2 Droin, P., Berger, G., and Pascal, L. (1998). Velocity dispersion of acoustic waves in cancellous bone. *IEEE Trans. Ultrason. Ferroelectr. Freq. Control* 45 (3): 581–592.
3 Jin, Y., Heo, H., Walker, E. et al. (2020). The effects of temperature and frequency dispersion on sound speed in bulk poly (vinyl alcohol) poly(*N*-isopropylacrylamide) hydrogels caused by the phase transition. *Ultrasonics* 104: 105931.

4 Krokhin, A.A., Arriaga, J., and Gumen, L.N. (2003). Speed of sound in periodic elastic composites. *Phys. Rev. Lett.* 91 (26): 264302.

5 Jin, Y., Walker, E., Krokhin, A. et al. (2019). Enhanced instantaneous elastography in tissues and hard materials using bulk modulus and density determined without externally applied material deformation. *IEEE Trans. Ultrason. Ferroelectr. Freq. Control* 67 (3): 624–634.

6 Jin, Y., Walker, E., Heo, H. et al. (2020). Nondestructive ultrasonic evaluation of fused deposition modeling based additively manufactured 3D-printed structures. *Smart Mater. Struct.* 29 (4): 045020.

7 Bai, X., Zhao, Y., Ma, J. et al. (2018). Grain-size distribution effects on the attenuation of laser-generated ultrasound in α-titanium alloy. *Materials* 12 (1): 102.

8 Wan, T., Naoe, T., Wakui, T. et al. (2017). Effects of grain size on ultrasonic attenuation in type 316L stainless steel. *Materials* 10 (7): 753.

9 Liu, C.R. and Barash, M.M. (1982). Variables governing patterns of mechanical residual stress in a machined surface. *J. Eng. Ind.* 257.

10 Staron, P., Kocak, M., Williams, S., and Wescott, A. (2004). Residual stress in friction stir-welded Al sheets. *Physica B* 350 (1-3): E491–E493.

11 An, K., Yuan, L., Dial, L. et al. (2017). Neutron residual stress measurement and numerical modeling in a curved thin-walled structure by laser powder bed fusion additive manufacturing. *Mater. Des.* 135: 122–132.

12 Oliveira, J.P., Fernandes, F.M.B., Miranda, R.M. et al. (2016). Residual stress analysis in laser welded NiTi sheets using synchrotron X-ray diffraction. *Mater. Des.* 100: 180–187.

13 Pantawane, M.V., Yang, T., Jin, Y. et al. (2020). Crystallographic texture dependent bulk anisotropic elastic response of additively manufactured Ti6Al4V. *Sci. Rep.* 11 (1): 1–10.

14 Neogi, A., Walker, E., and Jin, Y. (2020). Nondestructive ultrasonic elastographic imaging for evaluation of materials. WO, Patent No. 2020242994A1.

15 Neogi, A., Jin, Y., and Yang, T. (2022). Ultrasonic sensor based in-situ diagnostics for at least one of additive manufacturing and 3d printers. U.S., Patent No. 17/677,592.

16 Yang, T., Jin, Y., Squires, B. et al. (2021). In-situ monitoring and ex-situ elasticity mapping of laser induced metal melting pool using ultrasound: numerical and experimental approaches. *J. Manuf. Processes* 71: 178.

17 Ambrosio, D., Dessein, G., Wagner, V. et al. (2022). On the potential applications of acoustic emission in friction stir welding. *J. Manuf. Processes* 75: 461.

18 Jin, Y., Wang, T., Krokhin, A. et al. (2022). Ultrasonic elastography for nondestructive evaluation of dissimilar material joints. *J. Mater. Process. Technol.* 299: 117301.

19 Mishra, D., Gupta, A., Raj, P. et al. (2020). Real time monitoring and control of friction stir welding process using multiple sensors. *CIRP J. Manuf. Sci. Technol.* 30: 1–11.

20 Hartl, R., Landgraf, J., Spahl, J. et al. (2019). Automated visual inspection of friction stir welds: a deep learning approach. *Multi. Sens. Technol. Appl.* 11059: 1105909.

21 Sudhagar, S., Sakthivel, M., and Ganeshkumar, P. (2019). Monitoring of friction stir welding based on vision system coupled with Machine learning algorithm. *Measurement* 144: 135–143.

22 Jin, Y., Wang, X., Fox, E.A. et al. (2022). Numerically trained ultrasound ai for monitoring tool degradation. *Adv. Intell. Syst.* 2100215.

23 Jin, Y., Yang, T., Wang, T. et al. (2023). Behavioral simulations and experimental evaluations of stress induced spatial nonuniformity of dynamic bulk modulus in additive friction stir deposited AA 6061. *J. Manuf. Processes* 94: 454.

16

Applications of Friction Stir Welding

Raja Gunasekaran[1], Velu Kaliyannan Gobinath[2], Kandasamy Suganeswaran[2], Nagarajan Nithyavathy[2], and Shanmugam Arun Kumar[2]

[1] Department of Mechanical Engineering, Velalar College of Engineering and Technology, Erode, Tamil Nadu, India
[2] Department of Mechatronics Engineering, Kongu Engineering College, Erode, Tamil Nadu, India

16.1 Introduction

Traditional fusion welding methods have difficulty in joining softer metal alloys, so these sectors are increasingly turning to solid-state welding and processing technologies. Thomas (1991) was the first to develop friction stir welding (FSW) at The Welding Institute (TWI), UK. It is one of the solid-state methods in which components were bonded to each other without achieving base material fusion. FSW is a comparatively emerging technique for welding solid phases which has both a high energy efficiency and a high degree of flexibility. FSW has seen widespread use in fields as diverse as transportation (cars, planes, ships, trains, etc.) as well as manufacturing. In comparison to conventional methods of welding, the FSW technique provides several advantages. The classification of friction stir techniques is illustrated in Figure 16.1. Both displacement-controlled and pressure-controlled machines are typically used in this technique. The main element of the FSW method is a tool which is nonconsumable. The system was constructed with shoulder and a pin. The type of metal being welded, its dimensions, the joint's configuration, and other necessary parameters will define the geometry and materials of the welding tool. Several investigators have detailed the history and development of FSW methods, as well as their designs, material choices, strengths, failure mechanisms, and methods to avoid failure [2]. The strength of FSW joints were adequately strong, nearly equal to the strength of the original material. This method is faster, more precise, and less expensive than the other welding methods.

In this method, friction is used to weld together two alloys of metals, whether they are same or different materials. Since FSW does not use filler, the weight of the welded material is reduced significantly. Since its development, FSW has established itself as a forepart in combining aluminum alloys of maximum strength that can be used in the automotive and aircraft industries. It has also had some limited success in joining other metallic alloys [3–5]. According to Thomas et al. [6], the fundamental concept of FSW is that a turning tool together with contoured probe was embedded in the adjacent surfaces of the objects. A continual weld is created by a rectilinear motion throughout the welding axis, which simultaneously generates frictional heat between the contact surface of tool and workpiece and as a result it softens the base material. This method can

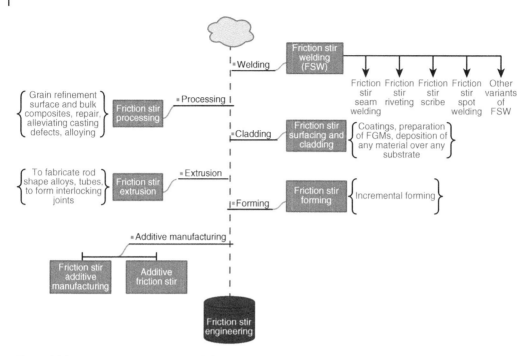

Figure 16.1 Classification of friction stirring techniques. Source: Rathee et al. [1]/with permission of Elsevier.

be used to join any kind of identical materials, including alloy, pure, and composite materials. It could be used to weld quickly dissimilar materials. Since FSW typically produces powerful forces while welding, the stiffness and force control are important considerations when designing an FSW machine.

FSW is a highly efficient method for welding metals and producing different aluminum composites. Furthermore, it offers exceptional mechanical characteristics in fatigue, tensile, and bending [7]. In early stages, FSW showed great promise in joining aluminum alloys [8], and newer studies have shown that it can join a broad range of dissimilar aluminum alloys. The schematic representation of FSW method is shown in Figure 16.2.

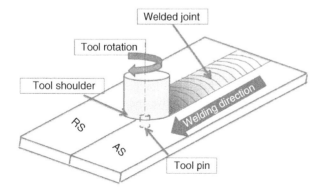

Figure 16.2 Schematic representation of FSW method. Source: El-Sayed et al. [9]/with permission of Elsevier.

In the aviation, aircraft, and automotive sectors, FSW can be used to join a wide range of sheet and plate parts. Several other arrangement and advanced manufacturing techniques also appear to be appealing, especially if cutting-edge FSW controller systems are widely available. While there are many advantages of FSW, such as excellent bond integrity, frequently good mechanical characteristics, low distortion, no need for surface treatment (cleaning), no porosity or related deformities, no fumes, no need for filler metal or shield gas, and lower input energy, there are also some fundamental issues that arise from microstructure changes that occur both inside the weld area and in the corresponding heat-affected areas, or stir-affected zones, just outside the actual FSW zone. Plastic materials are considered as the lightweight materials that are utilized in many industries due to their benefits in efficiency, chemically inert, and electrical properties. However, welding methods are frequently needed for large-scale and complex-shaped plastic components.

FSW is a low-heat-input technique that relies on heat formation due to friction between a turning weld tool and the substance. This makes it different from conventional welding processes which require minimum heat to complete the process. A vehicle's structural effectiveness can be increased by lowering its dry weight. The use of lighter components is one approach for weight reduction. Alloys made of aluminum were mainly used in the aircraft industry. Conventional fusion welding methods produced mechanically weak joints when used to combine the composite material. FSW has been shown to produce fault-free Al-Li 2195 joints with improved mechanical characteristics. Currently, FSW has primarily been applied to aluminum alloys [10, 11].

For large-scale plate or sheet-like materials, friction stir processing (FSP) appears to be the most dependable technique for particle refinement. In the aerospace industry, FSW uses high-strength aluminum alloys to create large containers for satellite launch vehicles, and several firms have been granted permission to utilize FSW in the production of lightweight aluminum aerospace parts for both military as well as for commercial aviation. With its numerous benefits, including improved mechanical characteristics (strength and wear), enhanced process strength, fewer problem related to health and environment, and reduces cost of operation, FSW has drawn significant attention in the automotive sector [12]. FSW is now used by the automobile industry to produce parts in large quantities, such as fuel tanks and lighter alloy wheels. FSW is widely utilized by the automotive and aircraft industries for welding large sheets of metals of the same or different types. It is easy to weld together the different kinds of materials, including metals with diverse mechanical properties. Shipyard and offshore, aircraft, automobiles, railways, robotics, personal computers, etc., are just some of the many fields that make use of FSW.

In this chapter, the importance of FSW in comparison with other metal welding processes was discussed. There was also discussion of recent advancements and uses of the FSW metal joining method in various sectors. The various industrial applications, such as aerospace, automobile, ship building, railways, and other sectors, were reviewed.

16.2 Application of FSW on Different Materials

16.2.1 FSW of Aluminum Alloys

Investigations were done how flow rate affected the structural and mechanical properties of AA5754 friction stir welded pieces by El Rayes et al. [13]. It is seen that increasing the feed rate resulted in fine particles of grain and greater low-angle structure. Therefore, increased tensile and yield strength at welded joints were obtained. Zhang et al. [14] evaluated the impact of the shoulderless FSW tool on the microstructure and mechanical characteristics of ultrathin AA1060 welded joints. It was found that using a shoulderless tool resulted in a smaller heat-affected zone (HAZ) and a narrower weld area. It was found that the nugget zone (NZ)'s hardness was increased than the BM's, and its maximum weld efficiency seemed to be 78.6% of the BM's tensile strength.

The effect of thermal behavior and mechanical characteristics of FSW factors on the optimum AA5083-O welded regions has been researched by El-Sayed et al. [15]. They determined that the welding optimum temperature was marginally affected by variations in tool pin profile and transverse motion. They also determined that welding at 50, 100, and 160 mm/min with a threaded tool pin resulted in faultless joints with increased tensile strength. Sun et al. [16] investigated the residual stress (Figure 16.3) that was obtained from thick AA7010 plates uses both traditional and stationary shoulder FSW technique. The weld stirred and heat-affected zones generated by the SSFSW method was narrower and very rigid across the plate's thickness. Furthermore, a "M" shaped distribution of residual stress was observed for both procedures. SSFSW, however, found that when the tensile zone was narrowed to its smallest possible breadth, its highest stresses dropped slightly. It was concluded that, when the travel speed increased, residual stress profile obtained was smaller and the maximum tensile residual stresses increased (Figure 16.3).

16.2.2 FSW of Magnesium Alloys

Mironov et al. [17] utilized EBSD and a digital image processing method to investigate the relationship between surface texture and strain displacement over the tensile test on AZ31 magnesium alloy. With optimal crystallographic configurations for the fundamental slipping, it was found that strain was concentrated in the SZ positions. Strain measurements obtained in the SZ's focal region indicated that the prism slip had been activated. In addition, the failure was found at the joint root because of double twinning, and strain measurements revealed that the front surface obtained more strain than the back surface as shown in Figure 16.4. Meanwhile, the "onion-ring" formation impacted the NZ. Othman et al. [18] used FSW method to weld a 2 mm AZ31 Mg alloy and examined the impact of a shoulder–pin diameter ratio and was found to be 2.25–5.5. According to the observations, a tensile strength of 91% was obtained at a ratio of 3.33, while the optimum ratio was found to be 5.5. Eventually, the tool ratio of 3.33 showed smaller particle size in SZ region than any other tool ratios.

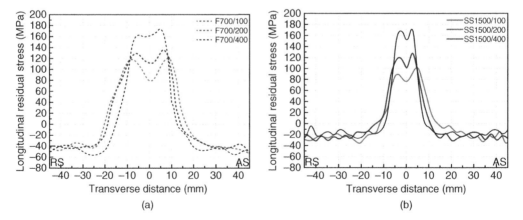

Figure 16.3 Longitudinal residual stress distributions measured across the welds' mid-plane for (a) the FSW and (b) the SSFSW samples, with increasing travel speed. Source: Sun et al. [16]/with permission of Elsevier.

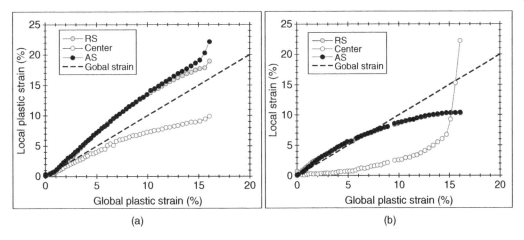

Figure 16.4 Evolution of local strain of the stir zone on the (a) front side and (b) back side. Source: Mironov et al. [17]/with permission of Elsevier.

16.2.3 FSW of Copper Alloys

Rao et al. [19] investigated how different FSW factors affected the physical characteristics of FSW joints made of 2200 copper alloys. The conclusions demonstrated that yield strength increases by increasing the rotational and welding speeds and drops while increasing the axial force. Furthermore, the threaded tool pin produced the best effects for mixing the plasticized material. Azizi et al. [20] reviewed the influence of traverse speed variation on the physical and material properties of thick pure copper joints. The observations demonstrated that higher welding speeds created joints with defects despite achieving finer grain sizes because of minimal heat input. More ductile fracture was observed at minimum traverse speeds, and it was found that increasing the traverse velocity has a significant impact on toughness but a negative impact on ductility.

16.2.4 FSW of Titanium Alloys

In order to examine the weld's mechanical and microstructure characteristics, Ramulu et al. [21] conducted FSW for three distinct alloys of titanium: TIMET-54M, Ti-6Al-4V, and ATI-425, Widmanstätten morphology was observed in the stirring zone of welded ATI-425 and TIMET-54M metals. Additionally, the HAZ processed bimodal particles and the TMAZ had a uniform grain texture in TIMET-54M and Ti-6Al-4V FS welded joints, while in ATI-425, the HAZ exhibited refined equiaxed infused with β. For TIMET-54M, Ti-6Al-4V, and ATI-425, the deformation appeared in the Stirring Zone, BM, and TMAZ/HAZ on the retreating side. The macrographs of butt welds of three different titanium alloys are shown in Figure 16.5. Using back heating as an aid, Ji et al. [22] analyzed the mechanical characteristics of FSW joints made of Ti-6Al-4V. Their observations revealed that, tearing-free welds could be made with conventional FSW at a speed of 100 rpm; whereas the back heating-assisted friction stir welding (BHAFSW) allowed for a greater variety of operating factors and created flawless joints by decreasing the thickness and difference in temperature of plate. While the efficiency of traditional FSW's joint reached at 98.9 percent, the BHAFSW's reached at 93.6% due to a greater amount of heat being applied. The breaking structures of the process joints show ductile fracture. Although the values of hardness obtained were greater than the BM and values obtained with BHAFSW were exceptional. It was proven that improved thermo-plasticization materials led to less tool wear during the BHAFSW technique.

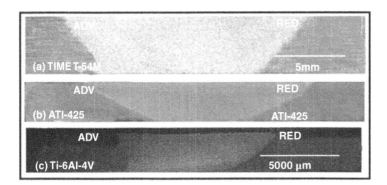

Figure 16.5 Macrographs of butt welds: (a) TIMET-54M; (b) ATI-425; (c) Ti-6Al-4V.
Source: Ramulu et al. [21]/with permission from Elsevier.

16.2.5 FSW of Steels

Using the neutron diffraction technique, Dawson et al. [23] stated that the different speeds affect the dispersion of residual stresses in oxide dispersion-strengthened (ODS) steel. As a result of maximum cooling rates, they concluded that the increased tensile stress measurement was 1200 MPa and it was due to optimum welding speed. Additionally, by reducing the welding pace, the optimum welding temperature increased while the TMAZ tensile residual stresses significantly decreased as shown in Figure 16.6. The values for the longitudinal residual stresses are smaller when compared to the values of traverse residual stresses, indicating that there is no effect on the latter with the varying welding speed.

Nelson and Rose [24] researched on how the input heat and the material of backing affected the joints' ability of HSLA steel. They concluded that the cooling rate of post-weld had a significant effect on the SZ microstructure across various welding factors, with the HAZ's hard zone which disappears at cooling rates less than 20 °C/S. Additionally, it was found that there is decrease in lath width from 1.11 to 0.59 μm as the cooling rate was increased from 7 °C/S to 30 °C/S, leading to hardness increment from 234 to 298 HV. It is also found that relationship between lath length and heat input was independent of the backing plate material.

16.2.6 FSW in Composite Materials

The matrix portion of the composite is composed of different alloys, organic substances, or ceramic particles, while the matrix portion is made up of lightweight metals or alloys. The most widely applied form of MMC is particulate-reinforced MMC. It is made of magnesium, aluminum, titanium, and other lightweight metals [25]. Oxides, carbides, bromides, nitrides, and silicate precipitates are used to strengthen these metals. Due to its superior strength to weight ratio, aluminum matrix composite (AMC) has gained the focus of different sectors such as aviation, automobile, maritime industries, and nuclear power sector. Conventional joining methods for MMCs promote an unfavorable interaction among the reinforcement phase and matrix; this leads to the development of brittle joints. The melting of these alloys during welding process must be avoided in order to prevent the development of brittle joints. Thus, the FSW method is suitable for welding composites. FSW was utilized to join a distinct range of metals, including the alloys of magnesium, copper, zinc, lead, lithium, titanium, aluminum, MMCs, and so on, and this has been the case for many decades [26, 27]. Rahmat et al. [28] succeeded in joining 7075 aluminum alloy with polycarbonate

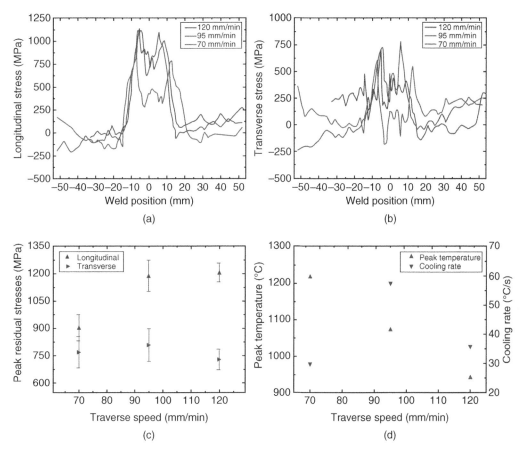

Figure 16.6 Line scan residual stress data at a depth of 1.4 mm from the top surface of the welds, in the (a) longitudinal and (b) transverse orientation; (c) peak tensile residual stress and (d) peak temperature and cooling rate as a function of traverse speed during welding. Source: Dawson et al. [23]/with permission of Elsevier.

sheet. The effect of welding speed on the weld interface was analyzed through several experiments. Adhesion of plasticized material on different tool pin shapes was evaluated by Mehta et al. [29]. The research focused in particular on the stresses produced in the tool's different pin profiles. It was determined that the profile with a greater number of sides will undergo less tension than the other profile.

16.2.7 FSW in Polymers

Thermoplastics, fiber-reinforced polymers, and aluminum-based alloys are increasingly used in a variety of fields because of their distinctive properties including the ratio of high strength to weight. While compared to other elements, polymers are lightweight materials that have experienced significant growth in almost all manufacturing sectors due to their superior formability, machinability, ease of manufacture, cheap cost of materials, etc. Some of the property of FSW includes low thermal characteristics, low melting point, and changing of volume during crystallization makes the conventional welding method more challenging to weld these types of polymer materials. FSW is a cost-effective and versatile welding method that produces a uniform weld area with minimal defect. Hence, FSW is an excellent choice for welding polymer composites. Huang et al. [30] performed the FSP of polymer composites made of polymer matrix and dissimilar metals, concluding that FSP is preferable for joining similar and dissimilar polymers and also for welding dissimilar metals to composites. An experimental study of the FSW for the materials made of high-density polymer was revealed by Moreno et al. [31], in which the tensile strength, toughness, crystalline nature, and thermal behavior of the welded joint were found to be affected by rotational and welding speed. Throughout many studies, the rotational speed was proved to be the greatest impact when compared with various processing parameters. With increasing rotational speed, weldment quality standards decreased.

16.2.8 FSW in Plastics

Metal FSW developments have made the utilization of FSW method more significant in thermoplastic composites. The profile of the stir pin was the primary focus of the tool's study in thermoplastic composites. Results from an FSW experiment conducted by Jaiganesh et al. [32] using cylindrical, square, and triangular threaded stir pin parameters to weld 5-mm poly propylene sheets revealed that the cylindrical tapered tool produced the strongest joints. The impact of tool-pin configurations on the joining surface and yield strength was examined by Ghasemi et al. [33]. They used FSW to weld poly propylene composite sheets with 30% glass fiber by weight. Three different types of pins were found in the tool: a taper pin, a cylindrical grooved pin, and a triangle pin. The conclusions revealed that tool-pin configuration significantly affected welding characteristics and joint surface. The tapered pin which contains a groove had the greatest weld surface and tensile strength, with a tensile measurement of 9 MPa.

16.3 Industrial Applications of FSW

16.3.1 FSW in Aerospace Industry

Because of its excellent durability and lightweight characteristics, FSW is used in the aircraft sector for manufacturing different parts and components, including airfoil, fuel containers, and airplane frames. Additionally, it can be used to fix the weak joints. The structures and parts used in aircraft are subjected to varying loading conditions, requiring the utilization of different materials in single part. It becomes effective for FSW to be able to build strong welds between different materials. Fuel tanks utilized by Air Bus, Boeing, and Lockheed Martin consist of AA2219 and were joined using FSW, which drastically reduced welding costs in comparison to tungsten inert gas (TIG) welding [34]. Boeing stated that using FSW reduced manufacturing costs for their Delta (II & IV) space launchers by 60%. NASA's Michoud Assembly Facility constructed the Orion spaceships using FSW for joining the bulkhead and nosecone [34]. FSW is used at NASA for joining the barrel parts, roofs of the space launch system (SLS), and rings of core stage. The fuel tank of SpaceX's Falcon 9 space launchers was completely friction stir-welded, becoming the largest aluminum–lithium alloy structure ever [35]. New Mexico's Eclipse Aviation Corporation used FSW to reduce assembly time and costs by replacing 7000 joints along with other components on their Eclipse 500 aircraft [34]. Additionally, it was stated that FSW is superior to traditional fastening methods in terms of fatigue life and strength. FSW has been used by various industries to reduce aircraft weight and enhance mechanical characteristics. Figure 16.7 exhibits the application of FSW in different sectors.

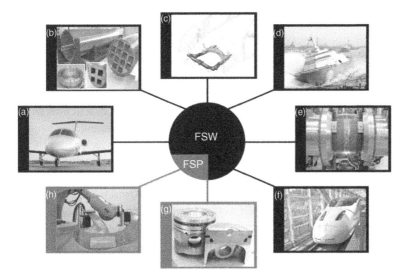

Figure 16.7 Examples of the industrial application of friction stir welding and processing: (a) the Eclipse 500 business jet, the first to use FSW, (b) 50 mm thick copper nuclear waste storage canisters, (c) dissimilar FSW of aluminum to steel in Honda front subframe, (d) deckhouse structure of Littoral Combat ship, (e) orbital FSW of steel pipes, and (f) floor panels of Shinkansen train. FSP surface modification of (g) automotive piston and (h) the Nibral propeller. Source: Heidarzadeh et al. [36]/with permission from Elsevier.

16.3.2 FSW in Automobile

Aluminum has become more prevalent in the automotive sector over the past decade, because of its lightweight characteristic and excellent strength-to-weight ratio. Vehicle weight reduced without endangering occupant safety. Due to FSW's ability to create strong welds with different aluminum alloys, the utilization of aluminum in the automotive industry has been a success. FSW was employed in the wheel rims, propellant carriers, tailored blanks, front of engines, and the frame of high-performance vehicles for automobiles. TWI recommended using FSW in 1998 to join aluminum sheet metals for use in car door frames. Then, FSW was employed in the massive production of automobile parts and also in the joining of different aluminum alloys and steel automotive components. Honda's mass-produced car suspension system utilized FSW process for welding dissimilar Al-Steel [37]. Press-formed steel and molded aluminum were welded in a lap arrangement to build the outer frame of the Honda Accord, which provided additional support for the engine and suspension components. Honda revealed that compared to a conventional steel subframe, their vehicle uses 50% less electricity and has a 25% lighter body [37]. When it came to mass-producing automobile door panels, Honda were among the first to use FSW to weld steel and aluminum [34]. Tower Automotive stated that using FSW instead of gas metal arc welding (GMAW) resulted in a weight reduction and is about 40% and a twofold increase in mechanical strength [34]. Friction stir spot welding was utilized to weld aluminum alloy sheets to galvanized steel brackets for the trunk lid of a Mazda MX-5. Mazda Motor Corporation stated that by using FSSW, they were able to save on production costs, energy consumption, and overall vehicle weight. When compared to GMAW, Ford stated that using FSW can improve joint strength and dimensional accuracy by 30% [34].

16.3.3 FSW in Ship Building and Marine Industry

Welding dissimilar aluminum alloys is one of the primary uses for FSW in the shipbuilding industry. The shipbuilding industry makes frequent use of aluminum alloys 5xxx and 6xxx for welding. FSW is used for a wide range of welding initiatives, including the construction of ship decks, bulkheads, sides, and floors; the interior surfaces of boats; the welding of aluminum hulls and superstructures; the construction of helicopter landing platforms; and the construction of ship bodies. In 1996, Sapa welded panels made from hollow aluminum extrusions were used in fishing vessels. In 1996, Marine Aluminium hired FSW to weld a sizable quantity of panels for the cruise ship "The World" in Haugesund. Mitsui Engineering and Shipbuilding (MES) constructed the "Super Liner Ogasawara," a passenger ship, with extensive use of FSW technique [38]. For 55-m long Inshore Patrol Vessels, which were built in 2004 by New Zealand and Australian naval, architects required the use of FSW for a large number of their welding requirements [39]. Inshore Patrol Vessels were constructed in Whangarie by the Donovan Group using FSW economically in the middle of 2005. In addition to these uses, in Japan, FSW is extensively utilized for manufacturing honeycomb and marine-grade corrosion-resistant panels. In addition to its extensive use in the construction of cruise ship hulls and offshore oil platforms, FSW is also utilized for a wide range of other commercial usages. Kawasaki Heavy Industries was able to effectively weld 50 mm thick plates of AA5083, which have a high resistance to corrosion from salt water, that are used in shipbuilding construction.

16.3.4 FSW in Railway Industry

As FSW has developed and the performance of weldments has increased, it has become more popular in the railway sector. Since 2001, SAPA's Munich subway trains have effectively utilized FSW for welding train exterior walls and surface portions. FSW replaced the traditionally used mechanized metal inert gas (MIG) method to join aluminum plates and its thickness is about 23 mm, resulting in the reduction of both weight and welding cost. Huge aluminum extrusion surfaces are widely used in the construction of modern railroad carriages. Particularly in Japan, FSW technique has been extensively used commercially by Hitachi. Arc welded aluminum sheets with lightweight did not require expensive straightening because of the low distortion of the welded sections. FSW was also used by Nippon Sharyo to produce surface panels for the new bullet trains. Additionally, Nippon Light Weight Metals used FSW in the Tokyo metro system's rolling stock [40]. FSW was also used by Kawasaki Heavy Industries to weld roof panels. They also constructed aluminum car body shells by utilizing the minimal distortion characteristic of FSW's thin sheets [41].

16.3.5 FSW in Rocket Tank Production

After extensive studies and investigations, the engineering implementation of rocket tank FSW on the tank portion, dome, and body was done [31]. FSW provides greater weld quality and geometry precision in aerospace applications compared to traditional fusion welding. With a very high First-Time Pass rate, the number of flaws that need to be fixed is decreased from $30 \sim 50$ to below 3. A 15% increase in tensile strength from 270–300 MPa to 320–350 MPa is achieved. The geometry precisions, including the perpendicularity of the tank to the axis and the slenderness of the tank's upper end surface, are improved by 30%. There has been a significant improvement in the accuracy and productivity of aircraft product.

16.3.6   FSW in Other Industries

FSW is widely used in the aircraft, railroad, and automotive sector, but it has also found use in other sectors like architectural and electrical industries. Aluminum, copper, and titanium facade panels, as well as window frames, pipe fabrications, and the construction of enormous structures like aluminum bridges and nuclear power plant reactors, improved by the use of FSW. However, FSW cannot be widely implemented in construction sector, because of its lack of mobility. The electrical sector utilizes FSW for various applications, including the joining of electric motor housings, electrical connections, and the electronic product encapsulation.

16.4   Conclusion

FSW is a permanent solid-state welding method that has been employed to effectively weld both same and as well as different materials. It has been widely accepted that FSW is among the most significant advancements in welding methods in the last two decades. Since less energy is needed to generate the weld, FSW is considered as the most energy-efficient method. FSW is widely used in current manufacturing industries, particularly in the aviation, marine, spaceships, and automobiles sectors, due to its superior characteristics over fusion welding methods, i.e. reduces defects due to porosity, HAZ also reduced, decreased environmental impact, and lower distortion. FSW has been accepted to be an economical and useful method of combining highly conductive (thermally) materials that are challenging to join using traditional fusion-based processes. Dissimilar joints have a broad range of applications. Several aspects of FSW including tool geometry, tool speed of motion, rotational speed, and tool and workpiece material have been determined and can be improved to achieve the intended process and product outcomes. This method provides a number of opportunities in highly significant fields, like aircraft, and has a possibility for widespread application. This technology is utilized by every sector of the transportation industry, including shipping, offshore, railway, automobiles, and aircraft. Major factors of FSW include low distortion, good repeatability, inexpensive, high strength, and good mechanical characteristics. FSW of aluminum has developed into a reliable and effective method, and is now widely used in the production of essential parts. It has the ability to be utilized in welding applications involving titanium alloys and other materials. FSW has expanded the use of welding specific materials, particularly for joining 2xxx and 7xxx alloys for the aerospace sector. The method is widely accepted because of its inexpensive cost, low need for repairs, high quality, and high degree of consistency as a result of its complete automation. FSW has widespread acceptance as an economically applicable and environmentally friendly alternative for fusion welding. As a green technology of welding, FSW has the potential for future implementations for large and small size industries due to its various merits than fusion welding, particularly in joint quality and environmental considerations. Additional investigations and development are being conducted to evaluate current FSW joint designs, build additional mechanical and corrosion data, determine approaches for the FSW of titanium, steel, and other complex materials, and ultimately to enhance additional utilization of this significant method.

References

1 Rathee, S., Srivastava, M., Pandey, P.M. et al. (2021). Metal additive manufacturing using friction stir engineering: a review on microstructural evolution, tooling and design strategies. *CIRP J. Manuf. Sci. Technol.* 35: 560–588. https://doi.org/10.1016/j.cirpj.2021.08.003.

2 Maji, P., Karmakar, R., Nath, R.K., and Paul, P. (2022). An overview on friction stir welding/processing tools. *Mater. Today Proc.* https://doi.org/10.1016/j.matpr.2022.01.009.

3 Threadgill, P., Leonard, A., Shercliff, H., and Withers, P. (2009). Friction stir welding of aluminium alloys. *Int. Mater. Rev.* 54: 49–93. https://doi.org/10.1179/174328009X411136.

4 De, P. and Mishra, R. (2011). Friction stir welding of precipitation strengthened aluminium alloys: scope and challenges. *Sci. Technol. Weld. Joining* 16: 343–347. https://doi.org/10.1179/1362171811Y.0000000020.

5 Salih, O.S., Ou, H., Sun, W., and McCartney, D. (2015). A review of friction stir welding of aluminium matrix composites. *Mater. Des.* 86: 61–71. https://doi.org/10.1016/j.matdes.2015.07.071.

6 Thomas, W.M., Johnson, K.I., and Wiesner, C.S. (2003). Friction stir welding-recent developments in tool and process technologies. *Adv. Eng. Mater.* 5: 485–490. https://doi.org/10.1002/adem.200300355.

7 Naik, S., Panda, S., Padhye, S. et al. (2021). Parametric review on friction stir welding for under water and dissimilar metal joining applications. *Mater. Today Proc.* 47: 3117–3122. https://doi.org/10.1016/j.matpr.2021.06.168.

8 Calder, N. (2000). The place for rapid manufacturing in military airframe production. *Mater. Technol.* 15: 34–37. https://www.tandfonline.com/doi/abs/10.1080/10667857.2000.11752853?journalCode=ymte20.

9 El-Sayed, M.M., Shash, A., Abd-Rabou, M., and ElSherbiny, M.G. (2021). Welding and processing of metallic materials by using friction stir technique: a review. *J. Adv. Joining Processes* 3: 100059. https://doi.org/10.1016/j.jajp.2021.100059.

10 Inaniwa, S., Kurabe, Y., Miyashita, Y., and Hori, H. (2013). Application of friction stir welding for several plastic materials. In: *Proceedings of the 1st International Joint Symposium on Joining and Welding.* Elsevier https://doi.org/10.1533/978-1-78242-164-1.137.

11 Tanaka, S. and Kumagai, M. (2010). Effect of welding direction on joining dissimilar alloys between AA5083 and A6N01 by friction stir welding. *Weld. Int.* 24: 77–80. https://doi.org/10.1080/09507110902842935.

12 Dawes, C. (1995). Friction stir process welds aluminum alloys. *Welding J.* 36: 1.

13 El Rayes, M.M., Soliman, M.S., Abbas, A.T. et al. (2019, 2019). Effect of feed rate in FSW on the mechanical and microstructural properties of AA5754 joints. *Adv. Mater. Sci. Eng.* https://doi.org/10.1155/2019/4156176.

14 Yue, Y., Zhou, Z., Ji, S. et al. (2017). Effect of welding speed on joint feature and mechanical properties of friction stir lap welding assisted by external stationary shoulders. *Int. J. Adv. Manuf. Technol.* 89: 1691–1698. https://doi.org/10.1007/s00170-016-9240-x.

15 El-Sayed, M.M., Shash, A.Y., Mahmoud, T.S., and Rabbou, M.A. (2018). Effect of friction stir welding parameters on the peak temperature and the mechanical properties of aluminum alloy 5083-O. In: *Improved Performance of Materials*', Design and Experimental Approaches, 11–25. https://doi.org/10.1007/978-3-319-59590-0_2.

16 Sun, T., Roy, M., Strong, D. et al. (2017). Comparison of residual stress distributions in conventional and stationary shoulder high-strength aluminum alloy friction stir welds. *J. Mater. Process. Technol.* 242: 92–100. https://doi.org/10.1016/j.jmatprotec.2016.11.015.

17 Mironov, S., Onuma, T., Sato, Y. et al. (2017). Tensile behavior of friction-stir welded AZ31 magnesium alloy. *Mater. Sci. Eng., A* 679: 272–281. https://doi.org/10.1016/j.msea .2016.10.036.

18 Othman, N., Ishak, M., and Shah, L. (2017). Effect of shoulder to pin ratio on magnesium alloy Friction Stir Welding. *IOP Conference Series: Materials Science and Engineering*. IOP Publishing. https://doi.org/10.1088/1757-899X/238/1/012008

19 Rao, A.N., Naik, L.S., and Srinivas, C. (2017). Evaluation and impacts of tool profile and rotational speed on mechanical properties of friction stir welded copper 2200 alloy. *Mater. Today Proc.* 4: 1225–1229. https://doi.org/10.1016/j.matpr.2017.01.141.

20 Azizi, A., Barenji, R.V., Barenji, A.V., and Hashemipour, M. (2016). Microstructure and mechanical properties of friction stir welded thick pure copper plates. *Int. J. Adv. Manuf. Technol.* 86: 1985–1995. https://doi.org/10.1007/s00170-015-8330-5.

21 Ramulu, M., Gangwar, K., Cantrell, A., and Laxminarayana, P. (2018). Study of microstructural characteristics and mechanical properties of friction stir welded three titanium alloys. *Mater. Today Proc.* 5: 1082–1092. https://doi.org/10.1016/j.matpr.2017.11.186.

22 Ji, S., Li, Z., Wang, Y., and Ma, L. (2017). Joint formation and mechanical properties of back heating assisted friction stir welded Ti-6Al-4V alloy. *Mater. Des.* 113: 37–46. https://doi.org/ 10.1016/j.matdes.2016.10.012.

23 Dawson, H., Serrano, M., Cater, S. et al. (2017). Residual stress distribution in friction stir welded ODS steel measured by neutron diffraction. *J. Mater. Process. Technol.* 246: 305–312. https://doi.org/ 10.1016/j.jmatprotec.2017.03.013.

24 Nelson, T.W. and Rose, S.A. (2016). Controlling hard zone formation in friction stir processed HSLA steel. *J. Mater. Process. Technol.* 231: 66–74. https://doi.org/10.1016/j.jmatprotec.2015.12.013.

25 Chawla, K.K. and Chawla, K.K. (1998). *Metal Matrix Composites*. Springer https://doi.org/10.1007/ 978-1-4757-2966-5_6.

26 Santos, T.F.d.A., Torres, E.A., and Ramirez, A.J. (2018). Friction stir welding of duplex stainless steels. *Weld. Int.* 32: 103–111. https://doi.org/10.1080/09507116.2017.1347323.

27 Manugula, V.L., Rajulapati, K.V., Reddy, G.M. et al. (2018). Friction stir welding of thick section reduced activation ferritic-martensitic steel. *Sci. Technol. Weld. Joining* 23: 666–676. https://doi.org/ 10.1080/13621718.2018.1467111.

28 Rahmat, S., Hamdi, M., Yusof, F., and Moshwan, R. (2014). Preliminary study on the feasibility of friction stir welding in 7075 aluminium alloy and polycarbonate sheet. *Mater. Res. Innovations* 18: S6-515-S6-519. https://doi.org/10.1179/1432891714Z.0000000001035.

29 Mehta, M., De, A., and DebRoy, T. (2014). Material adhesion and stresses on friction stir welding tool pins. *Sci. Technol. Weld. Joining* 19: 534–540. https://doi.org/10.1179/1362171814Y.0000000221.

30 Huang, Y., Meng, X., Xie, Y. et al. (2018). Friction stir welding/processing of polymers and polymer matrix composites. *Composites, Part A* 105: 235–257. https://doi.org/10.1016/j.compositesa .2017.12.005.

31 Moreno-Moreno, M., Macea Romero, Y., Rodríguez Zambrano, H. et al. (2018). Mechanical and thermal properties of friction-stir welded joints of high density polyethylene using a non-rotational shoulder tool. *Int. J. Adv. Manuf. Technol.* 97: 2489–2499. https://doi.org/10.1007/s00170-018-2102-y.

32 Jaiganesh, V., Maruthu, B., and Gopinath, E. (2014) 'Optimization of process parameters on friction stir welding of high density polypropylene plate', *Procedia Eng.*, 97, pp. 1957-1965. https:// doi.org/https://doi.org/10.1016/j.proeng.2014.12.350

33 Saeidi, M., Arab, N., and Ghasemi, F.A. (2009). The effect of pin geometry on mechanical properties of PP composite Friction Stir Welds. *IIW Int Congress Welding Joining*, Iran.

34 Bhardwaj, N., Narayanan, R.G., Dixit, U., and Hashmi, M. (2019) 'Recent developments in friction stir welding and resulting industrial practices', *Adv. Mater. Process. Technol.*, 5, pp. 461-496. https://doi.org/https://doi.org/10.1080/2374068X.2019.1631065

35 Fernández, L.A., Wiedemann, C., and Braun, V. (2022). Analysis of space launch vehicle failures and post-mission disposal statistics. *Aerotec. Missili Spazio* 101: 243–256. https://doi.org/10.1007/s42496-022-00118-5.

36 Heidarzadeh, A., Mironov, S., Kaibyshev, R. et al. (2021). Friction stir welding/processing of metals and alloys: a comprehensive review on microstructural evolution. *Prog. Mater Sci.* 117: 100752. https://doi.org/10.1016/j.pmatsci.2020.100752.

37 Worldwide, H. (2012). Honda Develops New Technology to Weld Together Steel and Aluminum and Achieves World's First Application to the Frame of a Massproduction Vehicle. September.

38 Khan, N.Z., Siddiquee, A.N., and Khan, Z.A. (2017). *Friction Stir Welding: Dissimilar Aluminum Alloys*. CRC Press https://doi.org/10.1201/9781315116815.

39 Delany, F. (2007). Friction stir welding of aluminium ships, *WELDING AND JOINING-HARBIN*, 5: 7.

40 Zubcak, M., Soltes, J., Zimina, M., Weinberger, T., and Enzinger, N. (2021) 'Investigation of Al-B4C metal matrix composites produced by friction stir additive processing', *Metals*, 11, pp. 2020. https://doi.org/10.3390/met11122020

41 Li, W., Fu, T., Hütsch, L. et al. (2014). Effects of tool rotational and welding speed on microstructure and mechanical properties of bobbin-tool friction-stir welded Mg AZ31. *Mater. Des.* 64: 714–720. https://doi.org/10.1016/j.matdes.2014.07.023.

17

Equipment Used During FSP

Kandasamy Suganeswaran[1], Nagarajan Nithyavathy[1], Palaniappan Muthukumar[2], Shanmugam Arunkumar[1], and Velu Kaliyannan Gobinath[1]

[1] *Department of Mechatronics Engineering, Kongu Engineering College, Erode, Tamil Nadu, India*
[2] *Department of Mechanical Engineering, Kongunadu College of Engineering and Technology, Thottiyam, Tamil Nadu, India*

17.1 Introduction

In recent scenario, aluminum metal matrix composites (AMMCs) play a vital role in manufacturing industries of aerospace and automotive owing to lightweight and elevated strength character. For fabricating this, liquid processing techniques and solid processing techniques are preferred.

17.1.1 Liquid Processing Techniques for Producing AMMCs

The following processing techniques are employed for producing AMMCs through liquid-state process.

17.1.1.1 Stir Casting

Ceramics or whiskers are mixed with a molten matrix through mechanical stirring. Then the resultant material are poured into the die for component manufacturing. This process is suitable for producing composites with 30% of volume fractions of reinforcement. The segregation of reinforcements by different process parameters poses a significant challenge in this technique, leading to an uneven distribution of particles.

17.1.1.2 Squeeze Casting

Squeeze casting is a method of producing metal that encourages solidification inside of a die by applying high hydrostatic pressure. With or without reinforcement, it produces high integrity technical components by combining permanent mold casting and die forging. Due to the high hydrostatic pressure, the porosity formed by gas or through shrinkage would be eliminated.

17.1.1.3 Compo Casting

Particulates, fibers, or whiskers are dispersed in semisolid slurry temperature of the alloy to form AMMCs. Reduction of particle sizes up to nanoscale is possible using this compo casting technique and it would improve the ductility of the composites fabricated. Intermetallic formation would be predominantly avoided in this process due to low operating temperature.

17.1.1.4 Spray Forming

Spray forming process involves the melting of the aluminum alloy in a furnace and forcing it into the orifice to mix it with reinforcement coming from the jet to form semisolid solution. It generates the microstructure along with fine and equiaxed grains. Moreover, the nature of spray and the thermal behavior of the droplets that are deposited on the substrate affect the microstructure formation.

17.1.2 Solid Processing Techniques for Producing AMMCs

The following processing techniques are employed for producing AMMCs through solid-state process.

17.1.2.1 Powder Metallurgy

Powder metallurgy is a multistep fabrication technique that involves the blending of powders and is compacted using the dies with close tolerance. Metallurgical bonding between the atoms would be obtained through sintering process and secondary operations are executed to obtain the close dimensional tolerance, good surface finish, and increased density. Due to the elevated cost of powder and expensive tooling, this process is unsuitable for mass production, although it enables the manufacturing of intricate shapes.

17.1.2.2 Diffusion Bonding

Diffusion bonding is a solid-state procedure that is used to combine two similar or dissimilar metals at high temperature (50–90% of the base metal melting temperature) by applying pressure over different time periods. It has an ability to produce sound joints with reduced flaws and porosity without having any loss of alloying elements. Metals, ceramics, alloys, and even powder-metallurgy materials are joined using this process and extensively used in electronics, nuclear, aerospace, and industries.

Liquid-state processing methods pave the way for the formation of detrimental phases and defects such as inclusions, shrinkage, and porosity that can be resolved by utilizing solid-state processing techniques. FSP is an alternative to the conventional processing methods which produce particle clusters and segregation. It is an eminent process that alters the microstructure and property of the composite material without melting the base substrate [1]. Further, the parameters can be varied to alter the microstructure, which is suited for particular applications.

17.1.3 Friction Stir Process

FSP employs a rotation tool with pin and shoulder. The tool plunges into the base material. This aforementioned action happens till the shoulder forms contact with the metal surface. During this process, the generation of heat caused due to friction softens and plasticizes the base metal. Plastic deformation through FSP modifies the base metal structure and results in significant grain refinement. Figure 17.1 shows the pictorial representation of FSP. The generated material flows from the front to back side of pin during the transverse action of FSP tool and forges on the retreating side (RS) through shoulder pressure and finally, it gets consolidated.

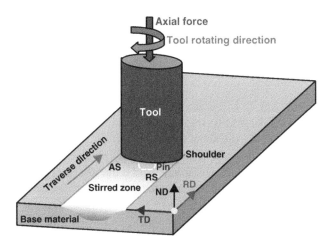

Figure 17.1 Schematic representation of FSP process.

Figure 17.2 Different zones of friction stir processed sample. Source: Mishra and Ma [1]/ with permission from Elsevier.

The combined thermal and mechanical action in FSP process results in three different zones such as like stir zone (SZ), Thermomechanical affected zone (TMAZ), and heat-affected zone (HAZ) and is shown in Figure 17.2. SZ is a fully recrystallized region primarily formed by the pin, which can be characterized with equiaxed and fine recrystallized grain formation. Based on some FSP parameters, the onion-ring structure is also seen in the NZ. TMAZ region, very next and adjacent to SZ region, is characterized by increased deformed structure with occurrence of null recrystallization in this zone. Next, to TMAZ, stays HAZ which experiences only thermal variation cycle and no encounters of plastic deformation with larger grain formation. Thus, the unaltered remaining portion of the material is called as the base material. An advancing side (AS) is a location where the tool travel direction and the deformed material flow are in similar direction. If the travel is in opposition to the other, then it is named as RS.

17.1.3.1 Process Parameters in FSP
Rotational Speed (N)
FSP tool can rotate either in a clockwise/counterclockwise movement along the line of processing. Faster tool rotation produces higher temperature changes owing to higher induced frictional heat. This caused the material to be mixed and stirred more vigorously. Because the coefficient of friction formed at the interface alters as the tool rotation rate increases [2].

Traverse Speed (V)

The traverse feed is a linear movement and an important process parameter of the tool. Reduced traversal speed results in more frictional heating. The agitated material is held within the tool shoulder by a proper tilt of the spindle in the trailing direction.

Plunge Depth

Another crucial aspect is the pin depth (target depth) of entry into the workpiece and is associated with the dimensions of the pin. The depth at which the shoulder is put into the workpiece is another way to describe the penetration depth [3]. If the insertion depth is too shallow, the tool shoulder does not come into contact with the surface of the workpiece. And if the insertion depth is exceeding the permissible limits, the excessive flash would be generated.

Tool Title Angle

The tool tilt angle or spindle angle is another crucial process parameter. The suitable tilt angle of the spindle enroute for the proper material consolidation. This confirms the tool-shoulder holding the stirred material and traverse the material capably from front to back end of pin [1]. Generally, the tool is tilted through at a small angle (ranging $1°–3°$). In such cases, the shoulder may enter the workpiece partially. The "heel" of the shoulder is the area where penetration occurs most frequently.

17.1.3.2 Forces Involved in FSP

During the FSP, the forces acting on the tool have a considerable impact. Axial force, also known as downforce, is the force that is exerted perpendicular to the tool axis (Z-direction). Traversing force is clear as the force exerted perpendicular to the X-axis direction of motion. The side force is the force that develops perpendicular to both the X and Z forces (Y-direction). The term "tool shoulder footprint" refers to the entire tool footprint on the workpiece surface [4].

17.1.3.3 Mechanisms in FSP

Continuous, discontinuous, and geometric dynamic recrystallization are the different processes available for the dynamic recrystallization process of Al alloys [1]. Discontinuous dynamic recrystallization is not observed due to fast recovery of aluminum alloy by its high stacking fault-energy, but it is seen in alloys, which have large secondary phase particles, i.e. $<0.6\,\mu m$. Jata and Semiatin [5] have identified the occurrence of continuous dynamic recrystallization. The low-angle grain boundaries in the parent metal are transformed to high-angle grain boundaries as a result of continuous rotation during the process. When there is limited boundary migration, strain causes the rotation on the sub-grains. This transforms the grain boundary into high angle boundary.

17.1.3.4 Precipitate Dissolution and Coarsening During FSP

During the FSP process, the precipitate may dissolve or coarsen depending on the processing temperature. The earlier studies made by Liu et al. [6] identified that the formed precipitates are larger in SZ than the base AA6061-T6 alloy. Sato et al. [7] have reported that all the precipitates in the NZ get completely dissolved into the matrix. Overall, the possible processes that occur during the FSP process are coarsening, dissolution, and reprecipitation of precipitates.

17.1.4 Fabrication of Surface Hybrid Composites

Two types of approaches are followed to produce SHCs, i.e. Particle reinforced FSP specimens: (i) in situ technique and (ii) ex situ technique. In situ technique requires the incorporation of reinforcements during the synthesis of composites. Ex situ technique follows the procedure of adding the reinforcements into the parent matrix externally and further stirring and mixing with the base material. It involves the following procedures:

i) A groove is taken on the workpiece with respect to the center of tool pin to avoid splashing of reinforcing powders during the process.
ii) Reinforcement powders are filled in the groove based on the calculated weight fractions.
iii) Reinforced powders are tightly packed using the scriber tool.
iv) Sealing the groove using pin-less tool to avoid sputtering of powder during the process.
v) FSP process is performed using the combinational process parameters.

Different procedures are followed by many researchers for the fabrication of SHCs. One of them is filling the powder in the groove by mixing it with volatile liquids like ethanol or methanol to ensure the close packing of powder. This avoids the sputtering action during the process. Another method is to drill a series of holes on the work material along the direction of tool travel and stuffing them with reinforcement particles to provide uniform distribution during the FSP process. The loss of particles is restricted by the portion of shoulder during the processing stage. There is no clear idea of drilling the holes in terms of hole network design, dimensions, pitch distance, and location to ensure the homogenous distribution of reinforcement particles. Thermal spraying, plasma spraying, and cold spraying techniques are other processes involved in the fabrication of SHCs. This involves higher amount of tool wear. Another approach followed along with the above-mentioned techniques is to fill the reinforcement powders in the consumable tool and it produces the composite surface layer with varying thickness. Based on their applications, different reinforcements like B_4C, SiC, Al_2O_3, Graphite, TiC, and WC are used in processing and each one has a sole property.

17.1.5   Advantages of FSP

The major advantages of utilizing FSP process are:

i) Solid-state process which does not need flux, shielding gas, or any filler material.
ii) Improved safety due to non-release of toxic fumes.
iii) Excellent mechanical properties at processed zone achieved through modified microstructural behavior.
iv) Loss of alloy elements and the reduction of strength in parent material is minimal.
v) Lower power consumption due to the absence of external heating.
vi) No position or orientation limitations.
vii) Less occurrence of defects.
viii) Combinational process parameters would be selected based on the requirements.

17.1.6   Disadvantages of FSP

The disadvantages of FSP are:

i) High investment for tooling.
ii) Lack of tool penetration would cause kissing bonds, that would be difficult to detect using nondestructive techniques.
iii) The process ends with an exit hole.

iv) Heavy clamp is required to uphold the material during FSP process owing to higher downward axial forces.

v) Less flexible than manual processes.

17.2 FSP Experimental Setup

17.2.1 Conventional Machines

FSP is performed using specialized FSP machines and is possible to adapt conventional machines by modifying them to accommodate the rotating tool and the required process parameters. Here are some examples of conventional machines that can be used for FSP:

Milling machine: A milling machine can be used for FSP or FSW by attaching a tool holder and a rotating spindle as seen in Figure 17.3. The workpiece can be clamped onto the table and moved in three dimensions to facilitate the FSP process.

Lathe machine: A lathe machine can be used for FSP by modifying it to accommodate the rotating tool and the required process parameters. The workpiece is fitted in the chuck which rotates at a controlled speed and it is processed using the FSP tool [8].

Drill press: A drill press can be used for FSP by attaching a tool holder and a rotating spindle. The workpiece can be clamped onto the table and moved in three dimensions to facilitate the FSP process [9].

CNC machine: A computer numerical control (CNC) machine can be adapted for FSP by adding a specialized FSP tool and modifying the machine control software to control the tool speed, feed rate, and other process parameters.

However, it is important to note that conventional machines may not provide the same level of precision and control as specialized FSP machines. Additionally, modifying conventional machines for FSP requires careful calibration and optimization to ensure that the process parameters are accurately controlled.

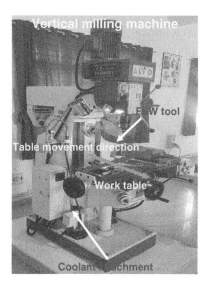

Figure 17.3 Conventional FSP machine.

17.2.2 Indigenous Machines

Indigenous machines for FSP are machines that are locally designed and fabricated using locally available materials and resources. These machines can be an affordable alternative to specialized FSP machines, particularly in developing countries where the cost of imported machines may be prohibitive.

Figure 17.4 illustrates the utilization of a 4-axis numerically controlled FSW machine for conducting FSP [10]. FSP experiments have been carried out by considering parameters such as spindle rotation speed and table travel speed. Their respective ranges are 10–2500 rpm and 20–800 mm/min. The spindle speed is controlled by the servo motor controller. Tool tilt angle was set by slanting the vertical spindle in the range of 2–3° toward trailing edge of the tool during experiments.

Here are some examples of indigenous machines that have been developed for FSP:

1) **Handheld FSP tool:** A handheld FSP tool is made using high-strength material like tungsten carbide or diamond. The tool can be attached to a drill press or similar machine and used to perform FSP on a workpiece. The different process parameters like tool speed, feed-rate, and pressure can be manually controlled by the operator [11].

2) **Modified lathe machine:** A lathe machine can be modified to perform FSP by attaching a specialized FSP tool and adding a motor to rotate the tool. The workpiece can be clamped onto the lathe and moved in three dimensions to facilitate the FSP process. The process parameters can be controlled manually or using a simple control system.

3) **Hydraulic press machine:** A hydraulic press machine can be adapted for FSP by adding a specialized FSP tool and modifying the control system to control the pressure, tool speed, and feed rate. The workpiece can be clamped onto the press and moved in three dimensions to facilitate the FSP process.

4) **Pneumatic machine:** A pneumatic machine can be designed and fabricated for FSP by using locally available materials such as steel pipes and valves. Throughout the FSP procedure, the machine can be utilized to regulate the pressure and tool speed.

It is important to note that indigenous machines may not provide the same level of precision and control as specialized FSP machines. Additionally, the design and fabrication of indigenous machines require a high level of technical expertise and careful calibration to ensure that the process parameters are accurately controlled.

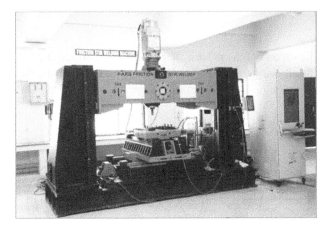

Figure 17.4 Pictorial representation of 4-axis friction stir welding machine.
Source: Muthukumar and Jerome [10]/with permission from Elsevier.

17.2.3 Tool Geometry

Design of FSP tool plays a significant role during the fabrication of composites as it affects the mechanical properties as well as the microstructural homogeneity. Selecting proper tool geometry also helps to produce a defect-free processed zone. Typically, a concave shoulder with a cylindrical pin that is threaded is employed. The pin and shoulder sizes are more important than other design elements. Moreover, the pin is crucial in reducing the grain size [12, 13]. FSP tool materials include high-speed steel, tungsten, high-carbon steel, and their selection depends upon the processed matrix material. The amount of the heat generated by the pin is seemed to be considerably smaller than that of the shoulder.

17.2.3.1 Tool Shoulder and Different Profiles

The purpose of the tool shoulder is to apply a lot of compressive stress and produce enough heat. The shoulder compressive stress reduces the likelihood of voids or pores developing in the consolidated metal. With a contact area up to three times bigger than the pin area, the shoulder diameter should be at least 50% larger than the pin diameter. The shoulder's surface profile determines how much heat it produces. According to the material being processed, the shoulder profile is designed as seen for different shoulder profile in Figure 17.5. This profile has the ability to alter the contact surface area and the quantity of heat produced. Also, it will alter the degree of deformation. As the tool spins, the shoulder comes into contact with the material, traps any profile shapes, and then carries them [1, 14]. It is possible to complete the operation with a zero tilt angle owing to the scroll kind of shoulder [15] without any flash formation.

17.2.3.2 Pin Designs

Single-piece tools: This tool has a simple design. The shoulder and pin are integrated to constitute a tool profile. The pin shape is often a smooth, cylindrical feature with flat ends. TWI created the Tri-flute tool. The dynamic to static volume ratio is larger for conventional tool than it is for the Skew-Stir tool. Tri-flute tool use increased traverse rate by 100% and reduced axial force by 20%, according to TWI investigations [16].

The pin profile of FSP tool is another important factor which affects the quality and properties of processed material. In addition, the pin profile determines the amount of heat generated, the amount of material that is displaced, and the material flow rate around the pin during the FSP process and some of the pin profiles like cylindrical pin and polygonal pin are shown in Figure 17.6.

17.2.4 Fixture Design

FSP fixture design is critical in ensuring the success of the process. The fixture must be designed to securely hold the workpiece in place, while allowing for precise movement and control during the FSP process. Here are some important considerations for FSP fixture design:

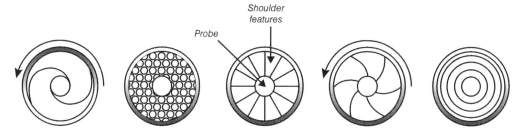

Figure 17.5 Different shoulder profiles. Source: Mishra and Ma [1]/with permission of Elsevier.

Figure 17.6 Various tool geometry employed like cylindrical pin and polygonal pin. Source: Mehta and Badheka [17]/with permission from Elsevier.

1) **Workpiece clamping:** The fixture must securely clamp the workpiece in place to prevent movement during the FSP process. The clamps should be positioned to minimize interference with the FSP tool and to allow for maximum access to the workpiece surface.
2) **Support structure:** The fixture must provide a stable support structure for the workpiece during the FSP process. The support structure should be designed to minimize vibration and ensure that the workpiece is held in a fixed position.
3) **Cooling system:** A cooling system should be incorporated into the fixture design to dissipate the generated heat during the FSP process. This is achieved through circulation of cooling fluid through the fixture, or by using a heat sink or other cooling mechanism.
4) **Motion control:** The fixture must allow for precise movement control of the FSP tool during the process. This can be achieved through the use of a computer-controlled motion system, or through manual control using precision positioning equipment.
5) **Accessibility:** The fixture must allow for easy access to the workpiece surface, particularly when changing the FSP tool or adjusting process parameters. The fixture should be designed to minimize interference with the FSP tool and to allow for easy removal and replacement of the workpiece.

The design of the FSP fixture should be customized to the specific requirements of the FSP process, taking into account the material being processed, the geometry of tool, and desired microstructural properties. The fixture design should be thoroughly tested and validated to ensure that it can effectively hold the workpiece in place and provide the necessary control and support for the FSP process.

17.3 Microstructural Characterization

17.3.1 Microstructural Behavior Based on Various Shoulder Profiles

The microscopic observations (Figure 17.7) depict the major defects like voids/cracks are not witnessed within the processed zone. This may be owing to the rapid deformation due to the pin design. A discussion on the effect of tool-pin on the microstructural characterization of structural steel (SS) specimen is considered. This represents the stirred zone of the FSP-processed SS under various pin profiles like triangular-pin, square-pin, conical-pin, and cylindrical-pin (Figure 17.7a–d). The results obtained by the author during the process clearly display the grain refinement owing to different recrystallization processes and severe plastic deformation. Besides, it is understood from the figure that during processing the excess plasticized material, amount of material flow and pulsating action causes consistent grain refinement with the pin contact surface and the processed material.

Figure 17.8 depicts the effects of three pin profiles like whorl pin, plain taper pin, and taper threaded pin on the macrostructure and microstructural characteristics of pure-grade aluminum (AA1100 H14 Al) and commercially available pure copper (Cu). Al–Cu elements are completely

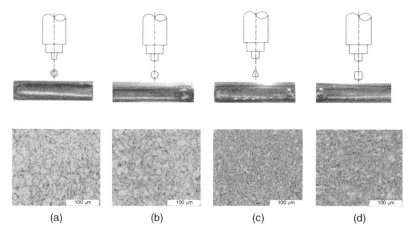

(a)   (b)   (c)   (d)

Figure 17.7 Optimal microstructure of FSPed structural steel by different pin profiles. (a) Conical pin profile. (b) Cylindrical pin profile. (c) Triangular pin profile. (d) Square pin profile. Source: Amirafshar and Pouraliakbar [18]/with permission from Elsevier.

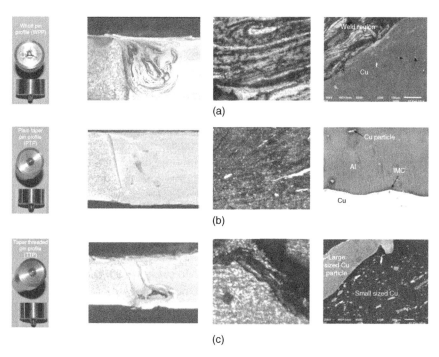

Figure 17.8 Different pin profiles of Al–Cu and their micro and macrostructural observations. Source: Muthu and Jayabalaan [19]/with permission from Elsevier. (a) Whorl pin profile: (i) photograph, (ii) macrostructure observation, (iii) optical microstructure of pin-influenced region, (iv) SEM image. (b) Plain taper pin profile: (i) photograph, (ii) macrostructure observation, (iii) optical microstructure of pin-influenced region, (iv) SEM image. (c) Taper threaded pin profile: (i) photograph, (ii) macrostructure observation, (iii) optical microstructure of pin-influenced region, (iv) SEM image.

mixed together in the whorl pin profile that is shown. Defect-free junctions are seen with low dispersion of Cu particles in the SZ when using plain taper pin profiles. In the pin-influenced zone of the threaded pin-profile structures, flaws are seen. Moreover, a partial ring-like material flow that is highly impacted by the pin motion is visible in the whorl profile (Figure 17.8a). As in SZ, small Cu particles of various sizes can be seen in the macrostructure of the plain taper pin profile (Figure 17.8b), as these can deform Cu particles that are oriented vertically toward the pin-influenced region. According to Figure 17.8c, threaded taper pin profile shape, the majority of the Cu particles are exhumed from the AS and swirled into the SZ.

Scanning electron microscope analysis, as shown in Figure 17.14, is used to thoroughly investigate the dispersion of Cu particles. Whorl pins, however, have more severely deformed Cu material particles than other pins in the SZ and exhibit severely distorted structure. A plain taper pin demonstrates the reinforcement of tiny Cu particles in the SZ to create a structure resembling the composites. Moreover, it offers a conduit for the plasticized material to flow from leading-edge of tool-pin to its trailing-edge. The profile of a threaded taper pin reveals distorted tiny and bulk Cu particles.

17.3.2 Microstructural Characterization Based on Various Shoulder Profiles

The macrostructural observations for the AA6061 as developed by straight cylindrical profile tool at three different shoulder diameters (15, 18, and 21 mm) are presented in Figure 17.9.

In the three macrostructures produced using straight cylindrical pin profile tool in AA6061 specimen (as seen in Figure 17.9), the specimen formed using 18 mm tool shoulder diameter is found to be defect free. However, the occurrence of other defects in the FSP zone (at 15 and 21 mm tool shoulder diameter) is a result of insufficient metal consolidation and inappropriate plastic flow. During the extrusion process, the metal is propelled after plastic deformation by the applied forces, and also the motion of tool pin owing to the action produced by the pin rotation are two factors that are understood to be influencing the material flow.

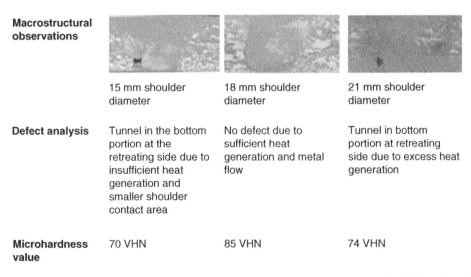

Macrostructural observations		
15 mm shoulder diameter	18 mm shoulder diameter	21 mm shoulder diameter

Defect analysis	Tunnel in the bottom portion at the retreating side due to insufficient heat generation and smaller shoulder contact area	No defect due to sufficient heat generation and metal flow	Tunnel in bottom portion at retreating side due to excess heat generation

Microhardness value	70 VHN	85 VHN	74 VHN

Figure 17.9 Macrostructure interpretations of the A6061 aluminum alloy developed through straight cylindrical profile tool at three different shoulder diameters (15, 18, and 21 mm). Source: Elangovan and Balasubramanian [20]/with permission from Elsevier.

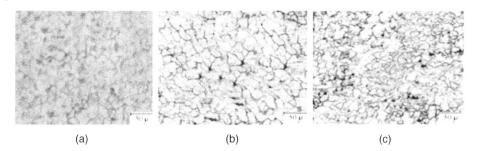

(a)   (b)   (c)

Figure 17.10 Microstructure analysis in AZ91/Al$_2$O$_3$ composite fabricated at 1250 rpm rotational speed and three different traverse speeds: (a) 20 mm/min, (b) 40 mm/min, and (c) 63 mm/min. Source: Khayyamin et.al. [21]/with permission from Elsevier.

17.3.3 Microstructural Characterization in Relation to Process Parameters

Khayyamin et al. studied the outcome of different process parameters on the mechanical properties of AZ91/Al$_2$O$_3$ composites. At constant rotational speed of 1250 rpm, three different traverse speeds (a) 20, (b) 40, and (c) 63 mm/min are selected on 8 mm thick AZ91 magnesium alloy. Figure 17.10 shows microstructures of the SZ and it is understandable that decrease in grain size is due to the increase in the transverse speed owing to the heat effect on grain growth.

In Figure 17.11, the optical micrographs depict the (as-cast) A356 and FSP samples created by standard threaded pin. The tool rotation speed is set at 900 rpm, and the traverse speed is 203 mm/min. The images clearly demonstrate a consistent rearrangement of the fractured Si particle within the Al matrix. It is evident that big, needle-like Si particles are broken up as a result of abrasive mixing and swirling action in the processed zone of the FSP A356. Additionally, this process resulted in the formation of dendrite structures and a uniform dispersion of Si particles over the Al matrix.

During FSP, the dynamic recrystallization in the NZ causes generation of fine-equiaxed grains. FSP parameters like tool-geometry, base matrix composition, temperature variations, active-cooling, and vertical-pressure employ important effect on the grain size. The pinhole defects, kissing bond, tunnel defects, cracks, and piping defects may form owing to improper metal flow and insufficient metal consolidation.

The macrostructure observations at three different locations with three different rotational speed is presented in Figure 17.12. The square pin profile tool at three different rotational speed establishes to be free from the basic defects. This may be due to two reasons on working of plasticized metal like sufficient working owing to the pulsating action of pin profile and excess working with

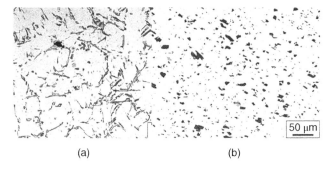

(a)   (b)

Figure 17.11 Optical micrographs showing the microstructure using standard threaded pin at rotational speed of 900 rpm. (a) As-cast A356. (b) FSPed A356. Source: Mishra and Ma [1]/with permission from Elsevier.

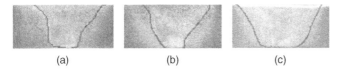

(a)   (b)   (c)

Figure 17.12 Influence of square pin profile tool with different rotational speed on the macrostructure behavior. (a) 1500 rpm. (b) 1600 rpm. (c) 1700 rpm. Source: Amirafshar and Pouraliakbar [22]/with permission from Elsevier.

broader FSP due to increased rotational speed (1700 rpm). From the figure, it is understood that the formation of defect-free FSP zone is due to the impact of both tool-profile and their rotational speed.

17.4 Mechanical Behavior of Composites Based on Various Tool Shapes

17.4.1 Mechanical Characterization

A predominant effect is recorded over surface modifications and mechanical properties of different specimens owing to different tool-pin shape Furthermore, the size of the tool shoulder diameter has a significant role in the generation of heat resulting from frictional forces. A fundamental observation is that a larger shoulder diameter leads to a greater amount of heat generation due to increased friction which are caused by the larger contact surface area.

17.4.2 Microhardness Measurements

Figure 17.13 displays the results of the microhardness measurements conducted by the author on the cross section of the ST14 specimen subjected to FSP. The material hardness depends majorly by its grain size variations like the lesser the grain size, greater the hardness. The plastic deformation formed during FSP surges the density of material dislocation. The presence of this structure along grain boundaries and sub-grain boundaries serves as a significant hindrance to the movement of dislocations. The figure reveals that the specimen with a square-pin profile exhibits a hardness of 320 VHN, while the triangular, cylindrical, and conical-pin specimens have hardness values of 311 VHN, 292 VHN, and 275 VHN respectively. This disparity can be attributed to the presence of

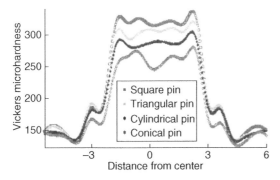

Figure 17.13 Microhardness variation values of ST14 structured steel for different tool pin design. Source: Amirafshar and Pouraliakbar [18]/with permission of Elsevier.

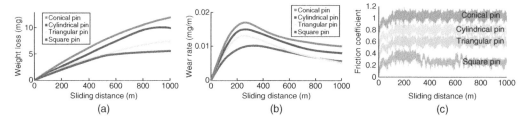

Figure 17.14 Image displaying the results of friction stir processed stainless steel using various pin designs for (a) weight loss, (b) wear rate, and (c) friction coefficient vs. sliding distance. Source: Amirafshar and Pouraliakbar [18]/Elsevier.

smaller grains in the processed region, which result from enhanced plastic deformation during dynamic recrystallization.

17.4.3 Tribological Characteristics

The tribological behavior of FSP-processed steel sheet (ST14) as observed in Figure 17.14 clearly elucidates the results taken through the pin-on-disk wear investigation technique [17]. The understandings from author show that enhancement in wear-resistance is correlated to high-hardness values owing to grain refinement and microstructural evolutions. On increasing the sliding distance, weight loss also increases (Figure 17.14a). The weight loss is minimum while using a conical pin than other pins like triangular cylindrical and conical-pin. The data indicates that the material processed by a square tool pin exhibits significantly lower wear rate, weight loss (Figure 17.14b), and friction coefficient (Figure 17.14c).

Additionally, friction coefficient with a square pin tool is less due to the process of confined severe plastic deformation in the dry sliding contact surfaces. The significant impact of tool profile and tool rotational speed on the tensile properties is apparent [20]. In comparison to other tool pin profiles, the square tool-profile specimen displayed improved tensile characteristics. Additionally, a tool with an 18 mm shoulder diameter displayed superior tensile properties compared to other shoulder diameters. Figure 17.13 supports this observation by showing variable microhardness values in the FSP zone created by the square pin profile at various tool shoulder diameters.

17.5 Mechanical behavior of Composites Based on Various Process Parameters

It is taken into consideration how several process parameters, such as rotational speed, traverse speed, and number of passes, affect the microhardness characteristics of the $AZ91/SiO_2$ composite (Figure 17.15a). Specifically, during the fabrication process, rotational speed is set at 1250 rpm, traverse speed is varied at 20, 40, and 63 mm/min for three passes. The results indicate that the engineered $AZ91/SiO_2$ composite layer exhibited a smaller grain size of 8 mm and higher hardness of 124 HV compared to the AZ91 alloy, which had a hardness of 65 HV. Tensile properties are assessed using a tensile tester for 1–3 passes, revealing an important improvement in tensile characteristics of AZ91 alloy due to FSP process (as depicted in Figure 17.15b). Increasing the traverse speed resulted in a reduction in grain size and rise in hardness, attributed to reduced contact time. Additionally, increased number of passes enables the $AZ91/SiO_2$ composite matrix to have a

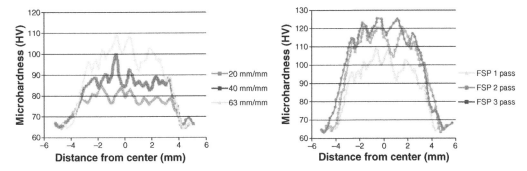

Figure 17.15 Hardness profile of specimens at rotational speed of 1250 rpm: (a) With traverse speeds of 20, 40, and 63 mm/min. (b) With traverse speed of 63 mm/min in 1, 2, and 3 pass. Source: Khayyamin et al. [21]/with permission of Elsevier.

homogeneous distribution of reinforcement particles, which enhanced the microhardness. The mechanical behavior was similarly affected by the increased traverse speed, which reduced the grain formation during the solidification process and thus increased the tensile strength.

17.6 Conclusion

In this chapter, FSP technique is applied for the fabrication of surface composites. Pin profiles like cylindrical, conical, square, triangular, whorl tool, tri-flute tool, and plain taper, taper threaded and skew-stir tools are utilized to modify the material flow characteristics. Furthermore, a comprehensive analysis is conducted to examine the connection between the material flow and mechanical properties like microhardness, tensile properties, and the wear characteristics. Additionally, the changes in microstructure are thoroughly investigated using optical microscopes and scanning electron microscopes. According to these characteristics, an indigenous setup made up of FSP tool profile with excellent mechanical qualities is effectively used to fabricate composites, particularly for automotive and aerospace applications.

References

1 Mishra, R.S. and Ma, Z. (2005). Friction stir welding and processing. *Mater. Sci. Eng.: R: Rep.* 50: 1–78.
2 Li, Z., Gao, S., Ji, S. et al. (2016). Effect of rotational speed on microstructure and mechanical properties of refill friction stir spot welded 2024 Al alloy. *J. Mater. Eng. Perform.* 25: 1673–1682.
3 Patel, V.V., Badheka, V.J., and Kumar, A. (2016). Influence of pin profile on the tool plunge stage in friction stir processing of Al–Zn–Mg–Cu alloy. *Trans. Indian Inst. Met.* 70: 1151–1158.
4 Banik, A., Saha Roy, B., Deb Barma, J., and Saha, S.C. (2018). An experimental investigation of torque and force generation for varying tool tilt angles and their effects on microstructure and mechanical properties: Friction stir welding of AA 6061-T6. *J. Manuf. Processes* 31: 395–404.
5 Jata, K. & Semiatin, S. 2000. Continuous dynamic recrystallization during friction stir welding of high strength aluminum alloys. Air force research lab wright-patterson afb oh materials and manufacturing . . . .

6 Liu, G., Murr, L., Niou, C. et al. (1997). Microstructural aspects of the friction-stir welding of 6061-T6 aluminum. *Scr. Mater.* 37: 355–361.

7 Sato, Y.S., Kokawa, H., Enomoto, M., and Jogan, S. (1999). Microstructural evolution of 6063 aluminum during friction-stir welding. *Metall. Mater. Trans. A* 30: 2429–2437.

8 Patel, P., Patel, S., and Shah, H. (2017). Design and experimental study of friction stir welding of AA6061-T6 Alloy for optimization of welding parameters by using lathe machine. *Int. Res. J. Eng. Technol.* 4: 26–32.

9 Banjare, P.N., Gadpale, V., and Manoj, M.K. (2018). Friction stir spot welding of commercial aluminum strip using a drill press. *MATS J. Eng. Appl. Sci* 3: 86–91.

10 Muthukumar, P. and Jerome, S. (2019). Surface coating (Al/Cu &Al/SiC) fabricated by direct particle injection tool for friction stir processing: Evolution of phases, microstructure and mechanical properties. *Surf. Coat. Technol.* 366: 190–198.

11 Murti, K., Kumar, C.L., Prasad, V., and Vanaja, T. (2015). Design and development of friction stir drilling and tapping. *Int. J. Sci. Res.* 78.

12 Kim, J.-R., Ahn, E.-Y., Das, H. et al. (2017). Effect of tool geometry and process parameters on mechanical properties of friction stir spot welded dissimilar aluminum alloys. *Int. J. Precis. Eng. Manuf.* 18: 445–452.

13 Shiraly, M., Shamanian, M., Toroghinejad, M.R. et al. (2017). The influence of tool geometry on the mechanical behaviour of FSSWed Al/Cu ARBed composite. *Trans. Indian Inst. Met.* 70: 2205–2211.

14 Bakavos, D., Chen, Y., Babout, L., and Prangnell, P. (2010). Material interactions in a novel pinless tool approach to friction stir spot welding thin aluminum sheet. *Metall. Mater. Trans. A* 42: 1266–1282.

15 Sevvel, P. and Jaiganesh, V. (2015). Effect of tool shoulder diameter to plate thickness ratio on mechanical properties and nugget zone characteristics during FSW of dissimilar Mg alloys. *Trans. Indian Inst. Met.* 68: 41–46.

16 Rai, R., De, A., Bhadeshia, H.K.D.H., and Debroy, T. (2011). Review: friction stir welding tools. *Sci. Technol. Weld. Joining* 16: 325–342.

17 Mehta, K.P. and Badheka, V.J. (2016). Effects of tool pin design on formation of defects in dissimilar friction stir welding. *Procedia Technol.* 23: 513–518.

18 Amirafshar, A. and Pouraliakbar, H. (2015). Effect of tool pin design on the microstructural evolutions and tribological characteristics of friction stir processed structural steel. *Measurement* 68: 111–116.

19 Muthu, M.F.X. and Jayabalan, V. (2016). Effect of pin profile and process parameters on microstructure and mechanical properties of friction stir welded Al–Cu joints. *Trans. Nonferrous Met. Soc. China* 26: 984–993.

20 Elangovan, K. and Balasubramanian, V. (2008). Influences of tool pin profile and tool shoulder diameter on the formation of friction stir processing zone in AA6061 aluminium alloy. *Mater. Des.* 29: 362–373.

21 Khayyamin, D., Mostafapour, A., and Keshmiri, R. (2013). The effect of process parameters on microstructural characteristics of AZ91/SiO$_2$ composite fabricated by FSP. *Mater. Sci. Eng., A* 559: 217–221.

22 Elangovan, K. and Balasubramanian, V. (2007). Influences of pin profile and rotational speed of the tool on the formation of friction stir processing zone in AA2219 aluminium alloy. *Mater. Sci. Eng., A* 459: 7–18.

18

Analysis of Friction Stir Welding Tool Using Various Threaded Pin Profiles: A Case Study

Bommana B. Abhignya[1], Ashish Yadav[1], Manu Srivastava[1], and Sandeep Rathee[2]

[1] *Mechanical Engineering Department, Hybrid additive manufacturing Laboratory, PDPM Indian Institute of Information Technology, Design & Manufacturing, Jabalpur, India*
[2] *Department of Mechanical Engineering, National Institute of Technology, Srinagar, J&K, India*

18.1 Introduction

Friction stir welding (FSW) has revolutionized the field of metal joining by enabling the production of high-quality, defect-free welds in a wide range of materials and applications [1, 2]. The FSW process involves a rotating tool that is plunged into the joint line between two workpieces, generating frictional heat and intense plastic deformation. As the tool moves along the weld line, it effectively stirs the material, causing it to consolidate and form a solid-state bond. The schematic diagram for the FSW process is presented in Figure 18.1.

There are various process parameters related to the tool that need attention while optimizing the performance of the FSW process as presented in Figure 18.2. These chiefly include parameters related to tool rotational speed, axial force on the tool, traverse speed of the tool, and the physical tool parameters itself [5, 6]. While the first three factors, i.e. axial force, traverse speed, and the tool rotational speed are more process/setup or machine-dependent, the last one which is the physical tool itself is totally controlled by the choice of tool material, geometry, and its dimensions as presented in Figures 18.2 and 18.3. Among the tool dimensions, the profile of the tool pin is an important factor that remarkably affects the weld quality and properties [8, 9].

In recent years, researchers have devoted considerable attention to investigating the influence of different threaded pin profiles on the FSW process. A variety of pin designs have been proposed, such as square, triangular, round, tapered, and concave profiles, each offering unique advantages and challenges. The threaded pin's geometry affects critical process parameters, such as heat generation, material mixing, and plastic deformation, which ultimately determines the weld quality and mechanical properties.

The main objective of this case study is to provide a comprehensive analysis of the effects of various pin profiles on the performance of the FSW tool. By systematically examining different pin designs, their geometrical features, and their impact on the welding process, the case study aims to gain deeper insights into the underlying mechanisms governing FSW. Additionally, this study will aid in optimizing the FSW process for different materials and applications, leading to improved weld quality and increased productivity.

Friction Stir Welding and Processing: Fundamentals to Advancements, First Edition.
Edited by Sandeep Rathee, Manu Srivastava, and J. Paulo Davim.
© 2024 John Wiley & Sons, Inc. Published 2024 by John Wiley & Sons, Inc.

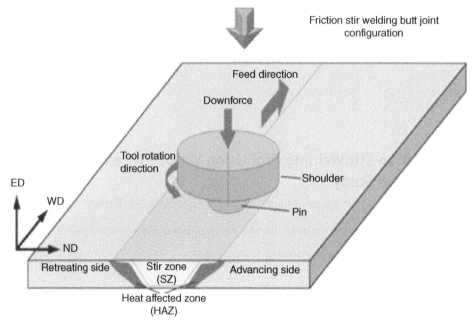

Figure 18.1 Schematic representation of the FSW process. Source: West et al. [3]/Elsevier/CC BY 4.0.

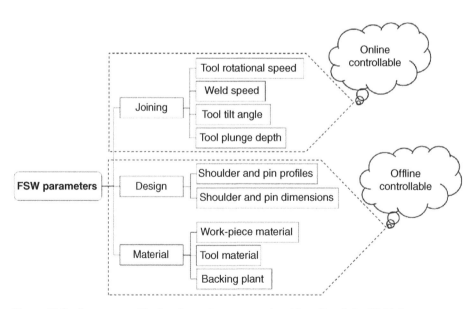

Figure 18.2 Parameters affecting the performance and weld quality of the FSW joints.
Source: Mishra et al. [4]/with permission of Elsevier.

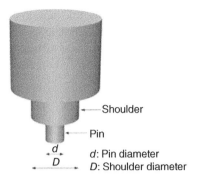

Figure 18.3 A labeled FSW tool. Source: Dehabadi et al. [7]/ Springer Nature.

18.2 Geometry Considered

In this case study, the stress analysis of FSW tool with three different pin profiles were considered as shown in Figure 18.4a–c, respectively, that depict the cylindrical, conical, and cylindrical pin profiles, respectively. The work piece's thermomechanical properties have an impact on the stress distribution of the tool pin in FSW. In this case study, three tools were created with threads incorporated into their profiles. The tools' pin shapes included conical, cylindrical, and cuboidal.

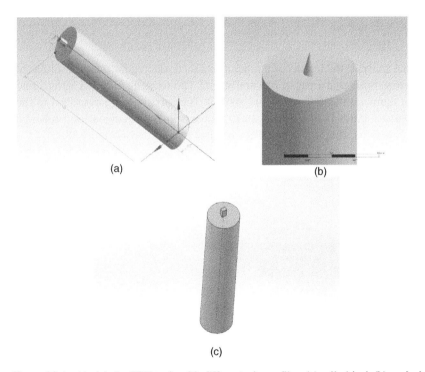

Figure 18.4 Models for FSW tools with different pin profiles: (a) cylindrical; (b) conical; (c) cuboidal.

The initial tool dimensions considered by taking into account the base material plate thickness and the induced structural stresses were examined to ensure that they were within allowable stress limits. The tools' modeling and analysis are done in ANSYS software to investigate stress distributions in the pin at various temperatures. The stress in the pin profiles are simulated by taking into account the frictional force between the tool shoulder and workpiece.

The dimensions considered for pin in finite element analysis (FEM) are:

- Cylindrical: pin diameter = 3 mm, pin height = 5.5 mm
- Conical: pin diameter = 3 mm, pin height = 5.5 mm
- Cuboidal: pin side = 3 mm, pin height = 5.5 mm

Constant shoulder dimensions considered are:

Shoulder height = 100 mm, shoulder diameter = 20 mm

The pin as well as shoulder dimensions are kept constant for all the three profiles to understand the exclusive impact of the pin profiles on the FSW process.

18.3 Results and Discussions – Analysis in ANSYS

On applying similar loading conditions, the stress distributions obtained for the cylindrical, conical, and cuboidal pin profiles are presented as Figures 18.5, 18.6, and 18.7, respectively.

The loading conditions selected for each pin are:

Axial load = 7000 N on tool pin
Tangential load = 2100 N on tool pin

In addition to the load analysis, the impact of temperatures is considered for the three different pin profiles. Temperatures considered are 600, 700, and 800 °C. The stress distribution is then analyzed at three different temperatures (600, 700, and 800 °C). The variations are described in the form of graph presented as Figures 18.8, 18.9, and 18.10 for the cylindrical, cubical, and cuboidal pins, respectively. A comparative chart for all the three profiles is presented as Figure 18.11.

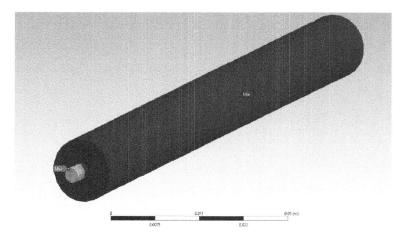

Figure 18.5 Stress distribution of FSW tool with cylindrical pin.

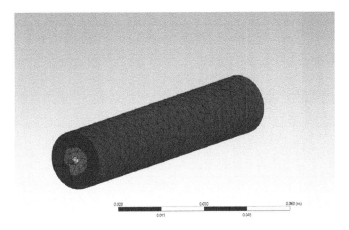

Figure 18.6   Stress distribution of FSW tool with conical pin.

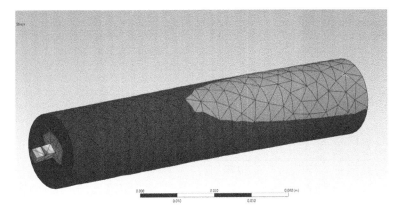

Figure 18.7   Stress distribution of FSW tool with cuboidal pin.

Based on the findings mentioned, it has been determined that among all the profiles, the tool with a cylindrical profile stands out as the superior for the chosen outputs. This particular profile exhibits an exceptionally small maximum stress distribution, indicating its excellent performance. Usually, profiles with threads display an increase in stress distribution as the temperature in the welding zone rises. The tool featuring a threaded conical profile, in particular, showcases the highest value of stress distribution.

18.4   Conclusion

The exclusive effect of different pin profiles of the FSW tool are analyzed in this case study keeping the remaining conditions same like pin length, loading conditions, temperatures, and shoulder diameter. It was established that cylindrical tool profile is best suited for the stress distribution and hence performance of the FSW tool. The cuboidal tool offered medium performance while the conical pins exhibited the highest stresses under same loading and temperature conditions and is therefore least preferred.

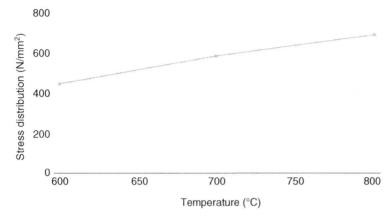

Figure 18.8 Variation of stress distribution with temperature in cylindrical pin profile tool.

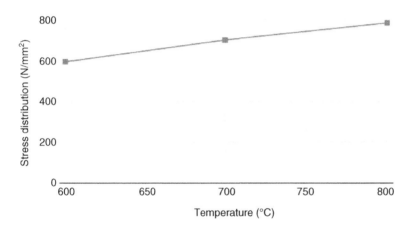

Figure 18.9 Variation of stress distribution with temperature in cubical pin profile tool.

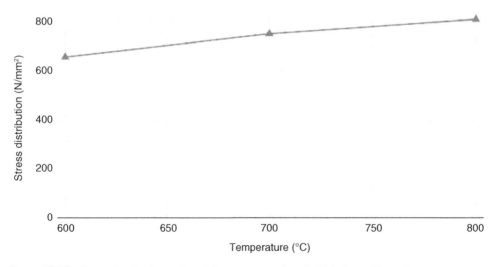

Figure 18.10 Stress distribution varies with temperature in cuboidal pin profile tool.

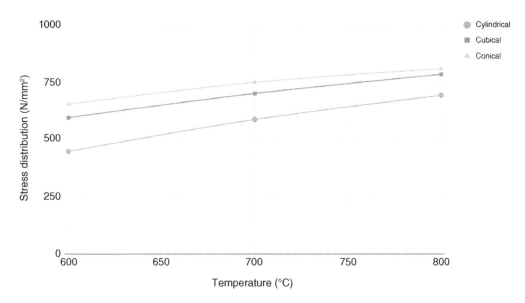

Figure 18.11 Comparative stress distribution variation with temperature in cylindrical, conical, and cuboidal pin tool profiles.

Further, the analysis can be extended by study regarding the effect of rotational speed on the maximum stress distribution. It is established that as the rotational speed is increased, these parameters do not undergo significant changes. These findings may highlight the stability and consistency of the stress distribution regardless of the rotational speed applied.

In conclusion, this study aims to contribute to the existing body of knowledge by presenting an exclusive analysis of various threaded pin profiles in the context of FSW. The findings from this study will not only deepen reader's understanding of the FSW process but also provide valuable insights for designing optimized pin profiles to enhance weld quality and process efficiency. Ultimately, this research has the potential to drive advancements in the field of FSW and facilitate its wider adoption in diverse industrial applications.

Acknowledgement

The authors thank the Science & Engineering Research Board (SERB) for its financial assistance under the project (vide sanction order no. SPG/2021/003383) to perform this work.

References

1 Srivastava, M. and Rathee, S. (2021). *A study on the effect of incorporation of SiC particles during friction stir welding of Al 5059 alloy. Silicon* **13** (7): 2209–2219.

2 Padhy, G.K., Wu, C.S., and Gao, S. (2018). *Friction stir based welding and processing technologies – processes, parameters, microstructures and applications: a review. J. Mater. Sci. Technol.* **34** (1): 1–38.

3 West, P., Shunmugasamy, V.C., Usman, C.A. et al. (2021). Part I: Friction stir welding of equiatomic nickel titanium shape memory alloy – microstructure, mechanical and corrosion behavior. *J. Adv. Joining Processes* **4**: 100071.

4 Mishra, D., Roy, R.B., Dutta, S. et al. (2018). A review on sensor based monitoring and control of friction stir welding process and a roadmap to Industry 4.0. *J. Manuf. Processes* **36**: 373–397.

5 Khan, N., Rathee, S., Srivastava, M., and Sharma, C. (2021). *Effect of tool rotational speed on weld quality of friction stir welded AA6061 alloys*. Mater. Today Proc. **47**: 7203–7207.

6 Khan, N., Rathee, S., and Srivastava, M. (2021). Parametric optimization of friction stir welding of Al-Mg-Si alloy: a case study. *Yugosl. J. Oper. Res.* **31** (2): 8.

7 Dehabadi, V.M., Ghorbanpour, S., and Azimi, G. (2016). Application of artificial neural network to predict Vickers microhardness of AA6061 friction stir welded sheets. *J. Cent. South Univ.* **23** (9): 2146–2155.

8 Khan, N., Rathee, S., and Srivastava, M. (2021). *Friction stir welding: an overview on effect of tool variables*. Mater. Today Proc. **47**: 7196–7202.

9 Rai, R. et al. (2011). *Review: friction stir welding tools*. Sci. Technol. Weld. Joining **16** (4): 325–342.

19

Static Structural and Thermal Analysis of Honeycomb Structure Fabricated by Friction Stir Processing Route: A Case Study

Nikhil Jaiswal[1], Umashankar Bharti[1], Ashish Yadav[1], Manu Srivastava[1], and Sandeep Rathee[2]

[1] *Hybrid Additive Manufacturing Lab, Mechanical Engineering Department, PDPM Indian Institute of Information Technology Design and Manufacturing, Jabalpur, India*
[2] *Department of Mechanical Engineering, National Institute of Technology, Srinagar, J&K, India*

19.1 Introduction

Honeycomb structures (HCS) have remarkable mechanical qualities and distinctive cellular architecture because of which they have a wide range of utility in the aerospace, automotive, and construction industries [1]. The hexagonal honeycomb design, as presented in Figure 19.1, provides higher strength and rigidity while minimizing weight. The advent of sophisticated fabrication methods in recent years has increased the potential for tailoring HCS to particular uses. Friction stir welding/processing (FSW/P) which incorporates localized heating and plastic deformation of materials using a rotating tool is one attractive example of fabrication and welding technique for HCS structures [2–4]. Due to its capability to produce fine-grained microstructures and alleviating flaws that are associated with traditional manufacturing methods, FSP has proven candidature to be particularly advantageous in fabricating, processing, and welding the HCS structures. It additionally offers a better level of control over the material flow during processing [1].

Various researchers have reported appreciable quantum of work related to the welding, processing, and fabricating HCS via the different routes. Arslan et al. [5] studied the experimental damage evaluation of honeycomb sandwich structures with Al/B_4C functionally graded material (FGM) face plates under high-velocity impact loads. Abhinav et al. [6] reviewed the origin of HCS and its sailing properties. Lin et al. [7] researched on the out-of-plane compression behavior of aluminum alloy large-scale super-stub cellular HCS. Ma et al. [8] reported progress in double-layer HCS as a special class of two-dimensional materials. Zhengxian et al. [9] studied the fabrication and mechanical behaviors of quartz fiber composite HCS with extremely low permittivity. Ciepielewski et al. [10] experimented on the static and dynamic response of aluminum sandwiched HCS structures, and so on. However, there is meager reported research on the analysis of HCS structures made via FSW/P route.

The goal of this case study is to investigate the static thermal and structural behavior of HCS fabricated via the FSP route. To fully comprehend the mechanical behavior of these structures under various loading circumstances, the analysis in this case study includes both experimental investigations

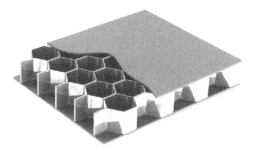

Figure 19.1 Honeycomb structure.

and numerical simulations. Internal heat distribution and temperature gradients of HCS during the FSP process are the main subjects of the thermal analysis. This research helps to comprehend how process variables affect the material's thermal history and the subsequent microstructural changes, which in turn influence the mechanical properties of the honeycomb structure. The results of this case study will help comprehend the potential of the FSP manufacturing approach to fabricate high-performance HCS. The findings can offer useful information for enhancing the design and production of strong, lightweight honeycomb components for a variety of engineering applications.

19.2 Modeling Details

This section provides details of the modeling including design methodology, dimensions of the hexagonal cell considered for the analysis, dimensions of the chosen panel, properties, and statistics of the structure.

19.2.1 Design Methodology

The initial stage of the design involves using simple tools to create a hexagonal cell structure and then performing extrusion. Following this step, a group of hexagonal cells are assembled for analysis in certain cases. Moving on to the next level, the design focuses on creating rectangular panels. Computer codes utilize finite difference methods or finite element methods, employing 1D, 2D, or 3D models to simulate various physical phenomena such as internal ballistics, fluid dynamics, continuum mechanics, and structural analysis. These methods enable accurate calculations and optimization, leading to the determination of the final geometry.

19.2.2 Dimensions of Hexagonal Cell

Chosen dimensions of the hexagonal cell considered for extruding, as presented by Figure 19.2, are:

- Edge length = 0.0035 m
- Radius = 0.007 m
- Depth = 0.023 m
- Thickness = 0.000068 m

19.2.3 Dimensions of Panel

The dimensions of the panel chosen are:

- Length = 0.100 m
- Width = 0.100 m
- Thickness = 0.0010 m

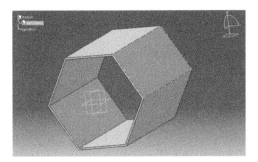

Figure 19.2 Hexagonal cell extrude.

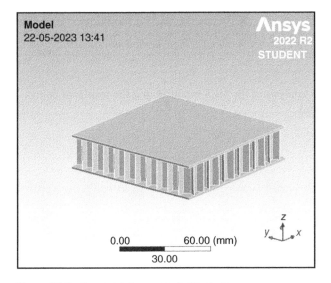

Figure 19.3 Sandwich honeycomb structure.

19.2.4 Properties

Volume of 50,445 mm^3, total mass of 0.39599 kg, and unity scale factor value is considered for the modeled HCS. The final model of sandwich HCS is presented as Figure 19.3.

19.2.5 Statistics

Basic statistics of the HCS model developed are presented in Table 19.1.

Table 19.1 Statistics of the HCS structure.

Bodies	3
Active bodies	3
Nodes	27,728
Element	3882
Mesh matric	None

19.3   Result and Analysis

This section presents details of static structural and thermal analysis of the HCS design proposed by FSP.

19.3.1   Static Structural Analysis

Force is applied in the Z direction by keeping degree of freedom (DOF) of the two opposite sides as zero. Resulting HCS deformation are generated using the above results.

As presented in Figure 19.4, the HCS structure is assumed as a fixed beam with point load, i.e. DOF for the right and left sides is taken as zero.

19.3.1.1   Input Definition

The input definition for the stress and strain analysis of the HCS is presented as Table 19.2.

19.3.1.2   Stress Analysis Results

Figure 19.5 presents the equivalent stress analysis of the HCS modeled and the results of equivalent stress analysis is tabulated in Table 19.3.

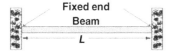

Figure 19.4   Fixed beam.

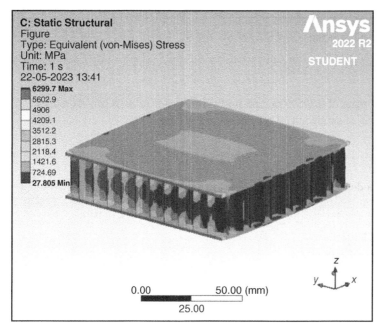

Figure 19.5   Equivalent stress distribution for the HCS model developed.

Table 19.2 Input parameter for stress and strain.

Type	Fixed support force
Defined by	Components
Applied by	Surface effect
Coordinate system	Global coordinate system
X, Y, Z	0, 0, −7000 N

Table 19.3 Stress analysis results.

S. no.	Minimum (Mpa)	Maximum (Mpa)	Average (Mpa)
1.	27.805	6299.7	716.18

Based on the static structural analysis of the structure, when a force of 7000N is applied in the negative z-direction with the degree of freedom fixed on two opposite sides, the following conclusions can be drawn regarding the stress distribution:

Maximum stress: The structure experiences a maximum stress of 6299.7 MPa. This indicates the highest stress concentration in the material which most likely occurs at specific points or regions where the applied force is being resisted.

Minimum stress: The minimum stress observed in the structure is 27.805 MPa. This indicates areas or regions where the stress levels are comparatively lower, potentially occurring in regions that are not directly affected by the applied force.

Average stress: The average stress across the structure is calculated to be 716.18 MPa. This value represents the overall stress distribution, taking into account both high- and low-stress regions.

These stress values provide insights into the structural behavior under the applied force. It is important to compare these stress values with the material's allowable stress limits or design criteria to ensure that the structure can withstand the applied load without compromising its integrity or safety. Further analysis and evaluation may be required to assess factors such as the factor of safety, material properties, and structural design considerations to determine the overall structural performance and adequacy.

19.3.1.3 Strain Analysis Results

Figure 19.6 presents the equivalent strain analysis of the HCS modeled and the results of equivalent stress analysis is tabulated in Table 19.4.

Based on the static structural analysis of the structure, when a force of 7000N is applied in the negative z-direction, the following conclusions can be drawn:

Maximum strain: The structure experiences a maximum strain of 3.1499e-002 mm/mm. This indicates the highest amount of deformation or elongation that occurs in the structure under the applied force.

Minimum strain: The structure exhibits a minimum strain of 5.6182e-004 mm/mm. This represents the least amount of deformation or elongation experienced in the structure.

Average strain: The average strain in the structure is calculated to be 4.2497e-003 mm/mm. This value provides an overall measure of the deformation across the structure due to the applied force.

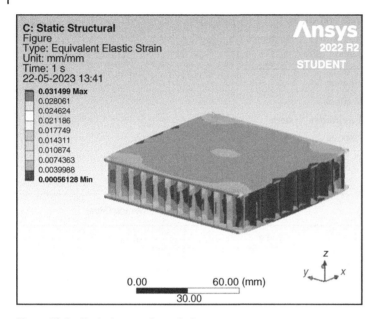

Figure 19.6 Equivalent strain analysis.

Table 19.4 Strain analysis results.

S. no.	Minimum (mm/mm)	Maximum (mm/mm)	Average (mm/mm)
1.	5.6182e-004	3.1499e-002	4.2497e-003

By analyzing the strain values, it can be determined how the structure responds to the applied load. Strain is a measure of deformation and indicates the relative change in length or shape of the structure. The maximum strain value indicates the region or point in the structure where the deformation is the highest, while the minimum strain value represents the area with the least deformation. The average strain provides an overall understanding of the deformation across the structure.

These strain values obtained from the static structural analysis help in assessing the structural integrity and performance of the component under the given loading conditions. It enables engineers to evaluate whether the structure remains within acceptable deformation limits and ensures that it can withstand the applied force without failure or excessive deformation. It is important to note that these conclusions are based on the specific analysis and assumptions made in the analysis process. The results should be interpreted in the context of the model, boundary conditions, and other relevant factors considered in the analysis.

19.3.2 Static Thermal Analysis

The static thermal analysis of the developed HCS model is considered in this section to understand its thermal behavior. The thermal stress and total heat flux were analyzed by applying a temperature of 500 °C on both plates.

19.3.2.1 Input Definition

The input definition for the static thermal analysis of the HCS is presented as Table 19.5.

19.3.2.2 Thermal Strain Analysis

The results of the thermal strain analysis of the developed HCS model are presented in Figure 19.7.

19.3.2.3 Total Heat Flux Analysis

The results of the total heat flux analysis of the developed HCS model are presented as Figure 19.8 and tabulated in Table 19.6.

Based on the analysis of heat flux, when a temperature of 500 °C is applied to both surfaces of the structure, the following conclusions can be drawn:

Minimum heat flux: The minimum heat flux observed is 2.4445e-014 W/mm². This indicates the lowest heat transfer rate in the structure, likely occurring in regions where the thermal energy is not being efficiently conducted or dissipated.

Maximum heat flux: The maximum heat flux observed is 2.4331e-011 W/mm². This represents the highest heat transfer rate in the structure, occurring in regions where there is efficient heat conduction or dissipation.

Table 19.5 Input for thermal analysis.

Type	Temperature
State	Fully defined
Scoping method	Geometry selection
Geometry	2 faces
Magnitude	500 °C
Suppressed	No

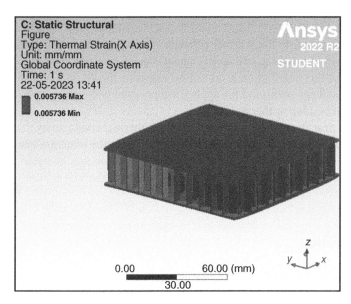

Figure 19.7 Thermal strain analysis.

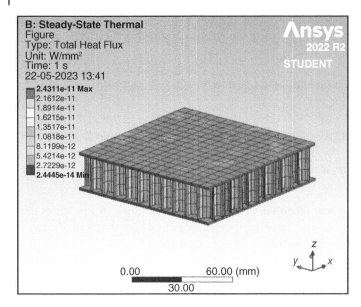

Figure 19.8 Total heat flux analysis.

Table 19.6 Heat flux result.

S. no.	Minimum (W/mm^2)	Maximum (W/mm^2)	Average (W/mm^2)
1.	2.4445e-014	2.4331e-011	1.4323e-011

Average heat flux: The average heat flux across the structure is calculated to be 1.4323e-011 W/mm^2. This value represents the overall heat transfer rate, taking into account both the minimum and maximum heat flux values.

These heat flux values provide insights into the thermal behavior of the structure when subjected to a temperature of 500 °C. They indicate the rate at which heat is being transferred through the material. Understanding the heat flux distribution is crucial for assessing the thermal performance of the structure, ensuring that it can withstand the applied temperature without any detrimental effects such as excessive heat buildup or thermal stress.

It is important to compare these heat flux values with the material's thermal conductivity, its expansion properties, and any thermal limitations or design criteria to ensure the structure can effectively manage and dissipate the applied heat. Further analysis and evaluation may be necessary to optimize the thermal design and address any potential concerns related to thermal management and structural integrity.

19.4 Scope of the Case Study

Objective of this case study is to conduct a static thermal and structural investigation of HCS made with the FSW/P route. The goal is to assess the HCS mechanical behavior and thermal properties under various loading scenarios. To acquire information and evaluate the performance of the developed HCS models, the research will involve running simulations and experiments. This will mainly involve fully understanding the following aspects:

Friction stir process: The principles, benefits, and drawbacks of the friction stir method will all be thoroughly examined. The study will look into the precise characteristics and methods employed in this procedure to create HCS.

Design of honeycomb structures: The research will focus on developing and perfecting HCS for particular uses. A number of geometric combinations, including cell size, cell shape, and core thickness will be taken into account. The design will seek to maximize the HCS's structural stability and thermal efficiency.

Validation Through experimentation: Experimentation will be used to confirm the mechanical and thermal characteristics of the fabricated HCS and to validate the simulation results. These testing could involve thermal strain, mechanical strain, and stress tests.

Performance assessment: Based on the simulation and experimental data, the research will assess how well the manufactured honeycomb structure performs. The analysis will evaluate the honeycomb structure's structural soundness, load-bearing ability, thermal efficiency, and other pertinent characteristics.

19.5 Conclusion

To conclude, this case study is focused on the static structural and thermal analysis of HCS fabricated using FSP. It aims to understand the mechanical behavior and thermal properties of these structures under various loading conditions. Both experimental investigations and numerical simulations were conducted to comprehensively analyze the honeycomb structure.

The results of the static structural analysis revealed the stress and strain distribution in the honeycomb structure when a force of 7000N was applied in the negative z-direction. The maximum stress was found to be 6299.7 MPa, while the minimum stress was 27.805 MPa. The average stress was calculated to be 716.18 MPa. The maximum strain observed was 3.1499e-002 mm/mm, while the minimum strain was 5.6182e-004 mm/mm. The average strain was determined to be 4.2497e-003 mm/mm. These findings provide valuable insights into the structural integrity and performance of the honeycomb structure under the applied load.

Additionally, the static thermal analysis investigated the heat flux distribution when a temperature of 500 °C was applied to both surfaces of the structure. The minimum heat flux was determined to be 2.4445e-014 W/mm^2, the maximum heat flux was 2.4331e-011 W/mm^2, and the average heat flux was 1.4323e-011 W/mm^2. These heat flux values help in assessing the thermal behavior of the structure and ensuring efficient heat transfer and dissipation.

The case study also discusses the design methodology for creating the hexagonal cell structure and the dimensions of the HCS panels. Finite element methods were used for modeling and simulation, allowing for accurate calculations and optimization of the design.

The study's objective was to enhance the understanding of the FSP manufacturing approach for fabricating high-performance HCS. The findings can be utilized to improve the design and production of lightweight, strong HC components for various engineering applications in industries such as aerospace, automotive, and construction. It contributes to the knowledge and advancement of HCS fabricated using the friction stir process. It aims to provide insights into their mechanical behavior and thermal properties, aiding in the optimization of design and manufacturing processes for these structures. Future research could explore additional aspects such as fatigue analysis, material selection, and further optimization techniques to continue improving the performance and applicability of HCS fabricated by FSP.

Acknowledgement

The authors, Dr Manu Srivastava and Dr Sandeep Rathee thank the Science & Engineering Research Board (SERB) for its financial assistance under the project (vide sanction order no. SPG/2021/003383) to perform this work.

References

1 Hara, D. and Özgen, G.O. (2016). Investigation of weight reduction of automotive body structures with the use of sandwich materials. *Transp. Res. Procedia* **14**: 1013–1020.

2 Kumar, A., Ganesh Narayanan, R., and Muthu, N. (2023). Friction stir spot welding of honeycomb core sandwich structure. In: *Low Cost Manufacturing Technologies*. Singapore: Springer Nature Singapore.

3 Rathee, S., Maheshwari, S., Noor Siddiquee, A., and Srivastava, M. (2018). A review of recent progress in solid state fabrication of composites and functionally graded systems via friction stir processing. *Crit. Rev. Solid State Mater. Sci.* **43** (4): 334–366.

4 Rathee, S., Maheshwari, S., and Noor Siddiquee, A. (2018). Issues and strategies in composite fabrication via friction stir processing: a review. *Mater. Manuf. Processes* **33** (3): 239–261.

5 Arslan, K. and Gunes, R. (2018). Experimental damage evaluation of honeycomb sandwich structures with Al/B4C FGM face plates under high velocity impact loads. *Compos. Struct.* **202**: 304–312.

6 Abhinav, S.N. and Budharaju, M.V. (2020). A review paper on origin of honeycomb structure and its sailing properties. *Int. J. Eng. Res. Technol.* 09(08).

7 Lin, S., Yuan, M., Zhao, B., and Li, B. (2023). Out-of-plane compression behaviour of aluminum alloy large-scale super-stub honeycomb cellular structures. *Materials* **16** (3): 1241.

8 Ma, M.-Y., Han, D., Chen, N.K. et al. (2022). Recent progress in double-layer honeycomb structure: a new type of two-dimensional material. *Materials* **15** (21): 7715.

9 Liu, Z., Zhao, W., Yu, G. et al. (2021). Fabrication and mechanical behaviors of quartz fiber composite honeycomb with extremely low permittivity. *Compos. Struct.* **271**: 114129.

10 Ciepielewski, R., Gieleta, R., and Miedzińska, D. (2022). Experimental study on static and dynamic response of aluminum honeycomb sandwich structures. *Materials* **15** (5): 1793.

20

Friction Stir-Based Additive Manufacturing

Ardula G. Rao[1] and Neelam Meena[2]

[1] *Naval Materials Research Laboratory, Defence Research and Development Organization, Thane, India*
[2] *Department of Metallurgical Engineering and Material Science, Indian Institute of Technology Bombay, India*

20.1 Additive Manufacturing: An Introduction

Additive manufacturing (AM), also known as 3D printing technology, is an innovative technology that employs a computer-generated design to create a three-dimensional (3D) object by melting and successively depositing layers of materials. This 3D model is composed of numerous two-dimensional layers that are fused together to form an intended 3D object. AM stands out from other manufacturing techniques, as it involves the gradual addition of material layer by layer, whereas subtractive manufacturing involves the removal of materials instead. The advantage of AM over conventional manufacturing processes lies in the precise engineering of complex geometries which are impossible to fabricate through conventional processes with minimal material wastage. It can be successfully employed at places where parts and components produced through conventional processing routes need replacement. AM can be applied to materials such as metals, ceramics, polymers, and composites. This technique was initially employed for creating prototypes for design purposes; however, now-a-days, even polymers and composite-based products are manufactured using AM. This technique has widely been utilized commercially by the fortune companies like Arcam, Concept lasers, General Electrics, and Siemens. AM reduces the hassle of assembling the parts of the components. Specifics part designs of the application-based components can be manufactured irrespective of the size and shape. Due to these advantages, the need and demand of AM-designed tools are increasingly gaining popularity in aerospace, automotive, medical, chemical, etc., industries [1].

This chapter provides an overview of solid-state AM processes, with a primary focus on friction stir additive manufacturing (FSAM). Subsequently, its application in various alloy systems is explored, emphasizing the alterations in microstructure and properties that can be achieved through FSAM.

20.1.1 Types of AM

According to ASTM standard (F2792-12a) [2], AM can be divided into seven categories, namely binder jetting, direct energy deposition (DED), material extrusion, material jetting, powder bed

Friction Stir Welding and Processing: Fundamentals to Advancements, First Edition.
Edited by Sandeep Rathee, Manu Srivastava, and J. Paulo Davim.

fusion (PBF), sheet laminates, and vat photo-polymerization. However, the most commercialized and evolved processes are PBF, DED, and sheet laminates (SL), these processes are discussed further.

PBF technique involves the use of laser or electron energy source to selectively melt the metal powders in a bed, layer by layer, to create a solid 3D object. The metal powder is spread on a build platform where first layer is melted using energy source. Following that, the platform is lowered, and a fresh layer of powder is introduced, repeating the procedure sequentially until the entire object is successfully built. The laser scanning strategy encompasses a range of parameters that determines the length, direction, speed, pattern, and spacing between the successive laser scans. The selected scan strategy plays a critical role in determining the final properties of the fabricated part, such as its density, mechanical behavior, and stress generation. It is crucial to monitor the residual stress generated in the microstructure to optimize the laser-based processes [3].

The DED process involves a focused beam or electric arc to melt metallic powder or wire feed-stock through layer-wise melting. Metal parts produced using DED processes are known to exhibit high cooling rates resulting in solidified microstructures. DED processes are highly versatile and can produce functionally graded (heterogeneous) parts by allowing the adjustment of material compositions at each layer. This is achieved by simply modifying feeding materials and process parameters, resulting in greater flexibility in the manufacturing process [3].

The SL process makes use principles of ultrasonic vibrations, diffusion bonding, laser, seam, and resistance welding to laminate/bond thin sheets of metals together. This process holds an advantage of joining two dissimilar sheets of metals. The bond forming at the interface of the sheet are weaker than within each layer which brings anisotropy in properties of the different layers of laminated sheets. This limits the use of SL process as compared to PBF and DED processes.

The PBF and DED processes have some advantages over other metal fabrication processes in terms of their direct nature. Both processes involve the use of energy sources, such as lasers or electron beams, to melt and fuse metal powders or wire feedstock into a solid part. One advantage of these processes is that they do not require an external processing step to complete the fabrication, as the part is built directly from the metal powder or wire. This can result in reduced lead times and potentially lower costs compared to other metal fabrication processes that require additional machining or finishing steps. Additionally, these processes offer the ability to create complex geometries and features that may not be possible with other fabrication methods. The ability to control the temperature and energy input during the process allows for precise control over the material properties and microstructure of the final part.

AM process is influenced by several variables, including the deposition rate, beam size and power, build environment, processing temperature, deposition mode, and scan strategy, which collectively determine the final properties of the product. The efficiency of which in turn is determined by several factors, including production times, the maximum size of components that can be fabricated, the ability to produce intricate parts, and the achieved dimensional accuracy during the manufacturing process which get influenced by the variables mentioned. These factors overall decide which process is viable for the use for bulk production. For example, powder-based feed material takes longer production time than wire-based which separates the powder-based material process to be used for smaller component fabrication and wire-based for large-size component fabrication. These processes involve use of laser and electron beam which melts the material and solidified microstructure is obtained. One common problem encountered during liquid-based AM processes are the issues of porosities, lack of fusion, and other solidification-based defect in structure. It prevails most when dissimilar metals are used for AM. Various alternatives such as use of nitrogen as shielding gas and maintaining high vacuum environment are suggested but cannot be adopted due to other associated limitations. A solid-state AM process, aside from the SL process, may offer a solution to circumvent these problems.

Selecting the right powder is critical to achieving the desired properties in the final product. The size and shape of the powder particles also play a significant role in the AM process. Powder size affects the surface quality and precision of the final product, while powder shape influences the flowability of the powder during the printing process. The consistency and purity of the powder material are essential to ensure consistent printing results and avoid defects. Contaminants or variations in particle size can cause issues like porosity, warping, or cracking in the final product [4].

20.2 Solid-State AM Processes

Solid-state AM process offers many advantages over other fusion-based techniques such as significant grain refinement and better structural homogenization/ Also, issues related to elemental segregation, porosities, and hot cracking are reduced. A versatile solid-state AM process is friction-based AM (FBAM). The original idea of joining metals layer by layer using friction was introduced and patented by White in 2002 [5]. The basic idea is to join the metal in solid state without melting. The peak temperature is always maintained below the solidus temperature of the alloy. The two material surfaces are joined with the principle of friction. The significant amount of heat generated at the interfaces increases the local temperature and softens the material and thus helps to consolidate [6]. A list of processes based on FBAM is shown in Figure 20.1.

In the rotary friction welding (RFW), the joining process is carried out either by keeping one part stationary and other rotating or both the parts rotates relative to each other. The axial pressure is increased upon contact and severe plastic deformation occurs which results in bond formation between the two parts. Sequentially, the successive layers are added and a bulk component can be fabricated. The process efficiency is highly dependent on rotating speed, axial force, and joining time of the parts. The limitation of this process lies where only cylindrical or round shape parts can be utilized for production. Due to dynamic recrystallization occurring during severe deformation, the resulting microstructure is very fine which results in better mechanical properties of the final product [7].

Similar to RFW process, joining and phase transition in friction deposition takes place in solid state. Friction deposition technique involves a consuming material to be used for deposition attached to the rotating spindle and brought in contact with the stationary plate. The material plasticizes and

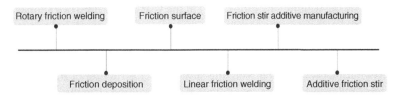

Figure-based additive manufacturing processes

Figure 20.1 Various AM processes categorized based on FBAM.

flows on the stationary plate and layer-by-layer deposition is performed. This method avoids the disparity in the microstructure based on different heat-affected zone as obtained in the RFW.

Friction surfacing method involves using a metallic consumable rod, the mechatrode, which is rotated and pressed against the substrate material with axial load. This process creates a viscoplastic boundary layer when the mechatrode tip reaches a temperature high enough to soften the material and generate bonding between the substrate and the softened material. With increasing dwell time, the plasticized material consolidates due to the higher rate of heat conduction through the substrate. This consolidation leads to the formation of multiple viscoplastic shearing interfaces, with each interface increasing the layer's thickness. Following the deposition of the first layer across and along the substrate, the second layer is deposited after machining the first layer [8].

In linear friction welding, one part is held stationary while the other is reciprocated back and forth under axial force. The reciprocating part rubs against the stationary part, generating frictional heat that melts and plasticizes the material. Once the interface material reaches the desired temperature and plastic state, the reciprocating part is held in place while the axial force is maintained. The heat generated by friction is dissipated through the stationary part, and the interface material solidifies and fuses the two parts together controlling the frequency, amplitude, and axial load allows for control over the heat input into the interface. This control is essential for achieving optimal conditions for the linear friction welding process [9].

FSAM involves creating a multilayered build by inserting a nonconsumable tool into the overlapping or abutting sheets or plates to be welded and then traversing the tool along the joint line. During this process, frictional heat softens the material, making it more malleable and allowing the tool to mix the layers and form a solid-state bond. Once the tool completes one pass along the joint line, it returns to the starting point to begin the next pass, building up the layers and forming the desired shape. FSAM is a highly precise manufacturing process that produces parts with excellent mechanical properties and dimensional accuracy [10]. Further details of the FSAM are discussed in the next section.

Additive friction stir (AFS) and friction stir deposition (FSD) processes are also based upon the principles of FSP/FSW. The AFS process involves the use of a nonconsumable rotating cylindrical tool to feed metal powder or solid rod, which generates heat and applies controlled pressure to plastically deform the material. This process builds successive layers upon a substrate. After each layer is built upon the substrate, the tool is lifted to allow for the placement of the next layer onto the previous layer. This creates a metallurgical bond between the layers. The feedstock material could be in the form of a rod or powder. The temperature of the processed reaches about 0.6–$0.9\,T_m$ of the melting temperature of the alloy. This process is advantageous to coat or repair the parts due to high deposition rates ($\sim 80\,cm^3/h$) [11].

20.2.1 How FSP was Derived from FSW for an AM Process

Based on the previous section, we will now shift our focus to the FSAM technique and discuss it in more detail. FSW was invented by the welding institute (TWI) in the year 1991 where two similar or dissimilar workpieces are joined using a rotating tool [12]. Over time, advancements have been made in the field of FSW, enabling the joining of materials with high melting points. Mishra et al. [13] later developed the friction stir processing (FSP) technique in the year 1991, which adopts the principles of FSW but focuses on modifying the surface microstructure of materials. Since welding and AM share similarities in their source nature, the resulting structures can be quite similar. However, there are differences between the two processes, particularly in the precision and speed at which AM operates. The relationship between process properties in welding is also of interest in AM of metals. Both are thermomechanical processes which share the similar microstructural field zone based on the amount of the heat dissipated.

The fundamentals of FSAM lies in the similar lines to friction stir welding/processing (FSW/FSP). However, their respective application functions are different. FSW or FSP as mentioned earlier is mainly used as the joining and surface modification technique, respectively. A schematic comparing the FSW and AFS deposition process is shown in Figure 20.2. The FSAM is performed by stacking the sheets. The central idea behind FSAM/deposition is quite straightforward like FSP – a specially designed rotating tool, featuring a customized pin and shoulder is utilized to join the overlapped sheets together by inserting it into the overlapping surfaces and moving it along the joint line. This tool is nonconsumable, meaning, it can be used repeatedly for subsequent layer deposition. During every succeeding layer addition, the tool height enables subsequent metal deposition, compelling the material to merge with preceding layer that ultimately results in strong interlayer metallurgical bond.

Similar to that of FSW/FSP, the material is pushed from advancing to retreating side. Microstructure zones developed in the material are also similar as of FSW/FSP where the length, breadth, and depth of the entire nugget zone depend on pin and shoulder dimensions. In contrast

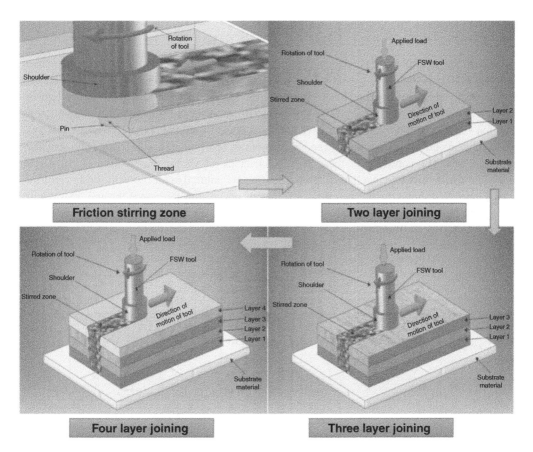

Figure 20.2 Schematic representation of FSAM process. Source: Kumar et al. [14]/with permission of Elsevier.

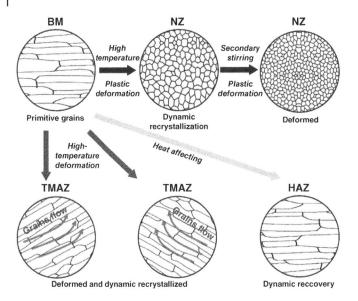

Figure 20.3 Schematic showing microstructure evolution in each heat-affected zone through FSAM process. Source: Jiang et al. [15]/with permission of Elsevier.

to conventional beam-based processes where solidification occurs, equiaxed microstructures develop in materials undergoing dynamic recrystallization as shown in Figure 20.3. This is crucial for achieving isotropic properties in the components. Microstructural kinetics is dependent on thermal cycle, strain, and strain rates. These factors in turn are controlled by variables such as rotational and traverse speed, tool geometry, and axial force. By adjusting these variables, it is possible to achieve optimal conditions for the FSAM process, resulting in a well-formed and structurally sound part. The control over these process variables allows for the creation of parts with unique microstructures and mechanical properties that are tailored to meet specific design requirements. Figure 20.4 shows the overall cross-section view of the alloy fabricated through FSAM with difference in the hardness variation from base to nugget zone. The high hardness is observed owing to recrystallized microstructure obtained in each layer following Hall–Petch relation.

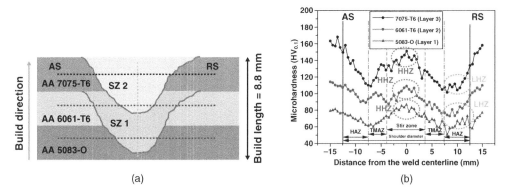

Figure 20.4 (a) Cross-section overview of the Al alloy composite fabricated through FSAM. (b) Hardness variation with each overlapped layer. Source: Kumar et al. [16]/with permission of Elsevier.

20.3 Case Studies of FSAM on Different Materials

20.3.1 Ferrous Alloys

Welding of iron-based alloys and steels has been successfully accomplished through FSW [17]. While limited research exists on FSAM of steels, Mishra et al. [10] have discussed the FSAM on P92 grade steel (9Cr–2W) and MA956 steel. Their study demonstrated that processing through FSAM enabled the overcoming of the strength-ductility trade-off, resulting in a two-fold increase in yield strength and similar elongation to the base metal. This improvement can be attributed to the finer microstructure and lath structure obtained after performing FSAM. Literature also reports on the successful FSAM performed on the bimetallic components with steels. One such study was reported on Fe–Cu-based bimetallic sheets obtained through FSAM process without any porosity or defect is shown in Figure 20.5 [18]. Under high temperature and strains, 2% of solid solution between Fe–Cu was possible to achieve where a wave-like interface between the two metals was seen which is considered better than a sharp interface which usually results in rapid failure of the components. Therefore, along with solid solution strengthening, grain boundary strengthening and deformation strengthening were attributed to the high shear as well as yield strength of the Cu–Steel part than alone base Cu sheet.

20.3.2 Nonferrous Alloys

Al-based alloys and Mg-based alloys have gained popularity in the automotive sector due to their lightweight properties. Al-based alloys are commonly used for manufacturing metal body parts in

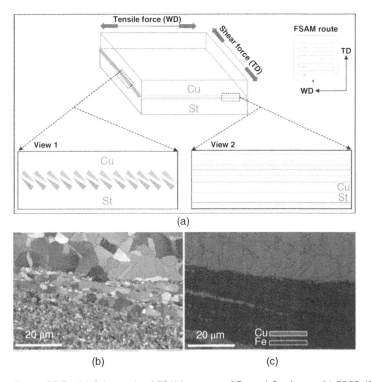

Figure 20.5 (a) Schematic of FSAM process of Fe and Cu sheets. (b) EBSD-IPF map and (c) phase map showing the interface between Fe–Cu sheet. Source: Guo et al. [18]/with permission of Elsevier.

aerospace and automobile applications due to their excellent strength to weight ratio and high corrosion resistance properties. However, fusion welding processes for these alloys are notoriously difficult. Fortunately, the FSP technique has enabled the solid-state weldability of Al alloys, making it easier to use them for AM of components through friction stir AM/deposition technique. Many studies are reported in the literature on the microstructure and property relationship on Al alloys fabricated using FSAM/deposition [19]. Al alloys mainly derive their properties from precipitation strengthening and dislocation strengthening mechanism. Precipitates present in the alloy try to retard the dislocation movement and strengthen the material. As different thermal regions in the stir zone of the processed part exhibits different microstructure, the precipitate that forms in the Al alloys with temperature variation also contributes vastly to the overall mechanical properties of the component. Due to rapid rotation speed and subsequent amount of heat generated, precipitates tend to dissolve, reprecipitate, coarsen, and their morphology and number fraction change from base metal to nugget zone. When Al–Li alloys are subjected to FSAM, random distribution of precipitates in different zones primarily govern the mechanical properties of the component. The T1 and θ' precipitates coarsening takes place in the heat-affected zone and precipitate-free zones are formed leading to reduced hardness value. Their fraction also varies within nugget zone from low/no precipitates to reprecipitation of T1 and θ' which increases the hardness value [20].

The FSAM technique has also demonstrated the feasibility of fabricating metal-composite materials [21]. The reinforcement particles can be incorporated into the sheets to be joined by creating holes in them. This allows for the introduction of the particles before the sheets are joined. Due to the pinning effect exerted by the reinforced particles, the grain growth is retarded at high temperatures achieved during FSAM and a fine recrystallized microstructure can be obtained. The incorporation of hard particles into the soft matrix results in an enhancement of the material's hardness.

Mg alloys on the other hand have also successfully shown the implementation of FSAM process. WE43 alloy that undergoes FSAM shows good strength and elongation than unprocessed part due to dissolution of intermetallic particles present in the nugget zone. Their coarsening and dissolution further are function of heat input and the strain experienced in the different regions. Therefore, tool speed plays a significant role in distributing the heat in the material during processing [22].

Inconel alloys are of great interest to the automotive and aerospace industry due to excellent mechanical properties, high temperature resistance, and corrosion resistance. One study revealed the successful manufacturing of IN625 alloy through AFS process in a HY80 matrix [23]. The microstructure developed in the build direction drastically reduced the grain size from 30 to ~1 μm in the part and to 0.25 μm at the interface region. An increase in strength from AFS process was achieved and when compared to the other AM process, it was found to be superior.

20.4 Advantages of FSAM Over Other AM Techniques

After reviewing the previous section, it has been observed that the FSAM process offers advantages over other AM techniques. This is because the FSAM process primarily does not involve melting the material, which can lead to defects in the microstructure. Because no melting is involved, the process is more energy-efficient and cost-effective. The process also does not restrict on the type of feedstock material. For the beam-based processes, material in powder form or a powder bed is required for carrying out the easy melting. Here, the material can be in rod shape or in powder form. Since the deposition is not constrained by the dimension of any powder bed or there is no

Table 20.1 Comparison of two solid-state AM process, USAM and FSAM [22].

USAM	FSAM
Build volume is constrained by the thickness of the material	Ability to fabricate large components
Only rolled foils of metal can be built leading to limited material selection	Faying surface contamination is not as vital as in UAM
Foil making is time consuming and costly	Variety of materials that are unable to build by fusion welding and UAM can be processed
Affected by linear weld density	Affected by process forces and tool geometry
Difficult to bond materials with high work hardening rate	High degree of reproducibility
Exhibit marginal fraction of the bulk material properties	Can produce variable microstructures for specific applications

restriction for any specific vacuum environment, this process is viable to produce large-size components. A comparison of FSAM process with other solid-state process of ultrasonic additive manufacturing (USAM), as taken from Palanivel et al.'s work [22], is shown in Table 20.1.

Other important advantages offered by the FSAM process are:

1) FSAM process offers advantage of greater production scale along with less material wastage over other AM processes.
2) It delivers superior structural integrity to the materials fabricated.
3) The FSAM process has the ability to address common issues found in fusion-based techniques, such as porosity, hot cracking, elemental segregation, and dilution. As material flows back and forth with high-speed rotating tool in the substrate, problem of elemental segregation is reduced.
4) The FSAM process can lead to grain refinement, which leads to more homogeneous and isotropic mechanical properties of the manufactured components.
5) Compared to other fusion-based processes, the FSAM technique results in lower levels of distortion in the manufactured parts and substrates. This is due to the fact that the deposited materials do not reach their melting point.
6) It is highly recommended to repair large components and to coat the material owing to its high deposition/build rates.
7) Problem of residual stresses in the material is less compared to other fusion-based processes.
8) In melting-based processes, issues related to oxidation of the material pertains which is completely eliminated in the friction stir-based AM process.

20.5 Advancements

MELD is a solid-state AM technology that has been recently developed by Aeroprobe Corporation [24, 25]. It is basically is used for manufacturing Al-based composite AM parts. Earlier, FSP was used successfully used to make Al composites by making holes/grooves in the plate and running a tool over having a pin which allows the reinforced particles to mix with the matrix metal [26]. A hollow tool is used to deposit the feedstock material. Dynamic contact friction, such as that which occurs at the tool–material interface and the material–substrate interface, generates heat.

Table 20.2 Summary of mechanical properties of different alloys fabricated through FSAM technique.

S. no.	Alloy	Mechanical properties	Properties	References
1	IN625	Tensile	Yield strength: 730 MPa	[23]
2	2205 Al–Li	Tensile	Tensile strength: 300 MPa (traverse direction) 399 MPa (longitudinal direction)	[20]
3	2060 Al–Li	Hardness	80–130 HV (single track) 90–110 HV (overlap track)	[15]
4	WE43	Tensile	Yield strength: 374 Tensile strength: 400 MPa	[22]
5	Cu–Steel bimetallic sheet	Tensile	Tensile strength: 435 MPa Shear strength: 345 MPa	[18]
6	AA2024-T4	Tensile	Tensile strength: 488 MPa	[28]
7	AA5083/AA6061/SiC composite	Tensile	Tensile strength: 291 MPa	[29]
8	AA5083/AA7075 Joints	Tensile	Tensile strength: Stir zone 1: 276 MPa Stir zone 2: 429 MPa Stir zone 3: 279 MPa	[30]
9	Al–Zn–Mg	Tensile	Tensile strength: Traverse direction: 400 MPa Building direction: 398 MPa	[31]
10	Al–Zn–Mg–Cu	Tensile	Tensile strength: 523 MPa	[32]

This heat can raise the temperature of the filler material, promoting plastic deformation. This plastic deformation, in turn, facilitates solid-state bonding between the filler material and the substrate. Due to high strain rates and high temperatures reached during processing, dynamic recrystallization results in finer grains. Al-based matrix composites (AMC) are successfully designed through MELD using the AMC in a rod form or loose particle form mixed with matrix powder in the hollow tool. The powders are consolidated in the tool before deposition with the action of dynamic friction contact that rises the temperatures inside the tool as it rotates. This provides better adhesion and homogeneous spread of the reinforcement particles. This technique is suitable to fabricate metal matrix composites on the principles of FSP/FSW [27]. Further, a comprehensive review on the mechanical properties achieved through FSAM technique is shown in Table 20.2.

20.6 Limitations

Despite significant advancements, every process or technique still has its unique niche. FSAM can be used to manufacture parts using materials such as aluminum and copper alloys, but it may not be suitable for materials with higher melting temperatures or materials that are difficult to process due to stringent requirement of tool material. Due to layer-by-layer deposition, the components require additional machining to give final finish. The size of parts that can be manufactured using FSAM is limited by the size of the build envelope of the machine. Large parts may need to be manufactured in multiple sections, which can lead to additional joining operations and potential quality issues. As far as MELD process is concerned, the limitation lies with the reinforced particles used which probably can hinder the material flow. Therefore, a check on size and nature of reinforcement particles is required before fabrication. The cost of FSAM equipment is relatively high compared to other AM technologies, which can limit its adoption by smaller companies or those with limited budgets [27].

20.7 Conclusions and Future prospectives

The scalability of any AM process depends on several critical factors, including ease of operation, energy efficiency, cost-effectiveness, high production rates, and sustainability. When all these criteria are met, the implementation of the process becomes more accessible. Effective defect mitigation techniques are required to produce sound components. The use of MELD process is required to be explored using different metals (other than Al alloys), nonmetals, and ceramics to work on making composite material-based AM components. Overall, the future prospects of FSAM processes are very promising, and as the technology continues to evolve, it could become a valuable tool for a wide range of industries.

References

1 Kok, Y., Tan, X.P., Wang, P. et al. (2018). Anisotropy and heterogeneity of microstructure and mechanical properties in metal additive manufacturing: a critical review. *Mater. Des.* 139: 565–586. https://doi.org/10.1016/j.matdes.2017.11.021.

2 Standard Terminology for Additive Manufacturing Technologies (2013). Standard Terminology for Additive Manufacturing Technologies (ASTMF2792 − 12a). *ASTM International*, pp. 10–12. https://doi.org/10.1520/F2792-12A.2.

3 DebRoy, T., Wei, H.L., Zuback, J.S. et al. (2018). Additive manufacturing of metallic components – Process, structure and properties. *Prog. Mater Sci.* 92: 112–224. https://doi.org/10.1016/j.pmatsci.2017.10.001.

4 Bourell, D., Kruth, J.P., Leu, M. et al. (2017). Materials for additive manufacturing. *CIRP Ann. Manuf. Technol.* 66 (2): 659–681. https://doi.org/10.1016/j.cirp.2017.05.009.

5 White, D. (n.d.). Object consolidation employing friction joining. *United States Patent*. https://patents.google.com/patent/US6457629B1/en.

6 Mishra, R.S. and Palanivel, S. (2017). Building without melting: a short review of friction-based additive manufacturing techniques. *Int. J. Addit. Subtract. Manuf.* 1 (1): 82. https://doi.org/10.1504/ijasmm.2017.10003956.

7 Kishore, V.A.N., Babu, N.K., and Krishna, K. (2021). A review on rotary and linear friction welding of inconel alloys. *Trans. Indian Inst. Met.* 74 (11): 2583–2598. https://doi.org/10.1007/s12666-021-02345-z.

8 Dilip, J.J.S., Babu, S., Rajan, S.V. et al. (2013). Use of friction surfacing for additive manufacturing. *Mater. Manuf. Processes* 6914 (28): 189–204. https://doi.org/10.1080/10426914.2012.677912.

9 Achilles, V. (2022). Linear friction welded titanium alloy joints : a brief review of microstructure evolution and mechanical properties. *Weld. Int.* 36 (11): 647–654. https://doi.org/10.1080/09507116 .2022.2149366.

10 Mishra, R.S., Haridas, R.S., and Agrawal, P. (2022). Friction stir-based additive manufacturing. *Sci. Technol. Weld. Joining* 27 (3): 141–165. https://doi.org/10.1080/13621718.2022.2027663.

11 Prabhakar, D.A.P., Shettigar, A.K., Herbert, M.A. et al. (2022). A comprehensive review of friction stir techniques in structural materials and alloys : challenges and trends Analysis of Variance. *J. Mater. Res. Technol.* 20: 3025–3060. https://doi.org/10.1016/j.jmrt.2022.08.034.

12 Liu, H. and Zhou, L. (2007). Progress in friction stir welding of high melting point materials. *Hanjie Xuebao/Trans. China Weld. Inst.* 28 (10): 101–104.

13 Mishra, R.S., Mahoney, M.W., McFadden, S.X. et al. (2000). High strain rate superplasticity in a friction stir processed 7075 Al Alloy. *Scr. Mater.* 42: 163–168.

14 Kumar, A., Kumar, N., and Rai, A. (2021). Friction stir additive manufacturing – an innovative tool to enhance mechanical and microstructural properties'. *Mater. Sci. Eng., B* 263 (June): 114832. https://doi.org/10.1016/j.mseb.2020.114832.

15 Jiang, T., Jiao, T., Dai, G. et al. (2023). Microstructure evolution and mechanical properties of 2060 Al-Li alloy via friction stir additive manufacturing. *J. Alloys Compd.* 935: 168020. https://doi.org/ 10.1016/j.jallcom.2022.168020.

16 Kumar Jha, K., Kesharwani, R., and Imam, M. (2022). Microstructural and micro-hardness study on the fabricated Al 5083-O/6061-T6/7075-T6 gradient composite component via a novel route of friction stir additive manufacturing. *Mater. Today Proc.* 56: 820–825. https://doi.org/10.1016/ j.matpr.2022.02.262.

17 Mohan, D.G. and Wu, C.S. (2021). A review on friction stir welding of steels. *Chin. J. Mech. Eng.* 34 (1): https://doi.org/10.1186/s10033-021-00655-3.

18 Guo, Y., Wu, X., Ren, G. et al. (2022). Microstructure and properties of copper-steel bimetallic sheets prepared by friction stir additive manufacturing. *J. Manuf. Processes* 82 (June): 689–699. https://doi.org/10.1016/j.jmapro.2022.08.022.

19 Phillips, B.J., Avery, D.Z., Liu, T. et al. (2020). Microstructure-deformation relationship of additive friction stir-deposition Al–Mg–Si. *Materialia* 7 (June): 100387. https://doi.org/10.1016/j.mtla .2020.100387.

20 Shen, Z., Chen, S., Cui, L., and Li, D. (2022, 2021). Local microstructure evolution and mechanical performance of friction stir additive manufactured 2205 Al-Li alloy'. *Mater. Charact.* 186: https:// doi.org/10.1016/j.matchar.2022.111818.

21 Srivastava, M. and Rathee, S. (2020). Microstructural and microhardness study on fabrication of Al 5059/SiC composite component via a novel route of friction stir additive manufacturing. *Mater. Today Proc.* 39: 1775–1780. https://doi.org/10.1016/j.matpr.2020.07.137.

22 Palanivel, S., Nelaturu, P., Glass, B., and Mishra, R.S. (2015). Friction stir additive manufacturing for high structural performance through microstructural control in an Mg based WE43 alloy. *Mater. Des.* 65: 934–952. https://doi.org/10.1016/j.matdes.2014.09.082.

23 Rivera, O. G. P.G. Allison, J.B. Jordon *et al.* (2017) 'Microstructures and mechanical behavior of Inconel 625 fabricated by solid-state additive manufacturing, *Mater. Sci. Eng., A*, 694 (October 2016), pp. 1–9. https://doi.org/10.1016/j.msea.2017.03.105.

24 Jeffrey, I., Schultz, P. and Mayberry, M. L. (2017). United States Patent Fabrication Tools For Exerting Normal Forces on Feedstock.

25 Mayberry, M. L. (2014). United States Patent Friction Stir Fabrication.

26 Sharma, A., Narsimhachary, D., Sharma, V.M. et al. (2020). Surface modification of Al6061-SiC surface composite through impregnation of graphene, graphite & carbon nanotubes via FSP: A tribological study. *Surf. Coat. Technol.* 368 (January): 175–201. https://doi.org/10.1016/j.surfcoat.2020.04.001.

27 Griffiths, R.J., Perry, M.E.J., Sietins, J.M. et al. (2020). A perspective on solid-state additive manufacturing of aluminum matrix composites using MELD. *J. Mater. Eng. Perform.* 28 (2): 648–656. https://doi.org/10.1007/s11665-018-3649-3.

28 Xiao, Y., Li, Y., Shi, L., and Wu, C.S. (2023, 2023). Experimental and numerical analysis of friction stir additive manufacturing of 2024 aluminium alloy. *Mater. Today Commun.* 35: 105639. https://doi.org/10.1016/j.mtcomm.2023.105639.

29 Jha, K.K., Kesharwani, R., and Imam, M. (2023). Microstructure, texture, and mechanical properties correlation of AA5083/AA6061/SiC composite fabricated by FSAM process. *Mater. Chem. Phys.* 296 (November): 127210. https://doi.org/10.1016/j.matchemphys.2022.127210.

30 Jha, K.K., Kesharwani, R., and Imam, M. (2023). Microstructure and mechanical properties correlation of FSAM employed AA5083/AA7075 joints. *Trans. Indian Inst. Met.* 76 (2): 323–333. https://doi.org/10.1007/s12666-022-02672-9.

31 He, C., Li, Y., Wei, J. et al. (2022). Enhancing the mechanical performance of Al–Zn–Mg alloy builds fabricated via underwater friction stir additive manufacturing and post-processing aging. *J. Mater. Sci. Technol.* 108: 26–36. https://doi.org/10.1016/j.jmst.2021.08.050.

32 Li, Y., He, C., Wei, J. et al. (2022). Effect of post-fabricated aging on microstructure and mechanical properties in underwater friction stir additive manufacturing of Al–Zn–Mg–Cu alloy. *Materials* 15 (9): 1–12. https://doi.org/10.3390/ma15093368.

Index

Friction Stir Welding and Processing: Fundamentals to Advancements, First Edition.
Edited by Sandeep Rathee, Manu Srivastava, and J. Paulo Davim.
© 2024 John Wiley & Sons, Inc. Published 2024 by John Wiley & Sons, Inc.

Printed and bound by CPI Group (UK) Ltd, Croydon, CR0 4YY

16/04/2025